ASTRONOMY FOR DEVELOPING COUNTRIES

SPECIAL SESSION OF THE XXIV GENERAL ASSEMBLY OF THE IAU

COVER ILLUSTRATION:

Histogram showing the distribution of countries by per capita GNP. For full description, see page 6.

THE ASTRONOMICAL SOCIETY OF THE PACIFIC
390 Ashton Avenue – San Francisco, California – USA 94112-1722
Phone: (415) 337-1100   E-Mail: catalog@aspsky.org
Fax: (415) 337-5205   Web Site: www.aspsky.org

Publisher

ASP CONFERENCE SERIES - EDITORIAL STAFF

Managing Editor: D. H. McNamara   LaTeX-Computer Consultant: T. J. Mahoney
Associate Managing Editor: J. W. Moody   Production Manager: Enid L. Livingston

PO Box 24453, 211 KMB, Brigham Young University, Provo, Utah, 84602-4463
Phone: (801) 378-2111   Fax: (801) 378-4049   E-Mail: pasp@byu.edu

ASP CONFERENCE SERIES PUBLICATION COMMITTEE:

Alexei V. Filippenko   Geoffrey Marcy
Ray Norris   Donald Terndrup
Frank X. Timmes   C. Megan Urry

A listing of all other ASP Conference Series Volumes and IAU Volumes
published by the ASP is cited at the back of this volume

INTERNATIONAL ASTRONOMICAL UNION
98bis, Bd Arago – F-75014 Paris – France
Tel: +33 1 4325 8358   E-mail: iau@iap.fr
Fax: +33 1 4325 2616   Web Site: www.iau.org

ASTRONOMY FOR DEVELOPING COUNTRIES

Proceedings of a Special Session of the XXIV General Assembly
of the International Astronomical Union
held at the Victoria University of Manchester
Manchester, United Kingdom
14-16 August 2000

Edited by

ALAN H. BATTEN
National Research Council of Canada
Herzberg Institute of Astrophysics
Dominion Astrophysical Observatory
Victoria, B.C., Canada

Library of Congress Cataloging in Publication Data
Main entry under title

Card Number: 2001087951
ISBN: 1-58381-067-6

IAU Publications - First Edition

Published on behalf of IAU by Astronomical Society of the Pacific

Printed in United States of America by Sheridan Books, Chelsea, Michigan

This volume is the record
of a meeting held in the
Victoria University of Manchester
where
Zdeněk Kopal (1914–1993),
Professor of Astronomy for thirty years,
gave their first opportunities to many
astronomers from developing countries.

Table of Contents

Preface xii
List of Major Contributors xv

Section 1: Survey Papers

Astronomy for Developing Countries
  Alan H. Batten 3
Critical Factors for a Successful Astronomical Research Program
in a Developing Country
  John B. Hearnshaw 15
Do Developing Countries Need Astronomy? (poster)
  Philippe Eenens 29

Section 2: The Current Situation in Astronomy Education

Astronomy Education: The Current Status in Zambia
  Geoffrey Munyeme and Peter C. Kalebwe 39
A Renewal of Astronomy Education in Vietnam
  Donat G. Wentzel 46
Revitalizing Astronomy in the Philippines
  Cynthia P. Celebre and Bernardo M. Soriano Jr 48
Astronomy Development in Morocco: A Challenge to Stimulate Science
and Education
  Khalil Chamcham 59

Experience in Developing an Astronomy Program in Paraguay
  Alexis E. Troche-Boggino 65
The Central American Master's Program in Astronomy and Astrophysics
  Maria C. Pineda de Carias 69
Astronomy Education in Universities of China
  Cheng Fang and Yuhua Tang 80

Section 3: Initiatives in Astronomy Education

Hands-On Astrophysics: Variable Stars for Astronomy Education
and Development
  Janet A. Mattei and John R. Percy 89
Exchange of Astronomy Teaching Experiences
  Rosa M. Ros 95

Public Education in Developing Countries on the Occasions of Eclipses
  Jay M. Pasachoff 101
Using Television for Astronomy Teaching
  Julieta Fierro 107
The Vatican Observatory Summer Schools in Observational Astronomy
and Astrophysics
  Christopher J. Corbally 110
Initiatives in Astronomy Education in South Africa
  Case L. Rijsdijk 117
Distance Education and Self-Study
  Barrie W. Jones 131
Principles for Tertiary-Level Astronomy Courses
  Derek McNally 141

SALT/HET Cooperation in Education and Public Outreach (poster)
  Mary Kay Hemenway and Sandra Preston 152
Problems Facing Promotion of Astronomy in Arab Countries (poster)
  Anas M.I.Osman 157
Astronomy in Cuba: Practice and Trends. An Effort to Develop a
Non-Formal Education Programme (poster abstract)
  Oscar A. Pomares 163
Astronomy Educational Activity in Jordan (poster abstract)
  Jack Baggaley 163
A Jump Start in Astronomy Education in Taiwan (poster abstract)
  Wen-Ping Chen 164
An Interactive Approach to Planetary Orbits at Secondary Level
(poster abstract)
  V.B. Bhatia 164
An Undergraduate Program for Astronomy in México (poster abstract)
  Hector Bravo-Alfaro et al 164
Astronomy in Romanian Universities (poster abstract)
  Mihail Barbosu 165
Astronomy Education in Thailand (poster abstract)
  Busaba Hutawarakorn et al 165
Astronomy Books in Spanish (poster abstract)
  Julieta Fierro 166
Teaching of Astronomy in India (poster abstract)
  Mandayam N. Anandaram 166
Summer Schools for European Teachers (poster abstract)
  Rosa M. Ros 166
The Leonids Observation Project by High-School Students all over
the World (poster abstract)
  B. Suzuki et al 167
New Student Laboratory Work about Pulsational Phenomena
in Astronomy (poster abstract)
  Salakhutdin N. Nuritdinov 167
Astronomy and Astronomical Education in the FSU (poster abstract)
  Nikolai G. Bochkarev 168

Conceptual Approach to Astronomy and Basic Science Education
(poster abstract)
    M. Melek 168
Research Oriented Astrophysics Course for Physics Students (poster title)
    Tapan K. Chatterjee and A. Pedroza-Mendelez 168

**Section 4: Current Status of Astronomy Research in Developing
Countries**
Astronomy in Algeria: Past and Present Developments
    Abdenour Irbah, Toufik Abdelatif and Hamid Sadsaoud 171
The Egyptian 1.88-m Telescope
    Anas M.I. Osman 179
Astronomical Research and Education in Tajikistan
    Pulat B. Babadzhanov 187
Work at Bosscha Observatory
    Moedji Raharto 197
Astronomy in Venezuela
    Patricia Rosenzweig 205
Astronomy Research in China
    Jingxiu Wang 210
SALT as an African Facility
    Peter Martinez 221

The Error in Solar-Diameter Measurements Induced by Atmospheric
Turbulence (poster abstract)
    Lyes Lakhal et al 230
Astronomy Research in Bolivia (poster abstract)
    Dmitry D. Polojentsev et al 230
Sofia Sky-Archive Data Center: Photographic Plate Collections for
Developing Countries (poster abstract)
    Milcho K. Tsvetkov et al 231
CCD Observations with Small Telesscopes of Moving Bodies (poster abstract)
    Oleg P. Bykov 231
Astronomy in Uzbekistan (poster abstract)
    Sabit P. Ilyasov et al 231
Astronomy in the Republic of Macedonia (poster abstract)
    Mijat Mijatovic 232
Astronomy without Astronomers? (poster abstract)
    Magdalena Stavinschi 232
A 100-Year Astronomical Data Bank: Collaboration Possibilities and
some Problems (poster abstract)
    Salakhutdin N. Nuritdinov 233
Galactic-Cluster Studies and Emission-Line Star Surveys with the Schmidt
Telescope at Bosscha Observatory (poster abstract)
    Suhardja D. Wiramihardja 233
Development of Astronomy in Ecuador (poster title)
    Ericson Lopez 233

Section 5: Small Telescopes or Internet Access?

The Choice of Small Telescopes
  David L. Crawford 237
What Can be Done with Small Telescopes?
  Boonraksar Soonthornthum 243
Simple Science, Quality Science
  John R. Percy 250
Simple Instruments in Radio Astronomy
  Nguyen Quang Rieu 255
Is Astronomical Research Appropriate for Developing Countries?
  Michael S. Snowden 266
Internet Resources for Astronomers Worldwide
  George Helou 276
Astronomy Research via the Internet
  Kavan U. Ratnatunga 279
The Role of Astronomical Catalogues in Modern Theory and Observation
  Oleg Yu. Malkov, Alexander V. Tutukov and Dana A. Kovaleva . . . 291

Design of a Small Automated Telescope for Indian Universities (poster)
  Mandayam N. Anandaram, B.A. Kagali and S.P. Bhatangar 303
Small Radio Telescopes for Education (poster)
  Koitiro Maeda and Noritaka Tokimasa 307
Cultural-Grant Aid in Astronomy for Developing Countries from the
Japanese Government (poster)
  Masatoshi Kitamura 312
Robotic Telescopes: A Link between Astronomically Developed and
Astronomically Developing Countries (poster abstract)
  Peter Martinez 314
The Four-College Automated Photoelectric Telescope (poster abstract)
  Saul J. Adelman et al 314
Developing an Astronomical Observatory in Paraguay (poster abstract)
  Alexis E. Troche-Boggino 315
Science with Small Observatory Instruments (poster abstract)
  Ron G. Samec et al 315

Section 6: Some Practical Matters

Some Cooperative Activities in East Asia
  Norio Kaifu 319
Third-World Astronomy Network
  Jayant V. Narlikar 324
Security of Equipment
  Lesley I. Onuora 329
Overcoming the English-Language Barrier
  Terry J. Mahoney 333

Practical Aid to Libraries in Developing Countries
    Peter D. Hingley 340
Preservation of Library Materials in Developing Countries
    Ethleen Lastovica 345
Electronic Access to Journals
    Helmut A. Abt 354

Making the Most of Publishing Software (poster)
    Terry J. Mahoney 357
The Physics-Astronomy-Mathematics Asia-Pacific Forum: A
Network for Librarians and Information Specialists (poster abstract)
    Jeanette Regan et al 363
ALA: Astronomy in Latin America (poster abstract)
    Philippe Eenens 363
The Working Group on Space Sciences in Africa (poster abstract)
    Peter Martinez et al 364
Pollution-Free Road Lighting (poster abstract)
    Duco A. Schreuder 364
Zimbabwe –The Place for Astronomy at the next Total Solar Eclipse
(poster abstract)
    Francis Podmore 364

Panel Discussion 369

Index of Countries 373
Subject Index 375

Only the first authors of multi-authored poster abstracts are given here. Names of co-authors will be found in the text.

Preface

The Special Session of the XXIV IAU General Assembly in Manchester originated as a proposal for a symposium devoted to the topic "Astronomy for Developing Countries" that was submitted by the Working Group for the Worldwide Development of Astronomy with the support and cooperation of Commissions 5, 38 and 46. The Executive Committee decided that the topic could be more suitably dealt with within a Special Session, the first of its kind, for which they allocated time during the General Assembly and made a financial commitment to provide travel support for the participants.

One of the stated aims of the original proposal was that the Proceedings, when published, would provide a manual for those who wish to begin (or to revive) astronomical studies in a developing country and for those in developed countries who may wish to help them. How far we have been successful in our aim is for readers of this volume to decide. Invited speakers were asked to keep the aim in mind when preparing their papers: some of them obviously have done so; others, equally obviously, have been motivated by other concerns. What we have done, however, is to bring together a wide variety of accounts from many different countries and a number of ideas about how we can strengthen astronomy research and education in those countries that want to create an astronomical tradition, either for its own sake or as a means of creating a broader scientific tradition. We have not been afraid even to include contributions by those who question whether a developing country should be trying to develop an astronomical tradition at all. I believe that this book does contain a representative picture of the worldwide state of astronomy in the year 2000.

One of the themes that emerges from the book, which is confirmed by my own experiences in many of the countries discussed, is the variety among developing countries. Perhaps, in one sense, the kind of manual we envisaged at the beginning cannot be produced because of these differences. There are small countries that have either never had an indigenous scientific tradition, or the one they had has lain dormant for at least a generation as the country concerns adjusts to the recovery of independence after a century or more of colonial rule. There are countries which, judged in terms of per capita wealth are poor but, because their populations are large, can afford a considerable investment in astronomical research. There are countries which had until recently a strong scientific tradition but, as a result of political and economic changes, have been thrown back into a status something like that of newly developing countries. Finally, there is the unique case of South Africa, with a strong but largely immigrant astronomical community, which has recently seen a profound, but fortunately peaceful, political change that has brought to power a government convinced of the value of continuing and broadening that astronomical community. It should hardly surprise us that many different approaches are needed in these different kinds of country and, perhaps, one of the important fruits of our Special Session has been to bring these differences out into the open.

The last paragraph makes clear that it is impossible to discuss the theme of astronomy in developing countries without reference to the political, economic, cultural and even religious differences between them. Neither is it possible to expunge all reference to such matters from the printed record of the Session. Where such matters are discussed, the opinions expressed should be regarded as the personal views of the respective authors. The Editor of this volume, himself, does not agree with all those views (some of which, indeed, are mutually contradictory); still less should they be taken as in any way being the expression of an official position of the IAU, which, of course, maintains a strict neutrality on topics that are not within the professional competence of astronomers, however familiar some pf its members may be with particular situations.

The arrangement of papers in this volume is similar to, but not identical with, the order in which they were presented during the 2.5 days of the Special Session. As anyone who has organized a meeting knows, the exigencies of the event inevitably mean that the actual order of presentation is never quite that envisaged beforehand. When the meeting is one of several parallel sessions to which many of the speakers also have commitments, it is even harder to keep to the intended schedule. I have, sometimes, restored the intended order, but I have also found, as I assembled the manuscripts, that some papers were more appropriate in other than the original positions. In particular, it became clear as the Session progressed, that there was considerable difference of opinion between the various participants as to the relative merits of astronomers in a developing country doing research with their own small telescopes or using data obtained from the Internet. We did not resolve this difference, possibly because, again, there are so many differences between countries that one solution is not necessarily right for all. We did not foresee this controversy and, perhaps, just the recognition of its existence is another important result from our Session. I have found it convenient to put all the papers that deal with this matter into one section of the book "Small Telescopes or Internet Access?" which does not correspond to a section of the original programme. In addition to these changes, I have endeavoured to interweave the poster papers that were presented in conjunction with the Session at the most appropriate places between the various invited, orally presented papers. A few selected poster papers are presented in full. Most are represented only by their abstracts. A few authors of posters neglected to sign the required copyright-assignment forms in Manchester and have not responded to subsequent appeals to do so. Their posters appear by title only. A final departure is that two papers are included here that were not presented in the Special Session. That by Abt (*Electronic Access to Journals*) was prepared by him at a time that he hoped to be present, but he was not actually able to come; the paper by Percy (*Simple Science, Quality Science*) was presented to a smaller group in Manchester two days before the Session began. It is so germane to our topic, however, that I invited Percy to submit a written version for inclusion in these Proceedings.

Participants in the General Assembly were free to come to and go from our Session as they chose, so it is difficult to give a list of those attending the Session. I have listed on the following pages, as contributors to the Proceedings, the names of those who gave invited papers and of those authors of poster papers who were subsequently invited to prepare a written version for this volume. At least one of the authors of each invited paper was, of course, present, usually most of the time. Similarly, at least one author of each poster paper was also usually present and the records of the discussions will show that others helped to form a core of regular attenders. The Index of Countries shows that about half the member states of the United Nations were considered, or at least alluded to, during the Session and a good fraction of those countries, and possibly some others, were represented among the participants. In particular, we had representation from seven African nations at the Session – probably a record for the IAU and a tribute to the activity of the UN sponsored Working Group for Space Sciences in Africa.

It remains to thank all those who assisted in the realization of this Session. First, those who helped to formulate the original proposal. Next the IAU Executive Committee for approving the proposal in a modified form and giving it the necessary financial support. I am grateful also to all those who served on the Scientific Organizing Committee and particularly those of them who acted as Chairpersons. The past and present General Secretaries, Johannes Andersen and Hans Rickman, gave help at many stages, not least in the disbursement of grants for attendance at the General Assembly which helped several who wanted to participate to be in Manchester. I also acknowledge the contribution made by the American Astronomical Society toward the cost of publishing the Proceedings. Finally, I thank Harold McNamara and his staff in the office of the ASP Conference Publications for their assistance in the production of this permanent record of the Special Session, and my long-time friend and colleague Murray Fletcher, without whose patient help and guidance I would never have mastered the intricacies of LaTeX

Alan H. Batten.

The eclipse maps on pp. 104-106 were prepared by F.G. Espenak of NASA.

Scientific Organizing Committee

Alan H. Batten (Chairman, Canada)
Olga Dluzhnevskaya (Russia)
Julieta Fierro (México)
John B. Hearnshaw (New Zealand)
Rajesh K. Kochhar (India)
Yoshihide Kozai (Japan)
Li QiBin (China)
Peter Martinez (South Africa)
Derek McNally (United Kingdom)
Lesley I. Onuora (United Kingdom/Nigeria)
Maria C. Pineda de Carias (Honduras)
Morton S. Roberts (United States of America)

Major Contributors

Toufik Abdelatif, Observatoire d'Alger-CRAAG, BP 63, Bouzaréah 1630, Alger, Algeria.

Helmut A. Abt, Kitt Peak National Observatory, Box 26732, Tucson, Arizona 85726-6732, U.S.A.

M.N. Anandaram, Department of Physics, Bangalore University, Bangalore 560 056, India.

Pulat B. Babadzhanov, Institute of Astrophysics, Dushanbe 734042, Tajikistan.

Alan H. Batten, Dominion Astrophysical Observatory, 5071, W. Saanich Rd, Victoria, B.C., Canada, V9E 2E7.

S.P. Bhatnagar, Department of Physics, Bhavnagar University, Bhavnagar, India.

Cynthia P. Celebre, PAGASA (Weather Bureau), 1424 ATB, Quezon Ave, 1104 Quezon City, Philippines.

Khalil Chamcham, King Hassan II University–Ain Chock, Faculty of Science, B.P. 5366, Maarif, Casablanca, Morocco.

Christopher J. Corbally, Vatican Observatory, University of Arizona, Tucson, Arizona 85721, U.S.A.

David L. Crawford, GNAT, Inc., Tucson, Arizona 85716, U.S.A.

Philippe Eenens, Department of Astronomy, University of Guanajuato, Apartado Postal 144, Guanajuato, CP 36000, México.

Cheng Fang, Department of Astronomy, Nanjing University, Nanjing, China.

Julieta Fierro, Instituto de Astronomía, UNAM, Apt 70 264, México DF 04510, México.

John B. Hearnshaw, Department of Physics and Astronomy, University of Canterbury, Christchurch, New Zealand.

George Helou, IPAC, Mail Code 100-22, California Institute of Technology, 1200 E. California Blvd, Pasadena, California 91125, U.S.A.

Mary Kay Hemenway, Department of Astronomy, University of Texas, Austin, Texas 78712-1083, U.S.A.

Peter D. Hingley, Royal Astronomical Society, Burlington House, Piccadilly, London W1V ONL, U.K.

Abendour Irbah, Observatoire d'Alger-CRAAG, B.P. 63, Bouzaréah 1630, Alger, Algeria.
Barrie W. Jones, The Open University, Milton Keynes MK7 6AA, U.K.
B.A. Kagali, Department of Physics, Bangalore University, Bangalore 560 056, India.
Norio Kaifu, National Astronomical University of Japan, Mitaka, Osawa 2-21-1, Tokyo, 181-8588, Japan.
Peter C. Kalebwe, Physics Department, University of Zambia, P.O. Box 32379, Lusaka, Zambia.
Masatoshi Kitamura, National Astronomical Observatory, Mitaka, Tokyo 181-8588, Japan.
D.A. Kovaleva, Institute of Astronomy, 48 Pyatnitskaya St., Moscow 109017, Russia.
Ethleen Lastovica, South African Astronomical Observatory, P.O. Box 9, Observatory, 7935, South Africa.
Koitiro Maeda, Department of Physics, Hyogo College of Medicine Nishinomiya, Hyogo 663-8501, Japan.
Oleg Yu. Malkov, Institute of Astronomy, 48 Pyatnitskaya St., Moscow 109017, Russia.
Terry J. Mahoney, Scientific Editorial Service, Research Division, Instituto de Astrofísica de Canarias, E-38205 La Laguna, Tenerife, Spain.
Peter Martinez, South African Astronomical Observatory, P.O. Box 9, Observatory 7935, South Africa.
Janet A. Mattei, AAVSO, 25 Birch Street, Cambridge, Massachusets 02138-1205, U.S.A.
Derek McNally, Department of Physical Sciences, University of Hertfordshire, College Lane, Hatfield, Herts., AL10 9AB, U.K.
Geoffrey Munyeme, Department of Physics, University of Zambia, P.O. Box 32379, Lusaka, Zambia.
Jayant V. Narlikar, Inter-University Centre for Astronomy and Astrophysics, Pune 411007, India.
Anas M.I. Osman, National Research Institute of Astronomy and Geophysics, Helwan, Egypt.
Lesley I. Onuora, Astronomy Centre, University of Sussex, Falmer, Brighton BN1 9QJ, U.K.
Jay M. Pasachoff, Hopkins Observatory, Williams College, Williamstown, Massachusets 01267, U.S.A.
John R. Percy, Erindale Campus, University of Toronto, Mississauga, Ontario, Canada L5L 1C6.
Maria C. Pineda de Carias, Central American Observatory of Suyapa, National Autonomus University of Honduras, Tegucigalpa MDC, Honduras.
Sandra Preston, Department of Astronomy, University of Texas, Austin, Texas 78712-1083, U.S.A.
Moedji Raharto, Bosscha Observatory, Lembang 40391, Java, Indonesia.
Kavan U. Ratnatunga, Department of Physics, Carnegie Mellon University, Pittsburgh, Pennsylvania 15213, U.S.A.
Case L. Rijsdijk, South African Astronomical Observatory, P.O. Box 9, Observatory, 7935, South Africa.

Nguyen Quang Rieu, Observatoire de Paris, Department DENIRM, 61 Avenue de l'Observatoire, 75014 Paris, France.

Rosa M. Ros, Department of Applied Mathematics IV, Technological University of Catalonia, Jordi Girona 1-3, Modul C3, 08034 Barcelona, Spain.

Patricia Rosenzweig, Universidad de Los Andes, Facultad de Ciencias, Departamento de Física, Grupo de Astrofísica Teórica (GAT), Mérida, Venezuela.

Hamid Sadsaoud, Observatoire d'Alger-CRAAG, B.P. 63, Bouzaréah 1630, Alger, Algeria.

Michael S. Snowden, P.O. Box 44, MacPherson Rd, Sinagapore 913402.

Boonraksar Soonthornthum, Sirindhorn Observatory, Department of Physics, Faculty of Science, Chiang Mai University, Chiang Mai 50200, Thailand.

Bernardo M. Soriano Jr, PAGASA (Weather Bureau), 1424 ATB, Quezon Ave, 1104 Quezon City, Philippines.

Yuhua Tang, Department of Astronomy, Nanjing University, Nanjing, China.

Noritaka Tokimasa, Nishi-harima Astronomical Observatory Sayocho, Hyogo 679-5313, Japan.

Alexis E. Troche-Boggino, Observatorio Astronomico "Alexis E. Troche-Boggino", Universidad Nacional de Asunción, Facultad Politécnica, 01 Agencia Postal Campus U.N.A., Central XI, Paraguay.

Alexander V. Tutukov, Institute of Astronomy, 48 Pyatnitskaya St., Moscow 109017, Russia.

Jingxiu Wang, National Astronomical Observatories, Chinese Academy of Sciences, Beijing 100012, China.

Donat G. Wentzel, University of Maryland, College Park, Maryland, 20742-2421, U.S.A.

Section 1: Survey Papers

Astronomy for Developing Countries

Alan H. Batten

National Research Council of Canada, Dominion Astrophysical Observatory, 5071, W. Saanich Rd, Victoria, B.C., Canada, V9E 2E7. e-mail: alan.batten@nrc.ca

Abstract. Developing countries have many claims on their limited resources and astronomy can expect only a small share of a small "pie". A useful rule of thumb is that a country's expenditure on astronomy is likely to be of the same order of magnitude as its per capita Gross National Product multiplied by the number of professional astronomers in the country. In the light of this, we consider how governments of developing countries can help their astronomers, how we can help them and how they can help themselves.

1. Introduction

When I tell friends and colleagues that I am working with astronomers in developing countries, I am frequently asked why such countries should be concerned with astronomy; do they not have many more urgent needs, for example, in the field of public health? I am sure that this is an experience shared by most of those who do similar work and that our colleagues actually working in developing countries are often posed the same question by the more tough-minded officials of their own governments with whom they have to deal. The question is a natural one and most of us must have pondered it ourselves and evolved our own answers to it. Some people point to the industrial spin-off of astronomy. For example, our colleagues in India, building large radio- telescopes, have greatly stimulated the steel and electronics industries of their country, with consequences far removed from their own immediate concerns. Don Wentzel likes to emphasize the versatility possessed by the holders of a modern Ph.D. in astronomy, so that they can make many different kinds of scientific contribution in their home countries. A former President of the IAU, Hanbury Brown, has often stressed the unity of science: you cannot foresee how developments in one area will affect those in quite different areas. For example, tomography is used both in modern medicine and modern astronomy. I am not sure in which field the process was first used but, undoubtedly, developments of it by astronomers have been of use to physicians who have also applied it for their own needs. My own response is to emphasize that most astronomers are also educators and, in that capacity, contribute to a very important aspect of the further development of any country. Most of all, however, the reasons for studying astronomy in a developing country are the same as those for studying astronomy at all. Astronomy is not one of the most obviously practical of human activities although,

3

historically, it led us to a greater understanding of the laws of terrestrial mechanics and Newton was, in an important sense, a precursor of the industrial revolution. Even more importantly, however, most of us feel that we are, in some sense, more fully human if we learn about the universe in which we live; this was brought out very clearly at another session of this Assembly by Julieta Fierro, drawing on her experiences of teaching elementary astronomy to the street children of México City. There is no reason why only those already privileged to live in wealthy countries should have the additional privilege of feeling more fully human.

2.   Some Facts of Life

If that sounds a somewhat idealistic note, let us turn to be quite realistic. We naturally ask "What can people living in developing countries hope to contribute to astronomy and how can the rest of us help them?" Perhaps a better question would be "What proportion of its resources should any country devote to the study of astronomy?". Let us begin by reminding ourselves how few astronomers there are. A recent ICSU document contains the estimate that there are somewhere between three million and ten million scientists in the world (Arber, 1999). Those figures exclude engineers, social scientists and people working in clinical medicine. If we take six million as a useful order-of- magnitude number, something like one person in every thousand is a natural scientist of some kind although, in developed countries, the ratio can be as high as one in a hundred. We all know that astronomers are a small fraction of the total number of scientists. The membership of the IAU is of the order of ten thousand. Even allowing for the fact that there are many astronomers not in the IAU, less than 1 per cent of all scientists, or less than 10^{-5} of the total world population, are professional astronomers. Yet, increasingly, we are asking for large and expensive instruments with which to pursue our research; instruments that cost, in total, billions of dollars. Even though the world's astronomers are concentrated in the wealthiest countries, few nations are now prepared to pay all the cost of these instruments from public funds and, unless private funds are found, only international consortia are building the world's largest instruments. Canada is about average in numbers of astronomers, in that about one person in every 10^5 is a professional astronomer, yet the instruments we have built, or our share in international instruments, have typically cost about 10^{-4} of our Gross National Product (GNP). Although I have deliberately focussed attention on professional astronomers, I recognize that in many countries – Canada included – there are amateurs who are making real contributions to the science, especially through such organizations as the AAVSO, who certainly support the devoting of a considerable share of national wealth to the study of astronomy and who can also play a role in helping astronomers in developing countries. In some of those countries, however, social and economic conditions prevent the emergence of a group with leisure enough to become amateur astronomers and it is difficult to make fair comparisons between developed and developing countries that take into account the contributions that many amateurs make.

   To return to budgetary matters, the capital cost of instruments is nearly always spread over several years; a closer examination of the astronomy budgets of many countries shows that the per capita GNP, multiplied by the number of professional astronomers in a country is a rough indication of the annual budget for public funding over a period of years. By "rough", I mean that the annual budget will not be more than a few times the product of the other two quantities. This is illustrated for a number of countries, including several in the G8, in a recent Canadian report (Pudritz, 1999). (That report more correctly uses the Gross Domestic Product – GDP). I have chosen to use GNP because the source available to me provides the per capita figure for that quantity for every country; the principle of the argument is not greatly affected by this choice.) Even so, several provisos are needed. In some countries it is difficult to separate astronomy budgets from space-science budgets; in others, it is unclear whether university astronomy is included in public funding of astronomy or in the general university budget. In all countries, of course, it is often uncertain just how many astronomers there are, since there are various ways of defining who is an "astronomer" and, as already mentioned, the role of amateurs differs widely from one country to another. Provided we are content with order-of-magnitude estimates, however, we will find that the product of per capita GNP and the number of astronomers will give some idea of what we might expect the annual expenditure on astronomy to be; rarely will the annual expenditure exceed ten times the product of the other two quantities. Whether or not that product really tells us how much a country should spend on astronomy, it at least provides a yardstick for estimating any given country's probable actual public funding for the field.

We immediately notice the shocking disparity in per capita GNP between the world's richest and the world's poorest countries; it is worse than appears from Figure 1, which condenses the wealthiest countries into one bin; the disparity is a factor of several hundred. Our colleagues in the developing countries suffer from two disadvantages: their per capita GNP is lower and so is the number of astronomers, if not absolutely then in relation to the total population. Let us take an imaginary country with a per capita GNP of U.S. $200 and employing five astronomers. I stress that this *is* an imaginary country, but it is not a fantastic one; some of the people who attended the Special Session in Manchester came from countries in which the situation is not very different. The formula given above implies that the annual budget for astronomy (including the salaries of the five) is going to be a few thousand U.S. dollars; if they are very lucky, perhaps as much as $10,000. Let us stop to think of the implications. Equipping each astronomer with a computer could absorb more than a whole year's budget! The smallest telescope with which useful research can be done (a 40-cm reflector) will likewise absorb more than one year's budget –even before any auxiliary instrumentation is bought or suitable housing for the telescope is constructed. Journal subscriptions are going to make a big hole in the budget, unless the journals themselves provide relief – as many do (Abt 1998). Only a few books can be bought and personal travel by the astronomers, whether to go to meetings or to observe, is almost out of the question. This should give those of us in wealthy countries some food for thought the next time we are tempted to complain about our budgetary restrictions! It is no use expecting people working in developing countries to pay for some of these things out of their salaries because, in fact, per capita GNP is quite a good indication of average family earnings. That does not mean that our colleagues live their daily lives in poverty, since basic needs

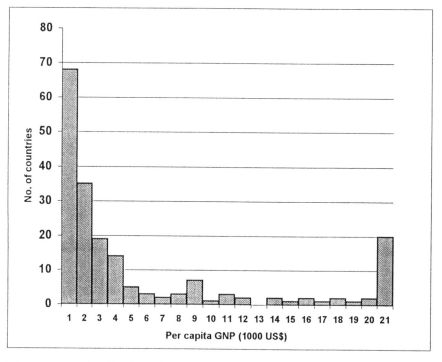

Figure 1.   Distribution of 193 countries by per capita Gross National Product in recent years. Abscissae are the per capita GNP (not necessarily for the same year) expressed in thousands of U.S. dollars; ordinates are the numbers of countries in each interval. For convenience, all countries with per capita GNP in excess of U.S. $20,000 are condensed in one bin. Source: *Encyclopaedia Britannica Book of the Year for 1999*.

provided within the country are cheap by the standards of developed countries, but it does mean that travel, journals, books, computers and instruments – all of which have to be paid for at the international "going rate" – are as out of reach for them personally as they are for their institutes. Possibly a country for which the proposed formula predicts an amount so small that no useful work in astronomy can be done at all should not encourage the study of that subject; in fact, few do, but the decision must remain one for the country itself to make, in the light of its circumstances and its own perception of its needs.

Compare the above with the situation of my colleagues in Canada. We have 300 to 400 active professional astronomers and a per capita GNP of around (U.S.) $20,000. We expect, then, that public funding in Canada is going to be of the order of 10^7 annually and that is indeed the case, when capital expenditures are included. Put another way, Canada is providing each of its astronomers with some tens of thousands of dollars annually (over and above their salaries). We are by no means the world's wealthiest country, and we have a reputation of being among the low spenders on scientific research (relative to GNP), in comparison with the other G8 countries. There are certainly countries whose astronomers are even better off.

3. What can Governments do?

In drawing attention to the disparities between countries, I am in no way trying to criticize the governments of developing countries or, indeed, of any country; rather, I am trying to convince those of us from the developed world how fortunate we are; something we can very easily forget, unless we make a conscious effort to remind ourselves. To avoid the appearance of criticizing particular regimes, I will, as far as possible avoid naming individual developing countries in this paper, but I remember one colleague, of about my own age, telling me that to work in the West, as he had briefly done, seemed to him like being in paradise! The situation has changed for the better, in many respects, in his country now, but those of us in wealthy countries are still immeasurably better off. I have no doubt that, if I were a government official in a developing country, I would have to draw the attention of would-be astronomers to the "facts of life" discussed in the previous section. Hard choices have to be made in the allocation of funds and we all must be realistic enough to accept that. On the other hand, there are non-financial constraints that governments could do something about if they wished. "Red tape" is often much more obvious in developing countries. For instance, on leaving one such country, I was required to show my passport four or five separate times at the airport. For a short-term visitor, such repeated checks are no more than a minor irritant and can, perhaps, even be amusing; those who are up against similar attitudes every day in their work must find them frustrating and, sometimes, infuriating. I have, for example, heard accounts of how access to university offices is restricted, or even forbidden, "out of hours". No doubt, government officials could offer convincing reasons for this kind of regulation. There are fears for the security of buildings and equipment (Onuora 2001, Snowden 2001). I know of at least one country in which the observatory is provided with an armed guard, twenty-four hours a day. There are also needs to create employment and, unfortunately, to forestall corruption by elaborate checks and counter-checks. The kind of bureaucratic attitude that these checks inevitably engender, however, is definitely inimical to an atmosphere in which good scientific work can flourish. Even worse, authoritarian governments will sometimes put their interests in controlling their populations above the free exchange of scientific information. I have been in at least one country in which I was told that provision for e-mail was in the astronomy budget but the astronomers could not have it, for political reasons. The regime in that country has changed since my visit, and the astronomers have e-mail now, but it is quite understandable that some governments should fear a medium of worldwide communication that they cannot censor. When such a fear results in a government deliberately restricting its own scientists from free communication with colleagues in other countries, however, those scientists are automatically condemned to second-rate status. Those of us who are more fortunate should take every opportunity to stress that governments that really want to encourage home-grown science can go a long way to achieving their end without much financial cost, if they will consciously try to create the conditions in which science flourishes, and, in particular, if they will relax restrictions on communication.

4.   What can Astronomers in Developed Countries do?

It is clear from all that we have considered so far that, if there is to be any astronomy in developing countries, there must be help from outside; financial help, of course, but, just as, if not even more important, the help of fellow astronomers who care about their isolated and less fortunate colleagues. This, I think, is clearly recognized in many quarters and the IAU has actually quite a good record of helping, considering its own limited financial resources. Those of you who took part in Joint Discussion 20 in Kyoto three years ago will have some idea of what the IAU has done, and is doing, along these lines; those of you who were not there can read about it in the account of that meeting (Batten 1998). We are not the only organization; the UN/ESA international workshops (also supported by The Planetary Society) have been very important and there are many who can testify to the help that they have received from the Vatican Observatory Summer Schools which, along with other similar initiatives, are described in more detail in this volume (Corbally 2001). It is particularly appropriate, in the record of a meeting in the University of Manchester, to mention that Zdeněk Kopal, for three decades Professor of Astronomy at that University, enabled many people from developing countries, particularly those in the Middle East, to go there to pursue studies up to the Ph.D. level and that most of those are now working in their home countries. He would, I believe, have been delighted that this meeting should have taken place there and that one of his earliest Manchester students should have organized it. I am sure that there are several other instances of particular individuals or institutes helping in such ways, but the majority of our colleagues in the wealthy countries do not seem greatly concerned about how their less fortunate colleagues might be integrated into the world community of astronomers and have not fully informed themselves about the conditions under which those colleagues work. One of the most important things we can do, and that I hope the meeting and its published proceedings will do, is to raise consciousness in the rest of the community about the needs of astronomers who work under much less ideal conditions than those of us in the most favoured countries enjoy.

   I have emphasized the lack of computers, instruments, books and journals in the preceding sections but, of course, there are people working to alleviate these conditions. For example, the Japanese Government's programme of donating telescopes of about 40-cm diameter to selected countries has already made a difference in several, as will also Querci's Project NORT programme as it develops (Querci 1995). There is a need, however, as is also brought out elsewhere in this volume (Snowden 2001, Wentzel 2001), for continuing cooperation with the international community to ensure that these generous gifts are used to their maximum capacity. Similarly, I know that the UN/ESA International Workshops have often been the means of introducing a number of computers into countries that otherwise would not have had them so soon. There are likewise several programmes for transferring books and journals to poorer countries (Hingley 2001). One problem about these programmes is that they usually rely on gifts of personal libraries made by retiring astronomers at the ends of their careers. One hesitates to criticize, even mildly, such generosity, but an inevitable consequence is that the books and journals sent are out of date. The crying need is for modern material; colleagues might well consider sending their personal issues of a journal to some institute in the developing world as soon as they have finished with them – or at least at the end of a calendar year.

Ultimately, however, the real problem faced by astronomers in developing countries is that of isolation – not only literally or geographically, but intellectually as well – and that is where any one of us can help. In a few countries, those which happen to have good potential observing sites, that isolation may be quite quickly broken down. Not many years ago, ESO had to renegotiate with Chile the contract under which it operates in that country (the anonymity of which cannot be preserved in this example), and the new contract had some very favourable terms for the local astronomical community written in (Mervis and Abbott 1993, Abbott 1993). However delicate and difficult those negotiations may have been at the time, that they were needed at all was a clear indication that ESO, and the other international observatories in Chile, had played a considerable role in reinvigorating the Chilean astronomical community and, indeed, those of other Latin-American countries. Similar developments could occur elsewhere. I have seen for myself the potential of the Atlas Mountains in North Africa and there may be similar opportunities in Central Asia, in countries that were formerly part of the Soviet Union (Babadzhanov 2001). That kind of development, however, is open only to a few naturally favoured countries. Many, if not most, developing countries do not have observing sites with the potential to attract the international consortia that build large telescopes.

Another route to minimize the effects of isolation is to encourage regional cooperation as, indeed, the IAU has done with its regional meetings. Of the two continuing IAU regions, however, one, Latin America, is reasonably compact, whereas the other, the Asian-Pacific region is rather large and amorphous. One can envisage several smaller regions in which astronomically advanced countries like Australia, New Zealand, Japan, China, India, Indonesia and Korea can act as centres for smaller nearby countries. This sort of thing is happening on a limited scale and should be encouraged (Kaifu 2001). Africa is a special case, but the establishment a few years ago of the UN Working Group on Space Sciences in Africa is an encouraging development. The complete political change in South Africa has removed barriers that once existed to cooperation between the internationally respected astronomical community in South Africa and communities in other African countries, but the great size of the continent and the difference in character between the countries on the Mediterranean coast and those south of the Sahara suggests that North-African countries should be more usefully linked across the Mediterranean, as is, in fact, beginning to happen. Southern Africa, on the other hand, is one of the regions where very favourable observing conditions are to be found and the planned construction of the Southern African Large Telescope (SALT) holds great promise for increasing regional cooperation on that continent, as is also described in this volume (Martinez 2001, Rijsdijk 2001, Hemenway and Preston 2001).

Grand schemes of regional cooperation have their place but, eventually everything depends on individual contacts, or at least on contacts between individual institutes. The great aim should be to enable the young astronomers in developing countries who have received a full modern education in astronomy to keep their contacts with the developed world and, above all, to take part in modern research, whether theoretical or observational. We all know that some modern research can, and should, be done with small instruments, but the young astronomer who has had a taste of working on a 4-m or larger giant telescope to obtain material for a Ph.D. thesis is likely to want to repeat the experience and to continue to work in fields in which such large instruments are necessary. Even among the astronomers in the developed world, it is now the exception for an individual to be awarded significant amounts of observing time on such instruments; rather, groups of half-a-dozen or more pool their expertise and compete for time with other similar-sized groups. We need a way for astronomers in developing countries to find acceptance in such groups and to take part in the reduction and analysis, if not in the actual collection, of the data. This means that astronomers who have studied in a developed country and returned home need to keep in touch with the respective institutes at which they studied (or perhaps held a post-doctoral fellowship). It means also that those former students must be able sometimes to return to those institutes for a period of months, to be refreshed by contacts with their former associates. Almost certainly, such return visits will be at the expense of the institute concerned. Are the more fortunate of us prepared to give up a few trips a year to make such return visits possible? In the end, we must give up something; I have tried to show how much more fortunate we are, and if we have to give up one meeting a year to bring one colleague over to us for a summer, the sacrifice is small but well worthwhile.

Many colleagues in the developed countries are convinced that the best way to help is by sharing the electronic data bases that we now have. Astronomers in developing countries, it is argued, can have access to these bases and work with modern data on modern problems. These possibilities are also elaborated in this volume (Helou 2001, Malkov, Tutukov and Kovaleva 2001, Ratnatunga 2001) and we should certainly be aware of them, but there are still practical difficulties. Having an e-mail address, or a fax number, do not in themselves guarantee access to the desired communication. Seven years ago I was in a country with a very inefficient telephone service. Calling outside the country was almost impossible, and, even to place a call across the principal city, one had to be prepared to try several times. While I was at a meeting there, a visiting representative of a European telecommunications company assured us that soon all would be well. A new kind of switch was to be imported very soon and everyone would enjoy complete access to the Web. Before the meeting in Manchester, I heard that the astronomers of that country were just getting e-mail but, as I learned in Manchester, access to it is at their own expense! More generally, e-mail connections seem to be less reliable in developing countries and fax machines often fail. I would not like to count the number of times I was frustrated in the organization of this meeting in attempts to reach would-be participants. Fax machines appear often to be switched off at night, and longitude differences then ensure that the machines are not accessible from North America at any reasonable hour. For a period of several weeks, colleagues in one country could not be reached because a lightning storm had knocked out the computer system on which they depended for e-mail. This could happen in any country, of course, especially in the tropics, but the damage is likely to be much more quickly repaired in a developed country – a point which is made very strongly in this volume (Onuora 2001). Even if all these problems are overcome, it is still necessary for the individual astronomers to have access to and to be able to use the sophisticated data-processing programs that are required to handle modern data properly – yet many colleagues presenting papers at this meeting told me that they could not make use of the LaTeX template provided for the preparation of their abstracts. Here again, there is need for a continuing relationship between individuals in developing countries and specific institutes in the developed countries; or at least for some cooperative organization in developing countries such as is envisaged in Narlikar's scheme for a Third-World Astronomy Network (Narlikar 2001).

5. What can Astronomers in Developing Countries do for Themselves?

It may seem surprising to suggest that astronomers placed in the sort of circumstances I have described can also help themselves, but I believe it to be true. One very simple thing that all astronomers in developing countries could do to help themselves, for example, is to be more punctilious about responding to the messages that do reach them. As mentioned above, I was often frustrated in the preparation of this meeting by unanswered queries and I do not believe that all these instances are to be blamed on the difficulties under which colleagues in developing countries often work and to which I have referred. Modern means of communication make possible effectively instantaneous communication around the world. On the west coast of Canada, for example, I can sometimes receive a reply from New Zealand within minutes of sending a question, even though (or perhaps because!) the New Zealanders' calendar date is day ahead of mine. Replies from most of Asia, of course, are delayed because their night-time hours coincide with my daylight ones, and none of us checks our e-mail every day. Even so, I often find a reply from Japan awaiting me on the morning after I send a query. This efficiency in communication becomes possible, however, only if the human being who uses the computer is prompt in replying.

There are other, perhaps more important, ways in which astronomers in developing countries can help themselves. As I have already implied, in many countries a necessary first step in the development of astronomy is sending promising students abroad for advanced studies. For various reasons, mostly either directly or indirectly the legacy of former colonialism, people of my generation in newly independent countries have often been cut off from modern astronomical research for much of their lives. They are unable, and in my experience freely admit that they are unable, to educate the new generation beyond imparting the very basic elements of the subject to them. I have also observed that the new generation is frequently impatient with their elders – which is an understandable reaction, not confined to developing countries, but particularly unfair in the context I have just described. The older generation, after all, did keep the torch burning, sometimes at great personal cost; they deserve a great deal of honour for doing what they could do in the situations in which they found themselves. Other papers in this volume discuss the potential of distance-learning for alleviating this situation (Jones 2001) but, sooner or later, graduate students need the personal contact with senior workers and fellow-students that can, as yet, come only through physical presence in a major institute. Some nations are understandably anxious about sending their most promising students abroad; will they return? This is a real problem; I received my graduate education in Manchester and, one year later, left for what I thought would be one year in Canada where, forty-one years later, I still live; so I know very well what the temptations are! The problem is perhaps not as great now as it was in my day, since all countries are having difficulty finding positions for their own science graduates, and the wealthy countries are not so ready now as they once were to "steal" graduates from poorer countries to make up their own shortfalls. The way for developing countries to encourage the return of graduates is not so much to attach conditions to their departure (which can often be got around by those determined to emigrate permanently) as to create conditions in which they will be able to use their newly obtained skills back in their own country. Most young people are both idealistic and patriotic and will work for, and in, their native land if they can see some chance of building up a scientific tradition there. While the provision of modern equipment may genuinely be beyond the present resources of many developing countries, as I have already said, much can be done to create working conditions congenial to scientists, at very little financial cost.

What has all this to do with how the scientific community in a developing country can help itself, without waiting for the politicians and bureaucrats to act? Just as the younger generation, in countries that have gone through decades of unrest or even outright civil war, should realize that their elders have often made great sacrifices to preserve any scientific tradition at all, those same elders must be ready to give their former students great freedom in choosing the fields in which they work and in their manner of doing that work. In the nineteenth century, an observatory director, in any part of the world, was "boss" – and could be a petty tyrant. All other scientists on his staff (and they *were* all male then!) were his assistants and his name usually appeared as first author of any paper that was published. When observatories were founded in countries that were then ruled as colonies, the same attitude was brought to them. In modern Europe and North America, this system is no longer in operation; it does not work in the modern world. You cannot give students all the education required to complete a modern Ph.D. and then expect them to act as assistants. They have been taught to think and to act independently. So, if a you, a professor in a developing country, have sent your most promising student to another country to get a modern astronomical education, then you must not expect that student to return simply to act as your assistant. If you try, the student will probably leave again at the earliest opportunity; and that is precisely the result that none of us wants.

It is also important for the few astronomers there are in any developing country to work together. I have seen instances of astronomy beginning (or at least reviving) in a country in which a small number of astronomers have coalesced around two or three institutes which then begin to compete for the very little funding there is. This can be fatal in any country. Even in Canada, a generation ago, astronomers had a fierce quarrel about the location of a major new instrument and the government of the day very quickly withdrew the promised funding, because we could not agree about what we wanted. If this can happen in a country with a tradition of astronomical research and, as I have shown, relatively generous funding of it, how much more easily can it happen in a country without such a tradition and with many competing claims on the public purse! In one of the countries that has been most successful in creating an indigenous tradition of astronomy after returning to independence, it has been a deliberate policy to build up only one centre, until that became well-established, before encouraging the growth of the subject in other centres. Thus there was no rivalry between institutes at the time at which the progress of astronomy would have been most vulnerable to disagreements among the astronomers.

The situation of astronomers in developing countries is likely to remain a difficult one for some years, possibly some decades, to come. Unfortunately, the poorer the country, the longer it is likely to be before any astronomers in it can hope for improvement. I have tried to make suggestions, however, of things that can be done to mitigate that plight – often at very little or no financial cost. Mitigation, however, does require of at least some of us, both in developing and developed countries, a change of attitude. Those of us who are fortunate enough to live in countries in which we can achieve a great deal of what we hope for need to give up the competitiveness which has, unfortunately, crept into astronomy in the last several decades and be prepared to achieve a little less if, by doing so, we can give a helping hand. Those of you who are struggling all the time with the problems of no money and intellectual isolation must also be prepared to cultivate attitudes of mind that, maybe, do not come naturally in the culture in which you have grown up. Governments of all kinds must come to accept that their job is not only to control their citizenry but also to serve it. If we all try to work together, there is still hope that astronomy will not only survive in developing countries, but also flourish.

References

Abbott, A. 1993, Nature, 364, 178.

Abt, H.A. 1998, in *Highlights of Astronomy*, 11B, ed. J. Andersen, pp. 929-30, Kluwer Publishing Company, Dordrecht, The Netherlands, p. 929.

Arber, W. 1999, Science International, Special Issue front cover and p. 1.

Babadzhanov, P.B. 2001, this volume pp. 187-196.

Batten, A.H. (ed.) 1998, *Enhancing Astronomy Research and Education in Developing Countries*, (J.D. 20) in *Highlights of Astronomy*, 11B, ed J. Andersen, Kluwer Publishing Co. Dordrecht, The Netherlands, pp.883-934.

Corbally, C.J. 2001, this volume, pp. 110-116.

Helou, G. 2001, this volume, pp. 276-278.

Hemenway, M.K. and Preston, S. 2001, this volume pp. 152-156.

Hingley, P.D. 2001, this volume, pp. 340-344.

Jones, B.W. 2001, this volume, pp. 131-140.

Kaifu, N. 2001, this volume, pp. 319-323.

Malkov, O. Yu., Tutukov, A.V. and Kovaleva, D.A. 2001, this volume, pp. 291-302.

Martinez, P. 2001, this volume, pp. 221-229.

Mervis, J. and Abbott, A. 1993, Nature, 363, 384.

Narlikar, J.V. 2001, this volume, pp. 324-328.

Onuora, L.I. 2001, this volume, pp. 329-332.

Pudritz, R.E. (ed.) 1999, *The Origins of Structure in the Universe* Report of the NRC-NSERC Long Range Planning Panel (Canada), p. 44.

Querci, F.R. 1995, in *Highlights of Astronomy*, 10, (ed. I. Appenzeller), Kluwer Publishing Co., Dordrecht, The Netherlands, pp. 677-9.

Rijsdijk, C.L. 2001, this volume, pp. 117-130.

Ratnatunga, K.U. 2001, this volume, pp. 279-290.

Snowden, M.S. 2001, this volume, pp. 266-275.

Wentzel, D.G. 2001, this volume, pp. 46-48.

Discussion

Dworetsky asked if there was documentation for the industrial spin-off of astronomy in India (apparently there is not). Dworetsky also wondered what argument one could present to a minister in a developing country for the support of astronomy. Martinez suggested that astronomy draws young people to careers in science and technology and that these people will eventually build up the science-and-technology base in developing countries. Ericson asked how could telescopes of 10-cm to 30-cm aperture produce useful results. He also wanted to know how we could fund temporary visits to developing countries by astronomers and astrophysicists from developed countries.

Astronomy for Developing Countries
IAU Special Session at the 24th General Assembly, 2001
Alan H. Batten, ed.

Critical Factors for a Successful Astronomical Research Program in a Developing Country

John B. Hearnshaw

Department of Physics and Astronomy, University of Canterbury, Christchurch, New Zealand. e-mail j.hearnshaw@phys.canterbury.ac.nz

Abstract.

I discuss the critical conditions for undertaking a successful research program in a developing country. There are many important factors, all or most of which have to be satisfied: funding, library holdings, computing access, Internet access (e-mail, WWW, ftp, telnet), collaboration with astronomers in developed countries, provision of proper offices for staff, supply of graduate students, access to travel for conferences, ability to publish in international journals, critical mass of researchers, access to a telescope (for observational astronomers), support from and interaction with national electronics, optics and precision engineering industries, a scientific culture backed by a national scientific academy, and lack of inter-institutional rivalry. I make a list of a total of 15 key factors and rank them in order of importance, and discuss the use of an astronomical research index (ARI) suitable for measuring the research potential of a given country or institution.

I also discuss whether astronomers in developing countries in principle fare better in a university or in the environment of a government national observatory or research institution, and topics such as the effect of the cost of page charges and journal subscriptions on developing countries. Finally I present some statistics on astronomy in developing countries and relate the numbers of astronomers to the size of the economy and population in each country.

1. Introduction

The purpose of this article is to consider those factors which contribute towards and are essential for a successful astronomical research program in any country, especially those countries which are developing economically. I have identified fifteen key factors which contribute to the viability of research. Most of them are essential if astronomical research is to be successfully pursued.

Adequate funding is obviously a key to most things, and although I identify this as the single most important factor, and although it is involved to some extent in all the factors that I have identified, it is too simple to assert that money is the only important thing for research. For example, the prevailing culture of scientific enquiry and the interactions with high-tech industry and with the teaching profession also play important roles.

Ideally, having identified fifteen separate factors, an astronomical research index (ARI) can be defined which measures the research potential of any given country (or institution). How the ARI should be defined in terms of the fifteen factors $n_1 \ldots n_{15}$ requires some care, so as to take account of two things: first, the relative importance of all fifteen factors is not the same – some (such as funding, computers and library holdings) feature far more prominently than others. Secondly, whereas a high score in all factors is conducive to research being successfully undertaken, a low score in just one of them may make effective research almost impossible, irrespective of the values of the other factors. This being the case, it is clear that the ARI must be a non-linear combination of the factors. In section 4 a logarithmic expression is proposed that takes this into account.

2.   Fifteen Key Factors for a Successful Research Program

I identify fifteen factors that underpin successful research. They are introduced in what I consider to be their relative order of importance. The list is of course subjective, and to some extent the relevant factors are inter-related, given that the boundaries separating them are not always clear-cut.

1. **Access to direct funding**
   Although most things in astronomy depend on funding, I consider here direct funding for astronomy, either in the form of individual research grants, or as institutional funding to observatories or to university astronomy departments, both for their operations and for particular research projects.

2. **Access to computers**
   Every astronomer today needs a computer for data reduction and analysis. This could be a PC, a mainframe computer in a local area network, or sometimes a powerful supercomputer, possibly in a national facility. In computer-rich nations, several computing options will be available 24 hours a day to all researchers, and many individuals will carry laptops and have private computers at home. Computers will be replaced every 3 or 4 years on average. In a developing country, one or two older PCs might be shared in a university department. In this section I do not include computers for Internet access – this is sufficiently important for a separate category (see No. 10 below).

3. **Access to an astronomical library**
   A good library is clearly essential if research of international standard is to be done. The major astronomical journals must all be subscribed to, and holdings need to go back several decades if research is not to be impaired. Journals must be available to all researchers, and be located at or near the place of their work. In addition to journals, a reference library, a good collection of relevant conference proceedings and a collection of review articles (e.g. *Annual Review of Astronomy & Astrophysics*) and of books (including text-books) are also required. Good libraries will also have a collection of observatory publications. The ability to browse in libraries is

important (access to the shelves is required), and prompt receipt of about a dozen major journals is also important.

Good libraries help to ensure that research topics are relevant and topical in a global context, and that proper standards of scholarship are maintained, including awareness of what has been accomplished by others in the past, and what is important now and for the future. Unfortunately, the cost of maintaining libraries is prohibitive for many developing countries. Not only are the journal subscriptions very high, but the true cost can be at least double the subscription, given the need to provide shelf space and a librarian. High inflation and the fact that subscriptions are usually quoted in U.S. dollars render many journals unobtainable in the developing world.

4. **The provision of office space**
Astronomers need to be housed in proper offices in which there is the normal office furniture, heating, lighting and storage space for books and papers. Offices need to be clustered together, so as to promote interactions between individuals. And offices need to be dedicated to one or at most a few individuals, and access should be provided on a continuous basis, seven days a week, and 24 hours a day.

5. **Having a critical mass of researchers**
Most astronomers learn their trade by interacting with others. Few can work in isolation. Isaac Newton certainly did for much of his scientific career, but even he needed the data of John Flamsteed in order to verify his theories! Today, astronomy is a vast web of knowledge, information, technology and know-how, which no individual could conceivably master unassisted. I consider a minimum critical mass of full-time active researchers interacting on a daily basis to be about four. Doubling this number to eight would however far more than double output.

6. **Access to a telescope (or to raw data)**
Telescope access is vital for observational astronomers, or for those whose research involves data analysis or its interpretation. Even theoretical astronomers need to have data to input into empirical models, and only the most abstract theoreticians can eschew data altogether. For many observational astronomers, data comes from telescopes; astronomers in many developing countries would consider that a telescope on home soil would not only greatly facilitate their research, but also act as a symbol of prestige. Unfortunately, not all telescopes in developing countries are adequately maintained, operated or equipped with instruments and detectors, so the mere presence of a telescope and willing astronomers is no guarantee of a flow of good-quality data. Today, in optical astronomy, a CCD detector is increasingly seen as an indispensable accessory to any telescope, yet the cost and necessary infrastructure (with liquid-nitrogen cryogenics and vacuum technology) are beyond the means of many developing countries.

If telescopes and associated instruments are not provided by the institution where astronomers are employed, then they need to be provided in a national or international facility that provides access to astronomers from developing countries. The outstanding example is Cerro Tololo Inter-american Observatory, that in principle provides access to South-American astronomers. Most astronomers in developing countries are not so fortunate.

7. **The presence of a teaching program, and the consequent supply of graduate students**
Graduate students are a huge bonus for any research team, and a university astronomy department with a ratio of full-time graduate students doing research to tenured academics of, say, two or three to one can enjoy the contribution from often talented young individuals, who are eager to work long hours for low pay and to provide ideas and enthusiasm. Government-funded observatories do not always have access to this resource, much to their detriment.

A flow of new graduate students every year in turn probably requires an active teaching program, especially at undergraduate levels, and which encompasses aspects of the research interests of the department. In universities the link between teaching and research is often asserted. Certainly some teaching broadens the research perspective of tenured lecturers, and undergraduate teaching stimulates interest in astronomy, thereby promoting the enrolment of graduate students and the willingness to consider astronomy as a career.

8. **Collaborations with astronomers from developed countries**
Astronomy is a global science, and collaborations provide another way of bringing in expertise, know-how, high-tech equipment and sound scientific judgement to a research project. Astronomers in developing countries, if they work in isolation, may not only lack the resources, but also the ideas to mount a successful research program. Collaboration is an ideal way of surmounting this. Those in astronomically favoured locations, such as Chile, certainly have a significant advantage. Those in moist tropical climates or inclement high-latitude ones may have less to offer, but some developing countries can exchange scientific benefits for cultural ones. In New Zealand we have a successful CCD microlensing project called MOA undertaken jointly between astronomers in New Zealand and Japan. Our observing site at Mt John is certainly not the world's best, although it offers some unique advantages (southern latitude, unique longitude); but in addition we offer our Japanese visitors a landscape and a lifestyle that contrast with and nicely complement those in Japan.

9. **Travel to conferences**
Travel to conferences is very much a part of the life of the scientist in today's world. The IAU General Assembly is testimony to that. Many astronomers in developed countries regularly participate in two or three international conferences annually. Conferences provide a further opportunity for personal interactions and exchange of ideas, vital for science to flourish in a modern environment. In the nineteenth century, conferences were largely conducted nationally through the meetings of bodies such as the Royal Astronomical Society, the Astronomische Gesellschaft and the

Académie des Sciences (to mention only three). Astronomers were also pioneers in promoting international conferences, and the meetings of the Astrographic Congress (that organized the Carte du Ciel) from 1887 and of the International Solar Union (in the early 20th century) were important predecessors to the foundation of the International Astronomical Union in 1920. Today there can easily be 50 or 60 astronomical conferences held around the world in any year. Participation is an important part of an astronomer's work, and conference papers may well constitute a majority of the papers published by any individual.

Participation at conferences on a regular annual basis requires not only funding, but also a lack of imposed travel restrictions and an absence of political interference in the scientific process in any country (both in the participants' home countries and in that of the conference host).

Countries like New Zealand are a long way from anywhere. This means that New Zealand astronomers only rarely host international conferences (few would come) and moreover, we must pay far more than many others to travel. Fortunately we are often able to make one or two international trips annually, usually to distant places.

10. **Good communications: internet access, fax, telephones, mail**
Communications are vital for astronomy. The Internet, including e-mail, ftp, telnet, and the WWW, is used by most astronomers on a daily basis. But astronomers also need fax machines, telephones in every office, and a reliable mail service. In developed countries, all these are taken for granted, but they may not exist at all in the developing world. The organizer of this Special Session, Alan Batten, has told me about the problems he had in communicating with home base in Canada from Nigeria. It is not always simple! Collaborations, data access and information all flow from good information technology. Some see the Internet as an alternative to more expensive libraries and telescopes; I doubt that it can ever be a complete substitute in the long run, but many developing countries may see the Internet as a possible replacement for libraries and telescopes in the interim.

11. **The ability to publish in international journals**
The whole question of the availability of the leading scientific journals for astronomers from developing countries to publish their papers is a thorny issue. For many, the page charges of the leading three American journals (*Astrophys. J, Astron. J, PASP*) and of *Astron. & Astrophys.* in Europe prohibit publication unless one has a collaborator as co-author in respectively North America or Europe. Some of these journals may waive page charges for authors from developing countries at the editor's discretion, which is a welcome but not widely invoked practice. Admittedly a few leading journals have no page charges (notably *MNRAS*), but high subscription costs then limit dissemination to libraries in the developing world.

Personally I believe that high but differentiated page charges (i.e. less or none for authors from developing countries) and very low subscriptions are the way forward, because then astronomers might curtail the huge volume of papers published, which in many cases few have time to read. The number of papers is at present driven by the "publish or perish" syndrome in the United States and other developed countries. We have a crisis in scientific publishing as a result. Electronic journals may force a solution on us, but if electronic publishing leads to lower costs to authors, it will not serve any great purpose, as lower costs will result in yet greater volumes of material that is read by few or even no-one. The solution is probably a very high cost to authors in the developed world for refereed electronic publications, publication costs being reduced or waived in the developing world, dissemination of articles via the Internet or CD-ROM, and low cost or free subscriptions. This would force astronomers to publish less, and enable scientists in the developing world better access to the literature. A much greater use of unrefereed web sites would be made to exchange information between specialists working in particular fields.

12. Employment opportunities

Many students are turned away from astronomy because they perceive there to be poor employment opportunities. This is especially true in developing countries, because astronomy is often regarded as not an essential part of national development goals. I note that the South African government's backing for SALT is a refreshing exception to this rule (Martinez 2001).

Students need identifiable career paths before them. That means a vigorous graduate program followed by a regular supply of postdoctoral fellowships, and at least occasional opportunities for tenured positions at universities and government observatories.

Postdoctoral positions are often a weak link, and this was certainly true in New Zealand until the mid-1990s. Fortunately both the government and the universities have heeded this weakness, and in recent years we have benefitted at Canterbury from a ratio of tenured astronomers to postdocs of about 1:1.

13. Lack of inter-institutional rivalry

Some may be surprised that I include this item; others who have experienced the destructive effects of rivalry between two institutions in a country, each vying for limited resources, will understand how damaging this is for the longer-term prospects of astronomy. Often this rivalry is between government-funded observatories and astronomy departments in the universities. Both the United Kingdom and New Zealand have had such rivalries in the past, both to their respective detriment. Fortunately these events are now history. I know that Australia has had them too, and I am aware of other serious cases elsewhere. In reality they are probably very common, and arise from the different ways in which universities and government-funded research institutions allocate their resources and set their priorities. Such squabbles can occur anywhere, and can cripple a developing country's effort in astronomy. Possibly the smaller the country and the scarcer the resources, often the more intense are the battles waged to assume power and control of whatever resources that can be allocated.

14. **Opportunities for interactions with high-tech industry**
    Astronomy is an expensive and high-tech science. For example, optical astronomers may well decide to build telescopes and instruments which will require well-developed optical, mechanical and electronic engineering capabilities as well as sophisticated computer control software. Partnerships between astronomers (whether in universities or government observatories) rarely occur in the developing world (such industries are generally absent). One only has to look at the enormous benefit to astronomy of the commercial and military development of CCD detectors and of infrared arrays to realize how astronomical progress is tied to developments in electronics, which are in turn driven by commercial and military interests in the developed world.

    But on a much smaller scale, such interactions can and do take place outside the G7 countries of North America, Europe and Japan. For example, in New Zealand at the University of Canterbury, we have had a marvellous relationship over 25 years with optical design and fabrication engineers based in the former DSIR Physics and Engineering Laboratory (now operated as a commercial company, IRL). This relationship continued with the individuals concerned after their retirement. It has enabled us to build the 1-m telescope and most of the instruments at Mt John University Observatory.

15. **The presence of a national scientific culture**
    For astronomy to flourish, science as a whole must flourish, and that needs a scientific culture to have developed. This may come from having a national academy at the helm; in New Zealand we have the Royal Society of N.Z., with an academy of fellows which can advise the government on science policy. However a scientific culture or respect for science probably starts much lower down, in the schools. In Japan, Singapore and South Korea, science is given a high priority in the schools, perhaps because science teaching is a well-respected profession. Students then want to be scientists (including astronomers) from an early age. The acceptance of astronomy as a taxpayer-funded endeavour requires such a climate for its national acceptance.

3. Can the Potential for Astronomical Research be Measured? – the Astronomy Research Index (ARI)

Devising an astronomical research index (ARI) as a measure of the potential of a country, or even an institution, to undertake astronomical research is a useful way of intercomparing countries or institutions within a country, and of identifying any weaknesses. My approach is as follows.

For each of the fifteen factors (in the order presented in section 2), a numerical quantity n_i is determined on a linear scale from 0 to 10. Zero represents a complete absence of that factor, while 10 represents the current best in the world (e.g. as represented by the United States). A value of about 5 for any factor would indicate a marginal value for mounting a successful astronomical research program. The assigning of values to these 15 factors is rather arbitrary, but generally developed countries will record 8, 9 or 10 for all of them.

Next, I have assigned weights (w_i) to the fifteen parameters to reflect the different level of importance they have for research. For simplicity, my weights are unity for n_1 to n_4 inclusive, $w_i = \frac{2}{3}$ for n_5 to n_{11}, and $w_i = \frac{1}{3}$ for n_{12} to n_{15}.

I note that a simple weighted mean of the n_i parameters would not be a meaningful ARI, because in practice a low value for any one parameter can completely nullify the entire research effort. For this reason, I define the index logarithmically:

$$\text{ARI} = \Sigma w_i \log n_i.$$

In the United States, where by definition the n_i values are all 10, one obtains ARI(US) = 10.0. In practice I consider ARI $\sim$ 8.0 as the minimum for a viable astronomical research program. Because this is a logarithmic scale, that represents about 1% of the resources being allocated to astronomy per astronomer that is allocated in the U.S. If, for example, all factors are individually $n_i = 5$, then ARI = 7.0, which, taken as a whole, is below my approximate critical value of the index for undertaking research successfully.

Table 1.   Table of factors conducive to research. The final column gives values of the factors for New Zealand astronomy at the University of Canterbury

i	Factor	Weight w_i	n_i(NZ)
1.	Funding	1	6
2.	Computers	1	7
3.	Library access	1	8
4.	Office space	1	9
5.	Critical mass of researchers	$\frac{2}{3}$	5
6.	Telescope (or data) access	$\frac{2}{3}$	6
7.	Teaching program; graduate students	$\frac{2}{3}$	9
8.	Collaborations	$\frac{2}{3}$	9
9.	Conference travel	$\frac{2}{3}$	$9\frac{1}{2}$
10.	Communications (incl. Internet etc.)	$\frac{2}{3}$	9
11.	Access to journals for publishing	$\frac{2}{3}$	6
12.	Employment opportunities in astronomy	$\frac{1}{3}$	6
13.	Lack of rivalry between institutions	$\frac{1}{3}$	8
14.	Interactions with high-tech industry	$\frac{1}{3}$	5
15.	Scientific culture in country	$\frac{1}{3}$	7
	Total	10	ARI(NZ) = 8.60

Table 1 lists the factors and their weights and gives, as an example, the estimated figures for the University of Canterbury, which is the principal place for astronomical research in New Zealand. Here ARI(NZ) = 8.60, and the weak links (with $n_i = 5$) are barely a critical mass of researchers (only four tenured astronomy academics), and few interactions with local high-tech industry. However, our strong points ($n_i \sim 9$) are good Internet and communications access;

good opportunities for collaboration (e.g. the MOA project , SALT); good provision of office space for astronomers; a good supply of talented graduate students and excellent opportunities for international travel.

4. Some Statistics on Astronomers, Demography and Economic Development

The IAU comprises 61 member countries with 8223 astronomers as individual IAU members (IAU Inf. Bull 82, 14 (1998)). In addition, there are 105 IAU members in a further 21 countries that do not adhere to the union. These statistics include the Central American Association for Astronomy, an association of six countries treated as a single member state. Of these 82 countries in total, which are either IAU member countries, or contain IAU individual members, all but five are also members of the International Monetary Fund (these five are China Taipei, the Vatican City State, Yugoslavia, North Korea and Cuba). I have used IMF data on the state of economic development in the remaining 77 countries; a useful single parameter is the IMF quota (in units of special drawing rights, SDR), which is based on national income, monetary reserves, trade balance and other economic indicators. IMF quotas have been taken from the Europa World Year Book (1998), as have total populations in each country (generally valid for about 1997). Table 2 lists these data, as well as the numbers of IAU members residing in each country.

Figure 1 shows there is a strong correlation between the IMF quota and the number of astronomers in a country. This supports the view that economic strength is essential for astronomical research. More instructive, however, is to plot the IAU astronomers/million of population against the IMF quota/million, as displayed in Figure 2. The latter is a measure of true wealth per head of population. Switzerland, on this basis, is the wealthiest country, while Belgium, Iceland, Saudi Arabia and Norway (in that order) are close behind. There is still a reasonably good correlation between astronomers/million and per capita wealth, though a few countries stand out. Estonia is a developing country, but the only nation with more than 14 astronomers/million. Its wealth is only 31.3 SDR units per million. Any value below about 50 units on this scale is what might be termed a developing country. On the other hand, Saudi Arabia (0.6 astronomers/million) and Norway (5.0 astronomers/million) evidently have invested few of their oil revenues into astronomy.

Figure 3 looks at more detail at those 44 countries which

(a) have IAU members or are member countries of the IAU,

(b) belong to the IMF,

(c) have a wealth index (SDR/million) less than 50.0 units.

The following comments can be made:

1. Astronomical activity as measured by the number of astronomers, varies from almost zero to some upper limit which increases linearly with wealth.

IAU members vs IMF quota (SDR)

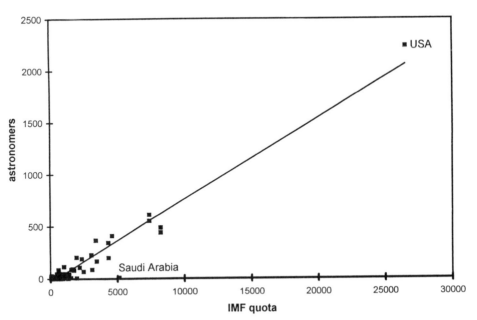

Figure 1.   IAU members versus IMF quota (SDR)

Astronomers/million vs wealth

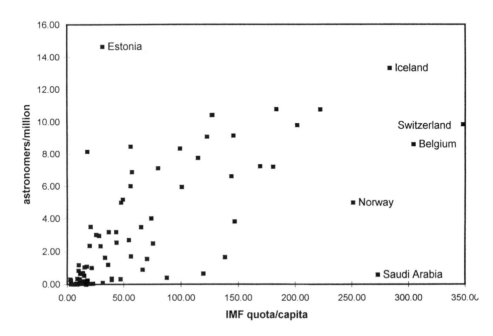

Figure 2.   Astronomers/million versus wealth per capita (SDR/million)

Developing countries: astronomers/million vs wealth/capita

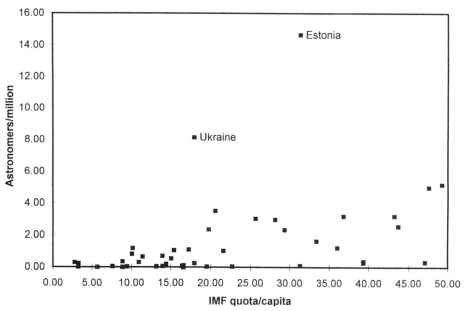

Figure 3.   Astronomers/million versus wealth per capita for developing countries

2. Just two developing countries stand out from this trend in having an unusually large number of astronomers/million (Estonia and Ukraine). Probably both these exceptions arose from political circumstances following the break-up of the Soviet Union, rather more than from astronomically enlightened governments, although I congratulate both countries on their good fortune in finding themselves astronomically well-endowed.

The last column of Table 2 gives astronomers/SDR, which is a measure of the number of astronomers per unit of economic activity in a country, and hence is an indicator of the fraction of the available economic resources that have been applied to astronomy. For 44 developing countries this is, on average, 0.066. This figure compares with 33 developed countries (those with SDR/million of population > 50.0), where astronomers/SDR = 0.051, actually less (though not significantly) than the number in developing countries. This shows that many developing countries are applying similar fractions of their available economic resources to astronomy as developed countries. Nevertheless, that conclusion is biased by the fact that only those developing countries with some astronomical activity are considered in Table 2. Two developing countries (Armenia and Estonia) have values of this index of about 0.46, much higher than any developed countries. The average value for North America, Western Europe and Japan is 0.064, and for 59 IAU member countries which are also IMF members, it is 0.074.

Table 2: Data on countries pertaining to economy, population and astronomers

Country	IMF quota (SDR)	Population (millions)	Wealth (SDR/million)	IAU members	astronomers (/million)	astronomers (/SDR) x 100
Albania	35	3.2	10.94	1	0.31	2.86
Algeria*	914	29.2	31.30	3	0.10	0.33
Argentina*	1537	35.2	43.66	90	2.56	5.86
Armenia*	68	3.8	17.89	31	8.16	45.59
Australia*	2333	18.3	127.49	191	10.44	8.19
Austria*	1188	8.1	146.67	31	3.83	2.61
Azerbaijan*	117	7.6	15.39	8	1.05	6.84
Bahrain	83	0.6	138.33	1	1.67	1.20
Belgium*	3102	10.2	304.12	88	8.63	2.84
Bolivia*	126	7.6	16.58	0	0.00	0.00
Brazil*	2171	155.8	13.93	109	0.70	5.02
Bulgaria*	465	8.3	56.02	50	6.02	10.75
Canada*	4320	30	144.00	199	6.63	4.61
Cent. Amer.*	740	32.6	22.70	2	0.06	0.27
Chile*	622	14.4	43.19	46	3.19	7.40
China PR*	3385	1223.9	2.77	368	0.30	10.87
China Taipei*		21.5	0.00	23	1.07	
Colombia	561	40.2	13.96	3	0.07	0.53
Croatia*	262	4.8	54.58	13	2.71	4.96
Cuba		11	0.00	5	0.45	
Czech Rep.*	590	10.3	57.28	71	6.89	12.03
Denmark*	1070	5.3	201.89	52	9.81	4.86
Egypt*	678	59.3	11.43	39	0.66	5.75
Estonia*	47	1.5	31.33	22	14.67	46.81
Finland*	862	5.1	169.02	37	7.25	4.29
France*	7415	58.4	126.97	609	10.43	8.21
Georgia*	111	5.4	20.56	19	3.52	17.12
Germany*	8242	81.8	100.76	488	5.97	5.92
Greece*	588	10.5	56.00	89	8.48	15.14
Hungary*	755	10.2	74.02	41	4.02	5.43
Iceland*	85	0.3	283.33	4	13.33	4.71
India*	3056	936	3.26	227	0.24	7.43
Indonesia*	1498	197.5	7.58	12	0.06	0.80
Iran*	1079	60.1	17.95	15	0.25	1.39
Iraq	865	22	39.32	8	0.36	0.92
Ireland*	525	3.6	145.83	33	9.17	6.29
Israel*	666	5.8	114.83	45	7.76	6.76
Italy*	4591	57.4	79.98	409	7.13	8.91
Japan*	8242	126.1	65.36	440	3.49	5.34
Kazakhstan	248	16.5	15.03	9	0.55	3.63
Latvia*	92	2.5	36.80	8	3.20	8.70
Lebanon	146	3.1	47.10	1	0.32	0.68
Lithuania*	104	3.7	28.11	11	2.97	10.58
Malaysia*	833	21.2	39.29	6	0.28	0.72
Mauritius	73	1.1	66.36	1	0.91	1.37
Mexico*	1753	81.2	21.59	83	1.02	4.73
Morocco	428	26.1	16.40	2	0.08	0.47
N. Korea		21.2	0.00	20	0.94	
Netherlands*	3444	15.5	222.19	167	10.77	4.85
New Zealand*	650	3.6	180.56	26	7.22	4.00
Nigeria	1282	97.2	13.19	4	0.04	0.31
Norway*	1105	4.4	251.14	22	5.00	1.99
Pakistan	758	134.1	5.65	1	0.01	0.13

Table 2 (cont.): Data on countries pertaining to economy, population and astronomers

Country	IMF quota (SDR)	Population (millions)	Wealth (SDR/million)	IAU members	astronomers (/million)	astronomers (/SDR) x 100
Paraguay	72	5	14.40	1	0.20	1.39
Peru*	466	23.9	19.50	1	0.04	0.21
Philippines	633	71.5	8.85	1	0.01	0.16
Poland*	989	38.6	25.62	117	3.03	11.83
Portugal*	558	9.9	56.36	17	1.72	3.05
Romania*	754	22.6	33.36	37	1.64	4.91
Russia*	4313	147.1	29.32	344	2.34	7.98
S. Korea*	800	46.4	17.24	51	1.10	6.38
Saudi Arabia*	5131	18.8	272.93	11	0.59	0.21
Singapore	358	3	119.33	2	0.67	0.56
Slovakia*	257	5.4	47.59	27	5.00	10.51
Slovenia	151	2	75.50	5	2.50	3.31
South Africa*	1365	37.9	36.02	46	1.21	3.37
Spain*	1935	39.3	49.24	204	5.19	10.54
Sri Lanka	304	18.3	16.61	2	0.11	0.66
Sweden*	1614	8.8	183.41	95	10.80	5.89
Switzerland*	2470	7.1	347.89	70	9.86	2.83
Tajikistan*	60	5.9	10.17	7	1.19	11.67
Thailand	574	60.8	9.44	3	0.05	0.52
Turkey*	642	63.7	10.08	53	0.83	8.26
UK *	7415	60.5	122.56	550	9.09	7.42
Ukraine*	997	50.5	19.74	119	2.36	11.94
Uruguay*	225	3.2	70.31	5	1.56	2.22
USA*	26527	267.6	99.13	2235	8.35	8.43
Uzbekistan	200	22.5	8.89	8	0.36	4.00
Vatican City*				5		
Venezuela*	1951	22.3	87.49	9	0.40	0.46
Vietnam	242	75.4	3.21	2	0.03	0.83
Yugoslavia		10.5	0.00	25	2.38	

* IAU member

5. Conclusions

In conclusion I note that I have not tried to correlate the astronomical research index (ARI), which measures research potential based on the prevailing conditions in a country, with IAU astronomers/million of population, which is a rough measure of actual astronomical activity. If that is a useful exercise, then astronomers from all these countries should first try to determine their ARI values, which requires careful assessment and local knowledge.

The potential benefits of such an exercise would be a more ready identification of the true weaknesses and impediments in each country's overall astronomical research effort.

References

IAU Information Bull. 82, 1998 (14, June).

Europa World Year Book, 1998, Europa Publications Ltd, London.

Martinez, P. 2001, this volume, pp. 221-229.

Discussion

Sterken suggested that in some countries with high per capita incomes there are several small astronomy research groups in small universities. Except for criterion 15 (national scientific culture), to which Hearnshaw had given the lowest weight, the ARI for such institutes (as opposed to the nation as a whole) would put those institutes below the level at which Hearnshaw considered effective research to be possible. Sterken expressed concern that, if governments were to apply the ARI to individual institutes, they might be led to wrong conclusions about funding. Hearnshaw expressed doubt that there were insitutuions in developed countries lacking the essential funding, resources and facilities for research and yet producing useful results. If, however, he was wrong about that, he believed that publication of his criteria was as likely to lead governments to increase the resources of such institutions as the reverse.

Martinez suggested that good communications (criterion 10) should rank higher than Hearnshaw had suggested – on a par with, or just below, access to good computers (criterion 2). Good communications would automatically take care of a number of Hearnshaw's other criteria (3: library access, 5: critical mass of researchers, 6: telescope or data access and 7: a teaching program). Hearnshaw agreed that good communications are vital and that the ordering of his criteria is partly subjective. He had no quarrel with Martinez' ordering but pointed out that, before about 1950, astronomers could communiucate only by ordinary mail and yet did much good research.

Do Developing Countries need Astronomy?

Philippe Eenens

*Department of Astronomy, University of Guanajuato, Apartado Postal
144, Guanajuato, CP 36000, México. e-mail: eenens@astro.ugto.mx*

Abstract. Astronomy can help, directly or indirectly, in the acquisition
and creation of new technologies, in attracting young people to scientific
careers, in providing a scientific education to the general public and in
fostering international collaborations. These and other benefits of professional astronomy are critically reviewed in the context of countries which
are facing urgent, basic needs. Several criteria are suggested for the best
implementation of astronomy in developing countries and the most efficient collaboration with industrialized countries.

1. The Question

What is the use of astronomy? This often-asked question becomes more acute in
developing countries, which are sometimes confronting many urgent problems,
such as hunger, diseases, illiteracy, lack of infrastructure, etc. Is it worth investing resources and people in an endeavor that will obviously not bring immediate
solutions to these and other basic needs? If astronomical research is to receive
a justification, it must take into account the complex situation of developing
countries. It is therefore important to make a critical assessment of the various
arguments presented to support astronomy. Such is the aim of this paper.

We will summarize the main arguments in favor of pursuing astronomical research in developing countries. These can be grouped in the following
categories: basic research, new technologies, scientific education, international
collaboration and broader cultural aspects. Each argument will be critically assessed; possible shortcomings will be pointed out and criteria to overcome them
will be suggested. research. Of course, the questions raised are also relevant to
industrialized countries, where astronomy can also be considered an expensive
luxury.

2. Argument No. 1: Basic Research

Astronomy deals with environments totally unlike those found in Earth-bound
laboratories, where it is impossible to reproduce the range of temperatures,
pressures and other physical conditions found in stars and galaxies. As such,

astronomical research opens the way to new theoretical discoveries, which may in turn lead to practical applications, although these may come only much later. One thinks for example of positional astronomy and Newton's law and their application to space flight, or possible future uses of the energy released in the nuclear fusion, as already studied in the center of the stars.

2.1. Caveats

Policies about basic research vary. One extreme position would be to cut all funding for basic research, on the ground that it does not produce immediate economical benefits. This would be a very short-sighted view, because long-term solutions should not be neglected. In the long run a total lack of skills regarding new scientific developments would hinder a nation from mastering new technologies and competing in the global market.

The other extreme would be to invest disproportionate amounts of resources to create a project that will be the "largest of its kind", motivated by the prestige it promises, in the belief that money alone will solve the deficiencies in training and infrastructure.

2.2. Recommendations

- Wherever possible, favor *low-cost alternatives*. Frontier research can be done with small telescopes. Observational data are also available in on-line archives. Fast computing is becoming affordable. Fast and cheap communication with collaborators, as well as video conferences and access to electronic publications are now possible through the internet.

- Promote a *limited number of key projects*. These could be selected on the basis of present expertise, of local interests, etc.

- Involve *amateur astronomers*. With a minimum of professional orientation, they can contribute to serious astronomical research.

- Spread the cost of larger projects by conducting them by means of *international cooperative efforts*.

3. Argument No. 2: New Technologies

Astronomical research is often presented as a way to:

- *Acquire new technologies:* e.g., the use of databases, analysis techniques, instrument control, computing, electronics, optics, space technology, etc.

- *Acquire new skills:* The instruments should not just sit and get rusty: astronomers and students should be trained to maintain and develop new astronomical instruments.

- *Create practical applications:* although not the main goal, these are side-effects of astronomical research (spin-off), which is thus believed to galvanize indirectly technological growth through large projects.

3.1. Caveats

Nonetheless, it must be conceeded that new technologies are not necessarily good.

- *Environmental problems* are a primary concern that come to mind, but one should also be concerned with other possible disruptions.

- The social fabric may also be damaged if new technologies bring about increased social and economic *inequalities.*

- The local economy may suffer if research efforts in developing countries lead to *debts and/or economical dependence* on the industrialized nations, thus benefiting the latter at the expense of the former.

- *Health* and general human well-being may deteriorate as a consequence of new technologies.

3.2. Recommendations

- Ensure that no debt or dependency are being created and that the users of the new technologies have full training and full control.

- Ensure that there is no damage to the natural, social and human environment.

4.   Argument No. 3: Scientific Education

The accumulation of capital and the attraction of investment depend less and less on the amount of a country's natural resources and labor. The key factor is the accumulation of technology based on the intensity of knowledge: science, technology and organization. Every nation needs people with good scientific training:

- to *understand* scientific discoveries made there or elsewhere

- to *apply* imported technologies

- to *fight* against pre-scientific attitudes such as astrology, reductionism in science, creationism and other prejudices

- to help leaders *make decisions* with global vision and objectivity

- to *compete* and to communicate with others on the international scene

5.   Astronomy as an Attractor

Astronomy has a special appeal to a large public, probably more than any other "exact" science. Given its popularity among the general public, astronomy can make a significant contribution to meeting the above goals in light of the following:

- Astronomy may *attract students* to scientific careers, thus providing a practical way to prepare good scientists and high-level technicians.

- Students in astronomy and space sciences acquire *valuable skills* of observation, analysis of data, use of scientific reasoning, writing, publication, etc.

- Such students can find work in *related fields*, including computer science, remote sensing, meteorology, communications, etc.

- Again, *amateur astronomers* should be considered and be given the opportunity to acquire these skills, for example through continuous education.

5.1. Caveats

- While trying to emulate the education level of industrialized countries, one must be careful not to create another dependence, a sort of *inferiority complex* that would set in the minds of students the impossibility of attaining the (perceived) level of other countries.

- A related problem is the *brain drain* (emigration of the best-trained people). Several reasons could explain this drain:
  – low salaries in developing countries
  – lack of good research conditions, such as telescopes, computers, libraries
  – the above-mentioned "inferiority complex" (doubts about the feasibility of frontier research in one's own country)
  – the lack of willingness to work toward strengthening astronomy research in the home country

5.2. Recommendations

- Perhaps the most important answer to these issues is the fostering of a *positive attitude* towards one's own country, both a realistic trust in its potential and a solid personal commitment to realize this potential. These should be the basic bricks of any educational program.

- As noted earlier, frontier research can be done with limited budgets, through the *creative use* of limited resources.

- Students from developing countries should be taught that helping to establish high-level astronomy at home can be as worthy a contribution to astronomy as doing pure research. During their studies, they must be encouraged to participate in the *development of local facilities*.

- Also, *expatriate researchers* have many opportunities to help their home country: remotely collaborating with astronomers from home, giving lectures during occasional visits home, supervising dissertations of students from their home country, inviting them to spend working visits at their institution, etc.

6. Argument No. 4: International collaboration

More than before, our world needs cross-cultural integration. Astronomy provides one of the most convincing *examples of collaboration between countries*:

- with peaceful goals

- without exploiting the poorer partner

- with coauthors frequently from different countries

- with free dissemination of discoveries, information, software (IRAF MIDAS), expertise (no one is surprised if the advisory panel for a new large telescope is made of the directors of the competing telescopes), etc.

Why is this so easy in astronomical circles? Probably because there is no direct economical benefit at stake.

It should also be noted that developing countries have *much to offer* to the larger astronomical community:

- their climate

- their territorial extension (e.g. more solar eclipses happen in developing countries than in rich countries)

- equatorial locations (e.g. for stationary satellites)

- access to southern-hemisphere objects (e.g. Chile)

- access to locations needed for all-Earth telescopes, large-baseline interferometry, etc.

- and finally, alternative views on old questions, based on different cultural backgrounds.

6.1. Caveats

- Collaboration should not merely use foreign land (cf. the case of Chile) but should also *benefit the host country* (astronomy research, technological spin-off, etc).

- Collaboration should help reinforce the *research groups* of local astronomers (cf. UN Workshops on Basic Science).

- Collaboration should allocate means for *teaching and outreach*.

The above also applies to the collaboration *between* developing countries (beyond national and regional rivalries).

6.2. Recommendations:

- International collaborations should seek to benefit the poorer partner.

- The benefits of the collaboration should not only be the installation of an astronomical instrument, but all that is necessary to build strong research groups: computing facilities, libraries, teaching programs, international interaction, etc.

- The collaboration should be concerned not only about initiating programs but also about the long-term growth of astronomical research in developing countries.

7. Argument No. 5: Cultural Impact

Is it enough to try to improve the economical sphere ("what we have") without taking care of *what we are*? This second sphere could be called "culture". What is meant here is our view of ourselves, our motivation for growth as individuals and as a society.

Could astronomy contribute to enrich humankind in that sense? in unexpected cultural and epistemological ways?

- Astronomy has deep roots in virtually every human culture. It has played a significant role in many civilizations of the past (e.g. Maya, Inca). Today it still holds a fascination for many people, as seen for example in the interest in eclipses and in the exploration of the solar system.

- Having no direct applicability, astronomy is well-placed to stimulate reflexions beyond the immediate.

- Astronomy explores the limits of physical reality and raises questions regarding:
  - the origin of the Earth and of the universe
  - our place in the cosmos and our destiny
  - our uniqueness in the universe.

- Astronomy discovers a universe that is always surprising, thus helping to transcend old views and find new solutions to old problems.

- Astronomy presents a universe extremely vast and complex; thus it helps put reality into perspective and encourages us to look for global, long-term solutions. Yet the majority of people remain unaware of the size of the universe, of the place of the Earth, of the distance to the stars and may even believe in astrology.

Astronomy is unique in that it addresses human questions from the standpoint of an exact science. Hence:
- It helps to formulate these questions.
- It helps to provide scientific evidence to confront theories about the origins, extension and evolution of our universe.
- It helps to integrate science into human concerns.

7.1. Recommendations:

• Promote the dialogue between exact sciences and thinking in the humanities.
• Astronomers must avoid the "guru" role: the road to the truth is a long and arduous one.

8. Conclusions

In this paper we have tried to show that many of the arguments usually put forward to justify astronomical research have a real value. Astronomy can contribute to the growth of developing countries. However the value of astronomy could be offset if attention is not paid to the way it is implemented as basic research, technology development, education, international cooperation and reflection on ultimate human questions. Therefore a thorough analysis of its implementation is necessary to ensure that it contributes to the intended goals we have set forth in the benefit of the society and individuals. It is hoped that the sketchy lines of thought presented here may foster further reflection and discussion of these important questions, in order to search for improved policies.

1. Astronomy can bring many benefits to developing countries, well beyond the scope of pure research:
   • technological spin-off
   • scientific training
   • international collaboration
   • changes to culture

2. However, criteria should be applied to ensure that the benefits offset the possible damages to the host country.

3. These criteria could be grouped under the following headings:
   • do not create dependency
   • respect the cultural identity of the host country
   • make provision for growth of local astronomy
   • educate the public
   • involve amateurs.

Acknowledgments. I am grateful to Richard L. Anderson for a careful reading of a previous version of this paper, and to the IAU for a travel grant to attend the General Assembly.

Section 2: The Current Situation in Astronomy Education

Astronomy for Developing Countries
IAU Special Session at the 24th General Assembly, 2001
Alan H. Batten, ed.

Astronomy Education: The Current Status in Zambia

Geoffrey Munyeme* and Peter C. Kalebwe

Physics Department, University of Zambia, P.O.Box 32379, Lusaka, Zambia. e-mail: g.munyeme@phys.uu.nl and PKalebwe@natsci.unza.zm

Abstract. There are many interlocking factors determining the introduction of astronomy education in Zambia. The process of infusing this new subject into an education system so centralised as that of Zambia is extremely complex. At school level the process is more complex than at university level, as all syllabuses are developed by a central body, Curriculum Development Centre (CDC) whose priorities are determined by perceived social and economic needs of the country. The prevailing notion in Zambia is that astronomy has no direct bearing on future employment needs. It is therefore not surprising that astronomy is at the bottom of the priority list among school subjects. The recent upsurge of interest in astronomy at the University of Zambia opens up the necessary background for developing astronomy in both school and university curricula. The University has recently formed the Astronomical Society and the Working Group on Space Science in Zambia. Coupled to this are exchange visits and collaborative work between the Physics Department of the University of Zambia and the South-African Astronomical Observatory. In this paper we present a review of the current activities in space science in Zambia and how they relate to the development of astronomy education.

*Current address: Debye Institute, University of Utrecht, Princetonplein 1, Utrecht 3508, The Netherlands.

1. Introduction

Astronomy is fundamentally an ancient science, which has played an important role in the general development of modern science and technology. At the XII IAU General Assembly held in Hamburg in 1964, Commission 46 (Teaching of Astronomy) was created (Blaauw 1994). The creation of this new commission made astronomy an educational tool, which has to date, found its way into the curricula of many universities worldwide. It is rather disappointing to note that such a treasured and developed science is non-existent in the curriculum of our University. What therefore is the missing link? Does it mean that astronomy has no direct relevance to the development of science and technology in Zambia or is it simply an unexciting science? An attempt (Munyeme 1997) to trace the many factors determining the missing link requires a detailed and lengthy analysis of the several factors claiming to influence the innovation of the physics/science curriculum in Zambia. It is not the intention of this paper to make this analysis, however we can easily point to the fact that science and technology are social phenomena which must be included in the social transformation and organization of a country. Both developed and developing countries follow this path, but with different degrees of success. In order to derive success in the application of science for social and economic development, the teaching of science must be broad-based. This has not been the case in Zambia, the emphasis on science and technology subjects has always been determined by: (1) the government, (2) the external agents, and (3) the interest of individuals or groups of scientists and teachers.

The above three factors have had strong influence on the past and present status of astronomy education in Zambia. We will address the combination of these factors with the current activities in space science in Zambia to explain their influence on the development of astronomy education.

2.   Curriculum Development

Zambia's educational practices are based on conceptions of education and science derived from colonial experience. Our history of colonial rule, combined with the post-independence government policy on science education and research has unfortunately resulted in low levels of scientific and technological progress (Kelly 1991, Goodson 1983). A central body currently known as Curriculum Development Centre (CDC) has always developed the school curriculum during both colonial and post colonial eras. CDC is a government body and its approach to curriculum development is largely determined by government policy on the social and economic needs of the country. CDC is responsible for developing curricula for all primary and secondary schools, pre-schools and preservice primary-school teacher-training programs.

2.1.   Space science in the school curriculum

Unfortunately, the centralized arrangement of developing the school curriculum makes it difficult to infuse new curricular materials and often leads to a very narrow selection of science topics. The current school curriculum addresses some elements of basic space science. A topic known as Universe I is taught in grade 5 of the primary-school syllabus. This topic introduces simple ideas of the solar system and features of the night sky. In grade 8 of the secondary-school syllabus is taught Universe II, which is a further study of the solar system and the position of the Earth in the universe. The contents include planets, comparison of Mars and Earth, stars, galaxies and the universe. Universe II marks the end of space-science topics at school level.

The great weakness of the school curriculum in Zambia involves assumptions that some kinds of areas of knowledge are more important than others and space science is among the less emphasized. Though appearing on the school curriculum, space-science topics have often been omitted by most of the schoolteachers. The common notion is that these topics have no direct bearing on the future career prospects of an ordinary Zambian. Space science is viewed by most of the schoolteachers as a subject for developed countries with capabilities in space technology (Stobie 1998). In our opinion the large percentage of teachers with inadequate understanding of space science contributes to the omission of these topics in their lessons. The findings of informal interviews conducted with some of the students admitted to the Physics Department of the University of Zambia show that at least 30% of the school leavers have not studied the Universe topics. We later discovered that these topics have been excluded from the school science-examination questions. The combination of this and the fact that space science is a non-career subject provides enough ground for pupils and teachers to shun it. The shunning of space science at school level has a detrimental effect on the promotion of astronomy education. It means that a lot of effort must be put into reversing the negative attitude that students and teachers have about space science.

2.2. The university curriculum

Schools and departments of the University have the statutory authority to formulate their own syllabuses. However there are many setbacks for the introduction of new material and courses into the existing curriculum. The constraints are mostly attributed to low funding levels and difficulties in securing high-level academic personnel. Recently the University of Zambia witnessed the emergence of new courses dictated by the rapidly changing world-wide educational system. The advent of the free market and democracy has brought in educational values that prepare young people for induction into today's competitive economy. Successfully introduced at the University of Zambia in preference to traditional courses are courses in human rights, democracy, computer studies, energy and environmental studies etc. The broadening of University curriculum reflects changes that can be made when suitable conditions are met.

3. Towards Astronomy Education

Having provided a general overview of curriculum development in Zambia, we shall now focus on the feasibility of developing astronomy education under the prevailing education conditions. From what has been said it might appear that there is no room for astronomy in the current educational system in Zambia. The question we must ask ourselves is: What type of instruments and methods do we have at our disposal to make astronomy compete favourably with courses dictated by the economic climate of Zambia? The answer to this question lies in the recent upsurge of interest in astronomy at the University of Zambia. There are three major initiatives in the Department of Physics, which are aimed at promoting astronomy in Zambia. These initiatives offer the necessary background and opportunity for infusing astronomy into both school and University curricula. In addition to this there are dedicated members of the Physics Department who are available and willing to spend a considerable part of their time and thought in organizing these activities. The three initiatives are:

1. The formation of the Astronomical Society of Zambia.

2. The formation of the Working Group on Space Science in Zambia.

3. The participation of the physics department in exchange visits and collaborative work with the South African Astronomical Observatory.

3.1. The Zambia Astronomical Society (ZAS)

The Zambia Astronomical Society was created in 1998 with the sole purpose of promoting the study of astronomy in Zambia. In its constitution, it is clearly stated that ZAS will provide the platform for popularizing astronomy among educational institutions and the general public through;

- Public lectures, seminars and exhibitions. ZAS will collect material like books, magazines, slides, movies, journals, etc. on astronomy and make them available to interested parties.

- Assisting the promotion and introduction of astronomy courses and activities in the school, college and university curricula.

- Encouraging co-ordinated research programmes and exchange visits between institutions within and outside Zambia.

Though different categories of membership were open to all those interested in the field of astronomy and other sciences, ZAS failed to command good membership. This did not come as a surprise to the organizers, since Zambia has no astronomical heritage and education. To many Zambians astronomy is a new vocabulary implying a new science in the history of Zambia. Fortunately a spectacular achievement of changing the image of astronomy was later facilitated by the formation of the Working Group on Space Sciences in Africa (WGSSA).

3.2. Working Group on Space Sciences in Zambia (WGSSZ)

The Working Group on Space Science in Zambia was formed on 27th May, 1999 as an affiliate of the Working Group on Space Sciences in Africa (WGSSA). African delegates to the 6th UN/ESA Workshop on Basic Space Science held in Bonn in 1996 founded WGSSA. It is an international and non-governmental organization whose long-term objective is to make possible the creation of an African Institute for Space Science. The immediate strategic objectives of the Working Group is to promote education in (a) astronomy and astrophysics (b) solar-terrestrial interaction and its influence on terrestrial climate (c) planetary and atmospheric studies and (d) the origin of life and exobiology.

In 1999 the Physics Department of the University of Zambia launched a membership drive for the WGSSA. The membership drive involved the distribution of the WGSSA membership forms obtained by the Department from the WGSSA co-ordinator Dr. Peter Martinez of the South-African Astronomical Observatory. The forms were distributed to various institutions involved in space science of one kind or another. The response was quite encouraging, a total of 38 members were registered and the Department forwarded their forms to the WGSSA co-ordinator. It was motivating for registered members to have received individual letters of acknowledgment from the co-ordinator of the WGSSA, urging them to support the WGSSA activities and its publication, *African Skies*, by contributing articles and communication news of developments in space science.

3.3. International cooperation

Zambia has no experience of astronomy education. In order to succeed effectively in introducing astronomy education, it is necessary to establish strong links with external institutions and individual scientists with wide experience of astronomy education and research. The links will promote skills which Zambian scientists are lacking and this will be beneficial to the general development of astronomy. The Physics Department of the University of Zambia has already established a link with the South-African Astronomical Observatory (SAAO). Two members of the department have visited South Africa and participated in the SAAO organized summer schools. In 1999, Dr. Peter Martinez of SAAO, who is the current coordinator of the WGSSA, visited Zambia during the inauguration of the Working Group on Space Sciences in Zambia. His visit was a landmark in the promotion of space science in Zambia and has strengthened the image of the Working Group in Zambia. Besides the SAAO link, the Physics Department has links with individual scientists who have assisted in the provision of materials such as journals, slides and bulletins.

4. Suggestions for the Way Forward

As already mentioned, the prime factors that influence curriculum change in Zambia are (a) the government (b) the external agents (donors) and (c) the interest of individuals or groups of scientists and teachers. The interplay of these factors is now possible through the current space-science activities at the University of Zambia. First and foremost, it is necessary to strengthen the existing structures. Either the Astronomical Society or the Working Group should be made to function thoroughly and efficiently so that it can be easier to advance the interests of promoting space science in Zambia. There is need to convince the government (in this case CDC), the teachers and the funding agencies that curricular changes responding to the needs of astronomy education should be introduced. However we should realize that the promotion of astronomy in Zambia is taking place against a background of social and economic change. This means that a systematic approach requiring dedication from the organizers is necessary for achieving the end results. The association should initiate programs aimed at:

- Influencing the integration of curriculum reform with an initial training of a small number of school teachers in space science. These teachers will transfer the acquired knowledge with enthusiasm to other teachers in schools and can reduce the shunning of the Universe topics at school level. The training can be accomplished through a series of workshops or summer schools.

- Circulation of information on space science through public lectures, slides, videos and newsletter.

- Initiation of a program of exchange of scientists for collaborative training and for R&D programs with Institutions like SAAO. This will enable high-level access for Zambian scientists to astronomical facilities.

- Establishing liaison between national institutions dealing with the components of space sciences.

- Assisting, where possible, the learning institutions to lobby funds for acquisition of equipment and facilities neccessary for teaching astronomy. As an example, in 1998 Prof. Donat Wentzel offered the University of Zambia U.S. $500 for the purchase of a telescope. Though a telescope costing this amount was found in the U.S.A., the Physics Department of the University of Zambia failed to meet the cost of transporting it from America. This would have been possible if there was a program dedicated to the acquisition of such instruments.

From the preceding discussions, it can be observed that the Physics Department of the University of Zambia has great interest in promoting and coordinating space-science activities in Zambia. The success of introducing astronomy in the physics curriculum of the University of Zambia is therefore very high. The constraints are attributed to lack of funding and difficulties in securing high-level academic personnel. In 1999 a proposal for the introduction of an undergraduate course in astronomy was written and circulated by the Physics Department to potential funding agencies. The project addressed among other things the need for capacity building in astronomy and international cooperation particularly with SAAO. Unfortunately the project has failed to attract positive response from potential funders. It is our feeling that funding will be hard to come by and other ways should be sought. The easiest is to fuse and to expand some astrophysics topics into the existing courses. In this case the Department can adopt the approach, and some contents, of the booklet *Astrophysics for University Physics Courses*, written by Prof. Donat Wentzel for the United Nations Office for Outer Space Affairs. The booklet presents an array of astrophysical problems, which can be selected and used within the existing physics courses. It covers topics on elementary mechanics, heat and radiation, kinetic theory, electrical currents etc. The other option is to look for voluteer scientists, particularly retired professors of astronomy, to assist in establishing the course for at least a year. Under this arrangement it is possible for the University of Zambia to arrange accommodation and local salary for the visiting scientist.

Acknowledgments. Financial support from the IAU for G.Munyeme to attend the 24th General Assembly in Manchester is gratefully acknowledged. The authors would also like to thank Alan Batten, Peter Martinez, Donat Wentzel, Julieta Fierro, Barrie Jones and J.V.Narlikar for their assistance in promoting the activities at the University of Zambia.

References

Blaauw, A. 1994 *History of the IAU* pp. 234, 237, Kluwer, Dordrecht, The Netherlands.

Goodson, I.F. 1983, *School Subjects and Curriculum Change*, Croom Helm, London, U.K.

Kelly, M.J. 1991, *Education in a Declining Economy, the Case of Zambia, 1975-1985*, EDI Analytical Case Studies No. 8.

Munyeme, G. 1997, *Constraints and Prospects of Education and Research for Basic Space Science Development in Zambia*, UN/ESA Workshops on Basic Space Science, Honduras.

Stobie, R.S. 1998, *Development of Astronomy within Africa*, IITAP reports.

Discussion

Orchiston asked if Zambian astronomers had joined with other scientists in their country to lobby for the establishment of a National Science Centre, which could lead to greater public interest in astronomy and thus to more interest among both high-school and university students.

Chambliss asked what plans the Zambian government had to enlighten its populace on the solar eclipse of 2001. In the past, in some countries (e.g. Indonesia) the emphasis seems to have been on dissuading people from observing an eclipse rather than on telling them how to look at it safely. Munyeme replied that the Physics Department of the University of Zambia had set up a Working Group in connection with the eclipse which they hoped would receive funds to work with the Zambia National Tourist Board on arrangements for accommodation and transport of visitors. This Group is also in touch with Jay Pasachoff. Pasachoff remarked that the IAU Working Group on Eclipses (Commissions 10 and 12) and the Program Group on the Public Understanding of Science (Commission 46) was in touch with Peter Kalebwe (Munyeme's co-author) and providing educational material, including instructions on how to observe the eclipse safely and describing how exciting a total eclipse is. Kalebwe plans to join Pasachoff and his students in observing the eclipse. Pasachoff also pointed out that convincing local authorities to advise the public correctly about eclipses is often difficult – and not only in developing countries. Finally, he mentioned that Zambians would see not only the total eclipse of June 2001 but also partial phases of the eclipse of December 2002. (*Editor's note;* see also the paper by Pasachoff on pp. 101-7 and the poster by Podmore on pp. 369-70.)

A Renewal of Astronomy Edcuation in Vietnam

Donat G. Wentzel

University of Maryland, College Park, Maryland, U.S.A. e-mail: wentzel@astro.umd.edu

Abstract. Vietnam was scientifically completely isolated for almost 30 years. With French help, several lecture courses have introduced modern astrophysics to Vietnamese physicists, and four students are studying abroad. The IAU program "Teaching for Astronomy Development" (TAD) has concentrated on modernizing the on-going astronomy course for students in the third year of the pedagogical universities. Three one-to-two week "Teachers' Workshops" have served to introduce selected up-to-date astronomical topics and a few modern teaching methods. The TAD program has also provided appropriate journals, books, a PC and educational software. A new text, *Astrophysics*, in Vietnamese and English on facing pages and with color pictures – apparently a first for any textbook in Vietnam – will first be used starting in September 2000.

Future additional activities: collaboration to plan a new astronomy course in the twelfth grade of the natural science branch of the secondary schools; collaboration so that the 41-cm telescope and astronomers in Hanoi can produce some simple quality science; helping the only planetarium in Vietnam to acquire a wider range of offerings; helping to create a set of B.Sc.-level astrophysics courses for three universities; and supporting the Vietnamese Astronomical Society in effective public outreach.

Vietnam was scientifically isolated for thirty years. Although astronomy has been a required subject for physics students in ten pedagogical universities for many years, the astronomy textbook used until this year is far out of date, and it has no photographs. Astronomy almost disappeared during thirty years of war. In general, science education is by lecture that is to be memorized (as is the case in many places); there are practically no hands-on laboratory experiments for students; and, during the thirty years of war, an entire generation of scientists has been trained without participating in the process of science, in exploring nature, in measuring things, in interpreting data. Therefore, the few senior Vietnamese astronomers have sought to re-introduce astronomy as a frontier science. With French help, several lecture courses have introduced modern astrophysics to Vietnamese physicists; and the first of several students studying in France has received a Ph.D. and will soon return to a position at the National University of Hanoi.

The IAU program "Teaching for Astronomy Development" (TAD) is designed to help scientifically developing countries that express an interest in enhancing astronomy in the country over the long term.

In Vietnam, the main task for TAD has been to train the teachers of the ongoing astronomy course in the pedagogical universities. Three 1-2-week "Teachers' Workshops" - one of which included nationally selected physics students - were organized with international faculty (N. Q. Rieu, J. C. White II, D. G. Wentzel). The workshops gradually introduced up-to-date astronomical topics and reviewed the associated physics. We used a small part of *Hands-on-Astrophysics* from AAVSO. The teachers have been delighted to learn more interactive ways of teaching than writing on a blackboard, and they practice at home what they learn at the workshops. Saunders College Publishing provided up-to-date textbooks, in English and containing many color photos. The TAD program has also provided appropriate journals, books, a PC, and educational software. We have written a new text, *Astrophysics*, printed in Vietnamese and English on facing pages. The IAU financial support made possible that the new text contains color pictures, which apparently is a first for any science textbook in Vietnam.

It is now time to expand the TAD activities. First, we plan to provide a support system for the university teachers in the semester when they first use the new text. Because most teachers do not yet have e-mail available, they will depend heavily on telephone to the three main universities which do have e-mail. Second, once the new course is established, we plan to help the Vietnamese teachers and astronomers to broaden the base of support for astronomy by sponsoring two more workshops. Third, we shall advise on a set of astrophysics courses that will be started in three universities so that a few students can study astronomy at a more advanced level. Fourth, we expect to support the Vietnamese Astronomical Society in seeking effective interdisciplinary public outreach.

The Vietnamese astronomers face a very difficult educational challenge. The ongoing astronomy course is given in pedagogical universities, but it focuses on the science of astronomy. It is not pedagogically oriented. Nevertheless, the university physics students taking this astronomy course will themselves teach astronomy courses in the schools. A twelfth-grade course on the solar system has been approved for the natural-sciences branch of secondary schools. Ideally, we would like the new course to become a seed for more effective science education nationwide. But the IAU cannot support the direct training of school teachers. That would overwhelm our resources. The Vietnamese will have to provide the teacher training. We can only help in the planning of the course and the textbook, we can only suggest that the new course include some learner activities and address frequent misconceptions, and so the educational seed we plant may be very small indeed.

There is also an observational challenge. The Pedagogical University of Hanoi received, from the Sumimoto Science Foundation of Japan, a 40-cm Meade Schmidt-Cassegrain telescope which is large enough that students, in principle, can experience the excitement of scientific inquiry by participating in observational research projects. However, the telescope has been used only for astrophotography. There is a CCD and a new computer, but the telescope is poorly mounted and there is not yet a capability to use the CCD and to analyze the data. There is no experience in what constitutes quality data. As a first step toward improving the observational capabilities, TAD sent a staff member to attend the UN/ESA workshop in Toulouse and to meet with experienced observers in Toulouse who use small telescopes.

The first planetarium in Vietnam has opened in Vinh City, with a GOTO projector received as a Cultural Grant-in-Aid from Japan. The audiences are both school classes and the public. The first show begins with local interests and then expands to survey all of astronomy. Two new shows (one donated) from Davis Planetarium, Baltimore, U.S.A. await translation and some cultural adjustment before they can be shown in Vietnam. A second planetarium is to be built in Hanoi, with support from France. We would like to see the planetaria used with the schools in a western-style class-participation mode, but this would require considerable adaptation of the local teachers and teaching system, far beyond the capacity of the current planetarium staff.

Acknowledgments. I want to thank my two long-term collaborators and advisors in this project, Professor Nguyen Quang Rieu from Paris Observatory and Professor Nguyen Dinh Huan, who is Rector of Vinh University and Vice-President of the Vietnam Astronomical Society. I thank James C. White II, Astronomical Society of the Pacific, for his advice and participation. Part of the TAD program was supported by grants to the IAU from ICSU.

Discussion

Rieu commented that, in addition to the optical telescope, there is a radio interferometer built at the Radio Astronomy Centre at Nançay (Paris Observatory) and offered to the University of Hanoi on the occasion of the solar eclipse of October, 1995, (*Editor's note*; see Rieu's paper, pp. 255-265.) He also said that part of the problem with the 40-cm telescope was vibration of the floor on which it is installed. Until that is solved, the telescope will not provide good images. Kozai emphasized that he himself took the telescope and CCD camera to Hanoi and assured himself that they worked well. (*Editor's note*: see also Kitamura's poster paper, p. 312-3.) Boice asked about science education in the primary and secondary schools of Viet Nam. Are students entering the undergraduate astronomy course well-enough prepared in mathematics and science to be successful? Wentzel replied that he did not know enough about Vietnamese schools to give a complete answer. The undergraduates appeared to be familiar with the appropriate principles and formulae but to lack the ability to think critically. In reply to a query from Celebre, Wentzel said that countries wishing to be consider4ed for a TAD program should, in the first instance, approach Batten.

Astronomy for Developing Countries
IAU Special Session at the 24th General Assembly, 2001
Alan H. Batten, ed.

Revitalizing Astronomy in the Philippines

Cynthia P. Celebre and Bernardo M. Soriano Jr

PAGASA (Weather Bureau), 1424 ATB, Quezon Ave, 1104 Quezon City, Philippines. e-mail: cynthia_celebre@hotmail.com and bolonging_soriano@yahoo.com

Abstract. Aware of the possibility that astronomy in the Philippines will remain as lethargic as it has been for a hundred years if drastic changes are not made, various revitalizing activities were planned in 1997 and some have been implemented. These activities were divided into five categories and included the promotion of astronomy throughout the country and the attendance of some personnel of the Atmospheric, Geophysical and Space-Sciences Branch at various international meetings. Project proposals were also prepared and submitted to various local and foreign institutions in order to acquire astronomical equipment. The Philippines also applied for, and received, associate mebership of the International Astronomical Union.

1. Historical Background

Work in astronomy in the Philippines started in 1897. It was one of the functions of the "Observatorio de Meteorologico de Manila" (OMM), which performed not only meteorological and astronomical services but also seismological and terrestrial magnetism services. Its astronomical activities were mostly limited to timekeeping and observation of solar and stellar phenomena.

The OMM began as a private institution in 1865 and became a government agency as the Weather Bureau in 1901 with its observatory in Manila as its central office. During the Second World War, the astronomical observatory was destroyed and a new observatory was constructed within the campus of the University of the Philippines in Quezon City in 1954. It remained there up to the present time, now under the Philippine Atmospheric, Geophysical and Astronomical Services Administration (PAGASA), as the only government observatory. From 1954, the observatory has not seen any major change up to the present time (1999). The construction of a planetarium in the PAGASA Science Garden in Quezon City in September 1977 is the only addition to the facilities of the agency.

Likewise, its activities are practically the same, except for the publication of astronomical data and the conduct of occasional telescope and stargazing sessions up to 1993. The revitalizing activities in astronomy in the Philippines began in 1997 with the celebration of the centennial of astronomy in the country. This paper will describe these activities.

2. Resources and Activities in Astronomy

At present, only PAGASA, the National Museum (NM) Planetarium, and the Manila Observatory (MO) undertake activities in astronomy in the Philippines. The NM Planetarium is under the Department of Education, Culture and Sports and is located at the Rizal Park in Luneta, Manila. The MO is a private institution in Quezon City that also conducts geophysical and astronomical observations and researches, some in cooperation with PAGASA. The succeeding paragraphs describe the resources and activities of PAGASA in astronomy. A section deals with the programs of the agency to revitalize astronomy in the Philippines.

2.1. Resources

PAGASA, now under the Department of Science and Technology (DOST), is composed of nine branches or divisions. Each branch/division is composed of sections. A section in astronomy is under the Atmospheric, Geophysical and Space Sciences Branch (AGSSB), which is basically the research and training arm of PAGASA. The Astronomy Research and Development Section (AsRDS) of AGSSB is staffed by 17 professional-level employees who hold college degrees, and by 17 sub-professional-level employees.

Since 1969, PAGASA has established a solar-radiation observation network that consists of 52 stations as of 1999. Ten of the stations are operated by the National Solar-Radiation Center (NSRC) under the AsRDS, AGSSB while the other divisions of PAGASA run the rest.

The NSRC maintains the instruments of, and produces the sunshine cards, for the network. The network records sunshine duration, and global, diffuse, direct, infrared and ultraviolet radiation. The solar-radiation data are sent to international solar-radiation centers for collation with other worldwide observations and for publication. The processed data are returned to the NSRC for its database.

As earlier mentioned, PAGASA has an astronomical observatory and a planetarium, which are both managed by the AsRDS, AGSSB. The observatory has twelve telescopes of various sizes, the largest five of which are the 30-cm reflector and four 25-cm reflectors. The other telescopes are of the refractor type; two each of apertures 15 cm and 10 cm, while the rest are of 7.5-cm aperture. One of the 15-cm telescopes is permanently installed at the Observatory. There are six other telescopes, which are also of the refractor type and have an aperture of 7.5 cm, that are distributed among the field meteorological stations of PAGASA. In addition, private individuals and institutions own 38 telescopes. The biggest of these is the 58-cm reflecting Newtonian telescope at the Science Centrum in Manila.

Procured in June 1998, only the first 25-cm Cassegrain telescope of PAGASA is equipped with a CCD camera, which is still not functional because a suitable air-conditioning system is not yet in place. A photographic camera was acquired in 1997 to record special astronomical observations like eclipses and the appearance of comets and meteors. A small video camera was procured in October 1998 for similar purposes.

In addition to telescopes, the astronomical observatory is equipped with a quartz clock for determining the time up to the nearest tenth of a second. The clock is at least twelve years old and needs to be replaced by modern time-keeping equipment. The transceiver that was used for transmission of time to the field meteorological stations has been inoperative for almost seven years and has not been replaced due to budgetary constraints. A global positioning system (GPS) was acquired in 1998 for time comparison with the quartz clock.

The Planetarium at the Science Garden can seat 88 people. AsRDS personnel give astronomical shows to visitors, who are mostly students and teachers in elementary and high schools. The visitors are charged a minimal entrance fee to help defray part of the maintenance expenses.

2.2. Activities of the astronomy section

The principal activities of the astronomy section of PAGASA (AsRDS) consist of sunspot and lunar-occultation observations and solar- radiation measurements. It also makes observations of the satellites of Jupiter and the transits of Mercury, of comets and of other planets. From 1980 to 1984, the Observatory made observations of variable stars. It observes astronomical phenomena that are seen in the country such as the three total solar eclipses in the century, whose paths of totality crossed the Philippines. The astronomy section receives and subscribes to international publications and, in turn, publishes data that are derived from computations based on the publications. Among its important and widely used publications are the *Almanac for Geodetic Engineers*, rising and setting tables for the Sun and Moon, for the Philippines and for selected fishing areas, a daylight-duration table, and calendar (Julian) data, and lunar data which are intended for the observance of the Ramadan by Filipino Muslims. The publications are one of the principal sources of income of the AsRDS.

The other important activity of the AsRDS is time-keeping. The agency was designated by law to be the official time-keeper of the Philippines. It used to perform this function through a radio transceiver. However, its radio equipment has become inoperative so that even the meteorological stations in the field have not been receiving time signals for nearly five years now. The only means of disseminating time at present is the telephone.

Lastly, PAGASA engages in the promotion of astronomy, including space science in the Philippines, through its planetarium shows and the publication of astronomical posters. Its staff conducts and/or serves as resource speakers in lectures and seminars on astronomy and stargazing and telescoping sessions in various parts of the country. It coordinates and collaborates with other agencies or institutions in this field, such as the organization of astronomical societies in the colleges or universities.

Currently, there are ten astronomical societies with a total membership of about 300 students. It should be stressed, however, that there is only one educational institution that offers a course related to astronomy or space science in the country. No college or university gives a full course in astronomy in the Philippines.

Personnel who do not have any formal education in astronomy perform all of the preceding activities. The knowledge of astronomy that they possess is obtained through the in-service training courses conducted by the agency and through the books that were procured, usually from overseas sources.

The training courses in astronomy in the Philippines are very infrequent mainly due to the lack of resource persons or lecturers in this field. Only two courses during the last 15 years have been held. The first of these was conducted by the Manila Observatory for the professional-level "astronomers" of the PAGASA in 1981 while the last one was held by the agency for astronomical observers in 1993. The rarity of training courses can likewise be attributed to the lack of opportunities in astronomy in the Philippines, as there are very limited staff positions in the AsRDS, the sole unit that deals in astronomy in the PAGASA.

2.3. Revitalizing activities

Promotion of astronomy: Celebration of National Astronomy Week. On 24 February 1993, Presidential Proclamation No. 130 declaring the third week of February of every year as the National Astronomy Week (NAW) was issued through the initiative of PAGASA and the Philippine Astronomical Society (PAS), Inc. It stressed the universal appeal of astronomy as it encompasses all fields of human interest and endeavor. It emphasized that the study of astronomy, especially by the youth, will afford a better understanding of our planet, and thereby encourage them to participate more actively in the conservation and preservation of the environment. Moreover, it underscored the need to focus the people's attention on the significant contributions of astronomy in the advancement of the other sciences such as mathematics, physics, chemistry, biology, etc.

To celebrate the first NAW, PAGASA declared all its field stations open to the public for special lectures in astronomy. Admission to its Planetarium was free of charge during the Week. Telescope and stargazing sessions were held in several key cities, participated in by hundreds of elementary and high school students and science teachers from various regional centers of the Philippines.

Total solar eclipse of 1995. Astronomy further gained nationwide interest among Filipinos when the total solar eclipse of 1995 was announced by PAGASA in November 1994. A national committee, led by the DOST and PAGASA, was created by the President of the Philippines to promote its observation and documentation. An extensive information campaign and lectures on the eclipse were conducted in many parts of the country, especially in Mindanao where the central path of the eclipse will cross the island province of Tawi-Tawi.

The event was highlighted by the attendance of then Vice President, later President Joseph Estrada, and many top government officials, including the Secretary of the DOST and the Director of PAGASA, to observe the total solar eclipse on 24 October 1995. Two local TV networks set up satellite transmitting equipment at Languyan, Tawi-Tawi, to beam the event nationwide. The TV networks also installed remote transmitting systems in several major cities in northern Luzon, central Visayas and southern Mindanao. Thousands of people, including members of astronomical societies and the radio and press, came to the site. Hundreds gathered in the PAGASA stations where a telescope was available for public viewing of the eclipse. A hundred eyepieces made of carton and mylar film were distributed by PAGASA to students and teachers in Tawi-Tawi and to TV crew and cameramen. The Philippine Long Distance Telephone Company set up a temporary satellite telecommunication facility in the area for the use of press reporters.

Discovery of Comet Hyakutake. The discovery of Comet Hyakutake on 30 January 1996, barely two months after the total solar eclipse of 1995, heightened the level of interest in astronomy in the Philippines even more. The public was made aware of the discovery through a press release issued by PAGASA in mid-February 1996. Increased concern by the public about the possibility of Comet Hyakutake crashing into Earth was perceived through numerous inquiries received by PAGASA as a result of the collision of Comet Shoemaker-Levy with Jupiter from 16 to 22 July 1994.

Celebration of the Centennial of Astronomy in the Philippines in 1997. To sustain the momentum gained during the past two years, PAGASA initiated the issuance of Presidential Proclamation No. 956, declaring 1997 as the Philippine astronomy centennial year and constituting a national committee for its celebration. The PAGASA/DOST was designated as the lead/chair agency with the National Museum of the Department of Education, Culture and Sports (NM/DECS) as vice chair, and the Philippine Information Agency and Manila Observatory as members. Among the activities implemented during the year are the conduct of: (i) information, education and communication (IEC) program in astronomy through press releases and radio/TV interviews; (ii) seminar/workshops for science teachers and students; (iii) symposium during the NAW; (iv) distribution of astronomical posters, grant of honor and recognition awards to five outstanding Filipino astronomers, and carrying out a contest on astronomy, termed as Astro Olympiad; and (v) preparation of a commemorative centennial publication.

The fortuitous discovery and appearance of Comet Hale-Bopp from 20 March to 11 April 1997 lent beneficial support to the celebration of the centennial of astronomy in the Philippines as the interest of the public was maintained at a high level from its discovery up to its disappearance. This momentous event elicited the remark that "even the heavens are celebrating with us" from the chief of the AGSSB.

Astro Olympiad. Astro Olympiad '98 was one of the activities in the DOST-GIA project, entitled "Promotion of Astronomy". The Olympiad was given full support by Dr. William G. Padolina, then secretary of the Department of Science and Technology, who wanted a science-oriented youth and gave encouragement to make the contest an annual activity.

It was a contest in astronomy for high-school and college students in the Metro Manila. The contest consisted of the elimination and final rounds, where 100 and 10 contestants in each level, respectively, were selected. The elimination round was a written examination held at the Philippine Science High School on 27 June 1998. The final round was an oral contest, held at PAGASA on 03 July 1998. A panel of five judges supervised the contest.

The 10 finalists and the winners for each level were chosen based on the highest scores. Ties in the finals were broken by additional questions for the tying contestants. Prizes were in cash and certificates of recognition for the finalists. Cash prizes (in Philippine pesos) amounted to P25,000 (U.S. $500) for the first prize, P20,000 (U.S. $400) for the second, P15,000 (U.S. $300) for the third and P5,000 (U.S. $100) each for the seven non-winners in the college level. In the high-school level, the cash prizes were P10,000 (U.S. $200), P5000 (U.S. $100), P2500 (U.S. $50) and P1000 (U.S. $20), respectively.

In 1999, the contest covered the entire Luzon area, which is composed of seven administrative regions. It applied the same contest rules used in Astro Olympiad '98. The only difference is the increase in prizes of the high-school winners which are P20,000, P10,000, and P5,000 for the first, second and third prizes, respectively.

Grant of Honor and Recognition Award.   During the awarding ceremony for the winners of the Astro Olympiad in 1998, plaques of honor and recognition were also distributed to five outstanding Filipinos who have made significant contributions in the field of astronomy. A screening committee led by PAGASA was formed which selected the awardees from those who were Filipino citizens at the time of their application/nomination and whose achievements in astronomy are exemplary. Filipino expatriates who have distinguished themaelves in astronomy were also considered for these awards.

The plaque was named the "Dr. Casimiro del Rosario Award", in honor of the first Filipino Director of the Weather Bureau after the Second World War, who pioneered in astronomy during his tenure. One of the awardees is the retired Director of the then National Geophysical and Astronomical Office, who was chosen because of his important contributions to the IGY Moonwatch Satellite Tracking Program. He also started the the publication of various astronomical data by PAGASA.

The second awardee is a priest who is presently Head of the Upper Atmosphere Division of the Manila Observatory. He is an active member of the American Geophysical Union and has been President of the Philippine Astronomical Society. He has written more than forty papers on astronomy and about the same number on physics and other various topics.

The third recipient of the award is presently the head of the National Museum Planetarium. His most significant accomplishment in astronomy is the promotion of that subject among young people through lectures and telescope and stargazing sessions.

The fourth and fifth awardees are a husband-and-wife team, now based in the United States. This couple's continuous efforts in observing led to the determination of the orbit of asteroid 6282, discovered by Carolyn Shoemaker in 1980, and were also recognized in 1995 by the IAU's Minor Planet Center. The husband is the first Asian to be an Associate Editor of *Sky and Telescope.*

A special (posthumous) award of recognition was also given to Dr Casimiro del Rosario for his significant contributions and pioneering work in astronomy.

Astronomical Diary.   In July 1998, PAGASA announced the issuance, on a monthly basis, of information about some significant astronomical events, in addition to the daily sunrise and sunset data. These publications, called the

Astronomical Diary, depict the phases and the positions of the Moon, the Sun and the stars and some of the planets and its satellites. The primary aim of the *Diary* is to enhance the interest and elevate the level of knowledge of Filipinos in science, in general, and in astronomy, in particular.

The *Diary* received enthusiastic responses from the media and the public, particularly the students. This interest was shown during the occurrence of the Leonids meteor shower in November. The meteor shower drew marked interest due to the predicted Leonids meteor storm.

Seminar/Workshop on Meteorology and Astronomy for Science Teachers. The seminar/workshop is a proposal which was submitted to the Department of Science and Technology for possible funding under its DOST GIA fund. The main objective of the project is to enhance the knowledge and skills in meteorology and astronomy of elementary-school and high-school science teachers to enable them to impart these effectively to their students. The two-week seminar/workshop will cover the entire country, which is composed of 15 regions. To be able to implement the project in these regions, it has a duration of 18 months.

2.4.   Attendance at Various International Meetings and Workshops

The participation of the Chief of the AGSSB in two workshops, jointly organized by the UN Office for Outer Space Affairs (OOSA) and the International Astronautical Federation (IAF), in Graz, Austria on 15-22 October 1993 and in Jerusalem, Israel on 10-14 October 1994, paved the way for increased interest in space sciences in the Philippines. His attendance at the UN/ESA Workshop on Basic Space Sciences: "From Small Telescopes to Space Missions" in Colombo, Sri Lanka on 11-13 January 1996, enabled the establishment of linkages with international space scientists and astronomers by the Philippines. This is considered as very important for the country as it enabled the Chief of the AsRDS to attend the subsequent workshops organized by the UN OOSA in Bonn, Germany, on 9-13 September, 1996, and by the International Astronimcal Union (IAU) in Kyoto, Japan, on 18-30 August, 1997, and the meeting of the Committee on Peaceful Use of Outer Space (COPUOS) in Vienna, Austria, on 03-14 June ,1996.

During these workshops, she was introduced to several prominent astronomers, who became sympathetic to assist the Philippines in strengthening its astronomical service after she presented a paper on the state of astronomy in the Philippines in the workshop in Bonn.

2.5.   Preparation of Project Proposals

Donation of a 45-cm telescope by the Japanese Government. A project proposal on the possible donation of a 45-cm astronomical telescope by the Japanese Government to the Philippines was prepared by the Chief, AsRDS, in 1997 as a result of her attendance in a workshop in Bonn in 1996 and the IAU GA in Kyoto in 1997. The proposal is aimed to upgrade the capability of the PAGASA Observatory to observe and to study astronomical objects as well as to enable its AsRDS staff to conduct astronomical researches in the future.

On 19-22 January 1998, Profs. Tomokazu Kogure and Masatoshi Kitamura, respectively, Director of Bisei Astronomical Observatory and Guest Worker of the National Astronomical Observatory of Japan, visited, in their personal capacity, the PAGASA Astronomical Observatory, the Department of Physics of the University of the Philippines, and the Manila Observatory. Their missions were not only to assist in the revision of the project proposal but also to assess the current status of research and education in astronomy in the Philippines. They also gave brief lectures in astronomy to some PAGASA astronomers. On 28 March 2000, the donation was formally approved in a ceremony held at the Department of Foreign Affairs (DFA) in Manila. The DFA Secretary and the Japanese Ambassador signed and exchanged diplomatic notes.

DOST-GIA Project Proposals. (i) Promotion of Astronomy: In May 1998, four 25-cm and one 18-cm telescopes, one video camera, and one CCD were procured through the DOST Grant-in-Aid (GIA) project "Promotion of Astronomy". The telescopes will be installed in four local regional centers and the CCD at the PAGASA Astronomical Observatory. The main objective of this activity is to promote astronomy in the countryside by allowing the public, particularly the students and the science teachers to observe the occurrences of important astronomical phenomena, thereby increasing their knowledge in this field, if not arousing their interest.

(ii) Upgrading of Time-keeping Equipment: PAGASA, by virtue of Presidential Decree No. 1149, is mandated by the government as the official time-keeper of the Philippine Standard Time (PST). The Time Service Unit (TSU), a unit under the AsRDS, performs this mandate through the maintenance and dissemination of the PST using its very old timing equipment.

To be able to perform its mandated task to the fullest, this project was prepared with a general objective of providing the national standard for the second of time and frequency to industry, science and technology and other users. The objectives of the proposal will be accomplished through the acquisition of a modern astronomical timekeeping system.

(iii) Three Portable Planetaria from the French Government: Since the two planetaria in the Philippines are located in the island of Luzon, PAGASA realized that the people in the islands of Visayas and Mindanao do not have easy access to the planetaria. To bring astronomy to these places, particularly the remote areas, the AsRDS requested from the French government, through its French Protocol Project in PAGASA, a donation of three (3) portable or mobile planetaria.

Aimed to promote astronomy in the countryside, these planetaria will be brought to these places upon their request on a first-come, first-served basis, at a very minimal fee. The planetariums will be called the "Roving Planetaria". When these materialize, the public can be educated in astronomy with the least time, money and effort.

Application to Different Foreign Training Courses and Fellowships in Astronomy
Since there is no university in the Philippines that offers a course in astronomy, some AsRDS personnel will apply for a Monbusho fellowship. They plan to take up a Master of Science degree in Astronomy.

Other AsRDS employees are also eagerly waiting for the International School for Young Astronomers, which is scheduled in Thailand in 2001. The AsRDS hopes that some of its personnel will be able to attend this prestigious training, to enable them to gain additional knowledge in astronomy, which may allow them to pursue research activities in this field.

In connection with the arrival of the 45-cm telescope donation from the Japanese Government in September 2000, the Chief, AsRDS undertook a familiarization study on modern telescopes at the Gunma Astronomical Observatory (GAO) of Japan on 06 - 12 March 2000 upon the invitation of Prof. Yoshihide Kozai, GAO Director. Although she has completed this course, she plans to stay at the Observatory for six months or more to undertake a training in astronomy, learn the basic techniques in astronomical observation and conduct an astronomical research. A request for support from the JICA office was made to enable her to realize the plan.

The AsRDS personnel are very optimistic that they can perform astronomical research if they can receive formal training in astronomy. Hence, these fellowships are the key factors that will hasten capacity-building in astronomy in the Philippines.

Application for Associate Membership in the IAU In June 1999, the Philippines through PAGASA submitted its letter of intent to the General Secretary of the IAU to apply for Associate Membership to the Union. The application was made in view of the possible donation of the 45-cm telescope from the Japanese Government. PAGASA believes that the Union can give assistance towards its new mission of revitalizing astronomy in the country and allow the AsRDS personnel to participate actively in regional and global cooperative activities in astronomy. The application was approved during the twenty-fourth General Assembly of the IAU in August, 2000.

3. Conclusion

Given the preceding information on the past and present resources and activities in astronomy in the Philippines, it is not difficult to make a projection of the status of the science in the near future. In 1996, it was foreseen that, in the next decade, the development of astronomy in the country will remain as lethargic as it has been for the past four decades, unless drastic positive changes were implemented.

Aware of this possibility, the Chief, AGSSB and the AsRDS staff, headed by its Chief, worked relentlessly since then and spent every effort to promote astronomy throughout the country. Likewise, it was very fortunate that the attendance of the Chief, AsRDS in the UN/ESA workshop in astronomy in Bonn, Germany paved the way towards the wider opening of the door of the international community for the country. She was able to meet some of the important people of the IAU, UN OOSA and the Japanese astronomers who are promoting the Cultural Grant-aid Program of their Government.

The project proposal on the possible donation of a 45-cm telescope from the Government of Japan is considered the best positive drastic action that was made that may lead towards a better prospect for astronomy in the Philippines. The installation of the computer-based telescope will signify a new beginning of astronomy, particularly in the field of research. To celebrate this, the Philippines will bid to host an international workshop on basic space science in the year

2001, coinciding with the inauguration of the new telescope in the PAGASA Astronomical Observatory.

Ultimately, the successful implementation of some of the revitalizing activities described in this paper and the approval of the application for associate membership of the country to the IAU, will enable the Filipinos to hope for a better and brighter future for astronomy and space science in the Philippines.

Discussion

Hidayat enquired about the role of the Jesuit Father Hayden. Celebre replied that he had indeed done a lot to promote astronomy in the Philippines and had been one of the lecturers in the 1981 in-house training course of PAGASA. In reply to Orchiston, who asked if interest in the Leonid meteor shower, as well as the solar eclipse and Comet Hyakutake, had been used in revitalizing astronomy, Celebre said that it had. Hingley asked if there were indigenous traditions and legends that could be used to spark interest in astronomy. Celebre replied that there are some, which are taught in primary schools, but she thought it would be difficult to relate them to a modern scientific approach.

Martinez said that, since there is not yet any astronomy at university level in The Philippines, before creating a department of astronomy they might consider introducing some astrophysics (10% to 20%) into physics courses. He pointed out that Wentzel had pioneered this approach with his *Astrophysics for Physics Courses* and might be able to offer some practical advice. Celebre expressed thanks for the suggestion.

Astronomy for Developing Countries
IAU Special Session at the 24th General Assembly, 2001
Alan H. Batten, ed.

Astronomy Development in Morocco: a Challenge to Stimulate Science and Education

Khalil Chamcham

King Hassan II University - Ain Chock, Faulty of Science, B.P.5366, Maarif, Casablanca, Morocco. e-mail: chamcham@star.cpes.susx.ac.uk

Abstract. From my experience in Morocco, I discuss the difficulties one can face while trying to set up projects in a country where astronomy is a forgotten science: everything has to be built from scratch and, at the same time, one is required to keep up the pace at the international level. But, on the other side, it is quite a relief to see the strong demand from students and the public. In these circumstances even professional astronomy cannot survive without feedback from the public and long-term investment in education at all levels.

1. Introduction:

In the early 1980s the Moroccan government officially encouraged the creation of an astronomy unit at the CNR, the Moroccan national institution coordinating scientific research. At that time, a small group of people trained in France was involved in setting up a network for solar oscillations, IRIS. One of the tasks of this group was to coordinate astronomy nationwide, in order to create other units in universities and engineering schools, and facilities were consequently provided to achieve this goal. Researchers from the university of Casablanca, coming from different areas in physics, established contacts to work in this context, but very soon the relationship between the coordinating group and its mother institution became very confrontational and the unit was diluted.

Also, from the beginning, the IAU expressed its support for Moroccan astronomy and Professor Jorge Sahade, acting as the IAU president, made an official visit to Morocco to express to the Moroccan authorities the support of the IAU for astronomy development. His visit was followed by the organization of an ISYA, in 1991, in Morocco, which was a very successful school. At the same time Japanese officials expressed their willingness to support Morocco in the development of astronomical facilities (i.e. a planetarium and an observatory).

This was a great period when Moroccan astronomy could have gone very far with this local and international encouragement had not some individuals in administrative positions failed to see the opportunity to invest in science and in the national coordination of astronomers scattered around the country. Unfortunately Moroccan astronomers lost this chance to organize themselves and to develop national facilities. From then on the above mentioned conflicts became a justification for unpleasant harassments.

2.   Present-Day Situation of Astronomy in Morocco

Despite this negative picture, some individuals carried on their intiatives and developed activities or units as part of the phyiscs department of their institution. These activites have been developed in the following places:

2.1.   Rabat

- At the CNR, two people carried on working on tomography and solar oscillations.

- One person moved from the CNR to the University of Rabat, but did not develop any activity. However he is mainly involved in amateur activities as part of a cultural association.

- The United Nations has designated the engineering school, EMI, as the home of the UN space Centre in Africa and a satellite project. This Centre is operational in principle, but it is not open to other people of the school. Also, the Moroccan Ministry of Research and Higher Education has initiated the creation of a national network for space research (RUSTE) wich is based at the same school. The activities of this network have been limited to local meetings and popularisation.

- Public Astronomy: a cultural association has built a small observatory to house a 50-cm telescope donated by the French government and a small planetarium. These facilities are in principle open to the public.

2.2.   Marrakesh

Another initiative was carried on by an individual at the University of Marrakesh, previously involved in the solar-oscillation project IRIS. He could include a course in astronomy as part of the high-energy physics group in the University. Also, he developed some amateur activity in the city of Marrakesh.

2.3.   Oujda

A couple of physicists have been working on astro-particles in collaboration with Italian groups. They have been active in developing research projects in connection with high-energy astrophysics and recently they submitted an M.Sc. project to the Moroccan authorities.

2.4.   Casablanca:

Since 1988, local astronomy meetings have organised at the University of Casablanca, as part of the activities of the nuclear and theoretical physics group and with the collaboration of French astrophysicists. These activities have been a stimulus, motivating students to come to astronomy and generating a gradual need to include astronomy as part of the physics curriculum. In fact, an undergraduate course and a postgraduate course have been developed and some of the students are fairly well advanced in their work.

There is now an Astronomy and Astrophysics Group structured as follows

- Two full professors,

- five Ph.D. students: one is now at SISSA and four are working between Casablanca and Italy with grants offered by Roma Observatory and the University of Pisa.

- Six postgraduate and seven undergraduate students.

- The curriculum focuses on fundamental astronomy, astrodynamics, stellar structure and evolution, galaxy formation, cosmology, numerical methods and English (i.e. the postgraduate course is attended by non-astronomy students from plasma physics, theoretical and particle physics).

- Research projects of the Group are concerned with celestial mechanics, CCD photometry, solar physics, stellar evolution, galaxy evolution and cosmology.

- The facilities available to the Group are: 1 PC, the IAU traveling telescope + CCD, the Edinburgh teaching package, *Hands On Astrophysics* and a collection of textbooks and proceedings. Most of this equipement has been acquired thanks to the support of the IAU, the UN Office for Outer Space Affairs, and a contribution from our faculty.

3. The Public Demand

The administration of academic institutions in Morocco is heavily bureaucratic and often appears to work for its own sake, as if the administrators have forgotten that they were nominated to serve the interest of students and staff. It sometimes seems as if science itself is looked upon as a marginal activity and that asking for budgeted money to be spent on the needs of students or the support of research is seen as a threat to the system, to be countered by personal harassment. These attitudes influence even some of the academics so that a small group trying to modernize teaching and research needs very strong international support. This is the situation of astonomy in Morocco and each step forward has been a real struggle.

Security is constant concern in our society and this also affects astronomy. For example, access to the Internet, seen as a source of subversive information, is strictly controlled. A telescope may be viewed as a device that can see far into a city (rather than deep into the sky) and take pictures: it may be prohibited for security reasons, although this may not be said explicitly.

Where science and culture cannot make progress, the irrational grows rapidly. On the other hand, people struggle for survival and want a good education, either for themselves or at least for their children. They want that education to include modern science and what the media show about the results of space research and other new discoveries in the developed world. From this perspective, astronomy benefits from wide public support and if, the administration has recognised the importance of astronomy, it is thanks to public pressure and the demand for astronomical activities from the students themselves.

4.   The Role of the IAU and International Collaboration

In the case of Morocco, the breakthrough came from the investigation of Alan Batten when he visited the country on behalf of the IAU. His understanding of the situation with all its ups and downs helped not only to give confidence to the astronomy community but also to prepare the administration to a more open position toward it.

Batten's visit allowed the preparation of a TAD contract between the IAU and the University of Casablanca, which was finalized during a visit by Donat Wentzel and Johannes Andersen. This had a great positive impact, moderating the attitude of the administration toward the astronomy community and establishing self-confidence and dynamism amongst students and the astronomers themselves. This can be understood if one knows that students who wished to follow the undergraduate astronomy course were pressured, or news was spread from unknown sources that the astronomy course was cancelled.

The other initiative which backed up astronomy teaching and research in the University of Casablanca is the kind support received from the Italian Astronomical Society (SAIt) and the Universites of Sussex and Glasgow. The SAIt provided travel grants and fellowships to students and staff members whereas the latter offered remote computing facilities and visiting grants. Particularly, the Osservatorio Astronomico di Roma has been very supportive in this respect and Pisa University is now donating computing facilities and providing training courses to all students. The Osservatorio Astronomico di Trieste has also been supporting research projects and the training of students. Thanks to these collaborations, we could set up research projects and strengthen our curriculum.

5.   The Importamce of Computing

Our teaching programmes are out of date and lack any sense of creativity. The undergraduate science programme is still based on the physics and mathematics of the early twentieth century and, in most advanced-level courses dates back to the early 1950s. When one sees that the table of contents of our official curriculum is nothing but a photocopy of the tables of contents of French textbooks, one can understand the effort involved to think of what kind of science we need to catch up with modern development and what kind of citizens we want to produce.

Considering these facts and what I have said above, it became clear to our Astronomy and Astrophysics Group that investment in computing is fundamental to back up students' projects and research. Also, in the absence of access to library facilities and journals, the Internet is now playing a major role in education and research by allowing people to access most of the necessary information they look for. It is essentially a way to introduce students to modern science without the need to go through the official channels, which in any case will never respond to any suggested reform of the curricula (modifying or replacing one single chapter of a course may require going through all the official institutions of the country, which is deterring).

The good thing about astronomy is that it is a science which fascinates people and excites their imaginations. This allows one to build strong relationships with people from all generations and social backgrounds. With a few pictures and some hand-waving, one can explain complex physics to "gran'ma" and even justify the technological and economical impact of astronomy on society, not excluding the cultural feedback.

In the case of Morocco, astronomy showed that even educated scientists who abandoned any hope of carrying on their scientific career could discover some joy by becoming interested in astronomy. The experience with the public also showed that people, at the end of a lecture or an open day, discover that the study of the sky provides a way to modernity and higher education, as long as they have the right to have a sky above their head.

6. Conclusion

It is exciting to carry this endeavour of bringing new hope and new light to individuals and a whole country through the mother of sciences, astronomy. I found it necessary to describe the social and ideological environment where astronomy used to be an honorable science, but is nowadays struggling for survival. This experience shows that societies do not always aim for the best, but a minority of individuals can aim for it and take the lead. The most positive aspect of this experience is the strong and permanent support amongst the astronomical community without which some astronomers could not survive in their own country. But wherever we go it is thanks to the taxpayers that things can change and improve, not only because of their money, but by their strong voice expressing what they want for their lives. I am glad that astronomy is teaching us that science is not isolated from the rest of our human experience.

Acknowledgment

I acknowledge the IAU financial support to attend the GA 2000 in Manchester and for the IAU-TAD programme in Morocco. I am very grateful for the permanent support of Alan Batten, Donat Wentzel, Derek McNally, John Percy and Hans Haubold. The AAVSO is kindly providing educational material.

Discussion

Hearnshaw noted that in many Moslem countries, particularly Malaysia, astronomers have successfully gained funds by using the link between Islam and astronomy. He wondered why Chamcham was reluctant to do this. Chamcham replied that he was aware of the Malaysian experience and had himself doe some work on Islamic astronomy. Funding applications to Moslem institutions, however, were compicated by bureaucratic procedures and often unsuccessful.

Chambliss asked which of Arabic, French or English was the language of science instruction in Morocco. Chamcham replied that the official language at high-school and college levels is Arabic and at university level it is French. Only a few people at his university spoke English, although he taught it as part of the postgraduate physics and required mature students who came to astronomy to take private tuition in English.

da Costa suggested that colleagues in Spain and Portugal could assist Moroccan astronomers. Chamcham replied that he had a few contacts with Spanish colleagues and IAU representatives had suggested that he send students to Spain for training in observational techniques. He had met Portugueses students in the U.K. but otherwise knew only da Costa himself. He would be happy to base future collabotration on these relationships.

Astronomy for Developing Countries
IAU Special Session at the 24th General Assembly, 2001
Alan H. Batten, ed.

Experience in Developing an Astronomy Program in Paraguay

Alexis E. Troche-Boggino

Observatorio Astronomico "Alexis Emilio Troche Boggino", Universidad Nacional de Asunción, Facultad Politécnica, 01 Agencia Postal Campus U.N.A., Central XI, Paraguay. e-mail: atroche@pol.com.py

Abstract. Encouraging IAU-ISYA schools were held in Argentina (1974) and Brazil (1977 and 1995). Contacts with IAU Commission 46 were soon established and the author became the responsible person for setting up a supporting framework for astronomy in Paraguay. A proposal for an IAU-VLP program was made and the program began in 1988, running until 1994. The IAU Traveling Telescope and another telescope were borrowed for hands-on astronomy. Three students went to the Brazil ISYA in 1995 and two others began to work as astronomers abroad. Others became high-school and university instructors. A small library was opened, thanks to donations from IAU members. The need to find jobs for these young astronomers led to the further development of an Astronomy Center around the National Astronomical Observatroy.

1. Introduction

Please, do not confuse Paraguay with its neighbor country Uruguay. Paraguay is a land-locked country between Argentina, Brazil and Bolivia. Its population is over five million inhabitants and its capital city is Asunción (latitude -25° 18', longitude 57° 35' W).

I became active in astronomy education 26 years ago. There was little astronomy taught in my country at that time, but I found opportunities for training high-school teachers. A new program for secondary schools was in a trial stage at that time. Basic astronomy topics were requested as a part of three high-school courses in natural science and as a part of the course in mathematics for the last year. Also, an elective astronomy course for junior students of physics were on schedule at National University of Asunción.

So, I used such opportunities to practice astronomy teaching as a university instructor and later I started to train high-school teachers in topics related to our science. I still do that and I have some former students in the same job at the Instituto Superior de Educación of the Ministry of Education and Culture.

Encouraging IAU-ISYA (International Schools for Young Astronomers) were held in Argentina in 1974 and later ones in Brazil in 1977 and 1995. Contacts with IAU Commission 46 were soon established and I learned what to do to start an astronomical supporting framework for my country.

Let me quote a motto from Dr. Cecylia Iwaniszewska, to whom I am grateful for continous kind assistance. "In order to reach each new generation with basic astronomy, we need to show its memebers the most outstanding facts, as well as to build a supporting frame".

What is the supporting frame's relevance? First, it means good reading materials, practical work, well-trained teachers, an astronomical observatory, a planetarium etc.

Second, with a poor framework it may happen that students may learn poorly and come to believe that they are experts. This means, in turn, instruction will become sketchy and few real experts will get good jobs. In my experience, one of the greatest difficulties in developing countries is that many people, both in government and universities, get their jobs not by their skill and knowledge, but through politics and personal connections. So many of our best students go abroad.

There are problems with the English language, which is less well-known the farther a country is from North America. This is bad for our students who cannot read some excellent astronomy pages on the Internet. What advice can you give to overcome this language barrier?

Among important visitors from the IAU astronomers were J. Sahade and R.M. West. These visits led to a Visiting Lecturers Program (VLP) for Paraguay, which began in 1988 and ran until 1994. I worked as local coordinator of six excellent astronomical courses with astronomers from Argentina, México and Italy. In particular, D.G. Wentzel's readiness to help, providing supporting material and visiting, improved the VLP.

Among the results of the IAU-VLP, two students went abroad for further study and practical work in astronomy. Another two attended the ISYA in Brazil in 1995. Two former VLP students are doing their doctoral work in physics abroad. Other former students are university and high-school instructors. A small astronomy library was has been formed with the help of donations from Canadian, American and other astronomers.

The IAU Traveling Telescope was lent for the last two courses in Astronomy. Mr. and Mrs. Eduardo Parini, local amateur astronomers, also lent us a 20-cm Celestron from their own private observatory. We consider the Parinis to be our "Maecenas" because of their generous assistance with their own equipment, maps etc. They acted as hosts, in their home, to many visiting professors and astronomers, and opened their observatory to students, amateurs and the public.

Natural events are of course useful. The visit of Halley's comet led to the formation of a first amateur astronomers' club in 1985 and two others were formed at the time of the total solar eclipse of November 3, 1994.

The president of one of these clubs, Dr Blas Severin founder of the Asociación de Amigos de la Astronomía, organizes regular star parties and gives lectures for beginning amateurs. Furthermore, he founded three amateur clubs at high schools.

A third group, the Sociedad de Estudios Astronómicos, publishes a bimonthly bulletin called *Antares*. Their members teach astronomy to the public and are active participants in international meetings. They are organizing a

Figure 1. The Astronomical Observatory of the Universidad Nacional de Asunción. The Observatory was inaugurated on June 6th, 2000 and named after the author of this paper.

meeting for Latin-American amateur astronomers next year. Thus, amateur astronomers help to build a strong framework for beginning an astronomy program.

However, not many of our physics students get really involved in astronomy activities because of the lack of jobs and low salaries. We needed an "attractor" for them. Our project to obtain instruments for a professional astronomical observatory began to be realized last year with the donation of a 45-cm Cassegrain reflector, a photoelectric photometer of high accuracy and CCD cameras and image-processing equipment by the Japanese Government O.D.A., last year.

The Astronomical Observatory of Universidad Nacional de Asunción was inaugurated last June 6th, at the University's Polytecnic School. A young physicist, Mr. Fredy Doncel was trained in Japan for six months in the use of the instruments. He is doing well in his new job.

We are looking for a senior veteran astronomer from abroad, to help us to work with photometry of binary stars and research on variable stars. Further, we welcome astronomers from the IAU who would like to assist by visiting us and giving astronomical lectures. We are particularly grateful to Drs D.G. Wentzel, R. Vicente and M. Kitamura for the many lectures they gave us few years ago.

Dr. J. R. Percy visited us for about a week in 1997. He did a remarkable job presenting several lectures, which were very useful, particularly for our astronomy instructors. Further he suggested to us the development of an Astronomy Center in Paraguay and he gave a strong support to the project of an astronomical observatory.

We would like to have your advice and help for further development of this project. Please, remember our motto: "Let all students who wish to learn astronomy do so as far as they are able and let us give them the opportunity as much as possible".

Discussion

Dworetsky commented that some papers are nearly incomprehensible in English; automatically translated versions may be no better! Troche-Boggino still thought it useful for people to have versions in their own languages. In reply to Rijsdijk he said that there were no charges for visiting the Observatory (on two days a week) but arrangements should be made in advance.

Astronomy for Developing Countries
IAU Special Session at the 24th General Assembly, 2001
Alan H. Batten, ed.

The Central American Master's Program in Astronomy and Astrophysics

Maria C. Pineda de Carias

Central American Astronomical Observatory of Suyapa, National Autonomous University of Honduras, Tegucigalpa MDC, Honduras. e-mail: mcarias@hondutel.hn http://www.astro.unah.hondunet.net

Abstract.

The Master's Program in Astronomy and Astrophysics for Central America arises as part of the project of the National Autonomous University of Honduras to contribute to the establishment of "Astronomy and Astrophysics" as an academic field within the region (Pineda de Carias 1993). In 1997, the same year that the Central American Suyapa Astronomical Observatory (CASAO) was officially inaugurated (within the frame of the VII UN/ESA Workshop on Basic Space Science), a degree course in astronomy and astrophysics at graduate level was approved. In 1998 the program was formally opened for Central American graduate students in physics, mathematics or engineering. In the year 2000, the first group of students is expected to finish their courses. In this document we present the main features of the Master's Program: the syllabus, resources, organization. A discussion of the results achieved and of future tendencies is also included, together with some recommendations about how the international community may contribute to the enhancement of this type of effort, and on how this model may be useful for developing countries.

1. Introduction

The Master's Program in Astronomy and Astrophysics for Central America arises as part of the project of the National Autonomous University of Honduras to contribute to the establishment of "Astronomy and Astrophysics" as an academic field within the region. In 1997, the same year that the Central American Suyapa Astronomical Observatory (CASAO) was officially inaugurated (within the frame of the VII UN/ESA Workshop on Basic Space Sciences, see Pineda de Carias 1997), a degree course in astronomy and astrophysics at graduate level was approved. In 1998, the program formally opened for Central American graduate students with a background in either physics, mathematics or engineering. Now, at the change of the millenium, the first group of students is expected to finish their courses, research projects and to obtain their degrees.

In this paper, a brief description is presented of what we have achieved during these past three years of intensive work. Seven components are briefly described: (i) students, (ii) staff, (iii) curriculum, (iv) research, (v) academic organization, (vi) cooperation and (vii) budgets.

A discussion of the results and future tendencies is included, together with some recommendations on how the international community may contribute to the enhancement of this type of effort and on how this model may be useful for the benefit of developing countries.

2.   Description

2.1.   Students

Registration. In August 1997, the Central American Astronomical Observatory of Suyapa announced for the first time its Master's Program in Astronomy and Astrophysics. At that time, 24 students from all the Central American countries, proposed by their national universities, showed interest, but at the time of registration for the propaedeutic course only ten of them formally applied. In 1998, the first group of graduate students (MAA-98) was formally accepted, four out of ten registered for the propaedeutic (see Table 1 – data for this and all tables come from Documents and Archives 2000).

Table 1.   Registration

Country	MAA-98				MAA-99
	1997	1998			1999
	Showed Interest	Propaedeutic Courses	MA-98	Showed Interest	MAA-99
Honduras	6	5	3	8	4
Guatemala	6	4	0	0	0
El Salvador	4	0	0	0	0
Nicaragua	2	1	1	0	0
Costa Rica	3	0	0	0	0
Panama	3	0	0	0	0
Total	24	10	4	8	4

The reason for this decrease 24:10:4 was very likely due to, first, their background. Astronomy and astrophysics are not for everybody and a non-solid background in mathematics or physics forced some of them to quit. Secondly, there are economic reasons. In Central America, a great number of students do not have enough money to pay for graduate studies and their families do not accept the idea of having a student member not earning money for a living.

In 1999, with the financial support of the Organization of American States (OAS –see Proyecto OEA-040/98, 1999), the CASAO invited a second group of students for the Master Program (MAA-99). This time, for the selection process, instead of a propaedeutic course we followed a different approach. A letter of intention, a paper on a topic in physics or astronomy and an interview were required. At the end, only four out of eight students remained.

Progress.   The Master's Program covers 20 courses plus a research project. The MAA-98 Group took six courses in 1998, another six in 1999 and this year (2000) they took the eight remaining courses. All of them are now working on their research projects (RP). The MAA-99 Group took seven courses last year (1999) and are taking nine courses in 2000; in 2001 they will take the last four courses and begin their research projects. In our grading scale, the students need to obtain at least 80% to be approved in a course.

Out of the total of four students who began in 1998 in Group MAA-98, only one abandoned the program. The same happened with the 1999 Group. In 1998, the one who failed in 1998 came from the Physics Department; the one who failed in 1999 came from the Mathematics Department. It seems that only those students related directly to the astronomical observatory are better prepared to succeed.

Table 2.   Progress

Group	20 Courses 1998	Research 1999	Project (RP) 2000	20001
MAA-98	6	6	8+RP	
MAA-99	0	7	9	4+RP
MAA-01				6
Total	6	13	17	10

Students spend most of the day at the Astronomical Observatory. They may also attend on weekends. Some of them live at the Observatory. On a regular day, we have graduate courses in the afternoon but during the morning our graduate students help with several sections of an undergraduate course *Introduction to Astronomy* for students from all departments of the university and a course on astronomy via the Internet. At the Astronomical Observatory, graduate students also have to conduct visits from elementary-school groups on Tuesdays, from secondary-school groups on Thursdays and from the general public on Fridays through an activity called "Astronomical Fridays". They prepare an *Astronomical Ephemeris* each month and collaborate at the library and at the computer center.

Financial Support.   As financial support for the graduate students, at CASAO we have offered six positions as Teaching Assistants: three for the Group MAA-98 and another three for Group MAA-99. Last year we had five scholarships from the Organization of American States (OAS). See Table 3.

2.2.   Staff

Only the Chair of the CASAO has a permanent position as Professor. The staff of the Master's Program is made up of visiting professors coming (mostly) from Argentina but also from Brazil, Spain, France and Costa Rica. The total numbers are given in Table 4. In 1998 two came from Brazil and one each from Honduras, Spain, France and Bolivia. In 1999, all five who taught the 1998 Group came from Argentina, while three from that country and one each from Honduras, Spain and Brazil taught the 1999 Group. In 2000, there were again

Table 3.   Financial support

Group	1998			1999			2000		
	SAO	ODU	OAS	SAO	ODU	OAS	SAO	ODU	OAS
MAA-98	3/4	1/4	0/4	3/3	0/3	3/3	3/3	0/3	0/3
MAA-99				3/4	1/4	2/4	3/3	0/3	0/3
Total	3/4	1/4	0/4	6/7	1/7	5/7	6/6	0/6	0/6

Notes: SAO = CASAO, ODU = Other Departments of the National Autonomous University of Honduras, OAS = Organization of American States. Figures give numbers of awards in each category/total numbers of students for each Group in the years 1998, 1999 and 2000, respectively.

five from Argentina and one from Costa Rica, who taught the 1998 Group and one each from Honduras, Brazil, Spain and Costa Rica taught the 1999 Group.

Table 4.   Staff

Group	Course	Staff			
		1998	1999	2000	Total
MAA-98	18/20	6	5	6	17
MAA-99	14/20	0	6	7	13
Total	32/40	6	11	13	30

Visiting professors come to Tegucigalpa for short visits from two up to twelve weeks. Each one is chosen on the basis of his or her field of interest, experience with graduate-school students and research projects. Almost all of them have enjoyed the experience and are willing to collaborate further.

Even though we have no permanent staff, we have the benefit of having experts on the topics we require; and since we have e-mail and Internet connections, students can keep in touch with the visiting professors as needed.

2.3.   Curriculum

The mission of the National Autonomous University of Honduras is to organize and to develop higher education in order to contribute to the knowledge of reality, the strengthening of national identity, promotion of science and technology, consolidation of ethical values and the development of a culture of quality. This is accomplished through the training of capable and sensitive professionals concerned with the transformation of society.

At CASAO we have the responsibilities of studying and contributing towards the development of astronomy and astrophysics, basic space science and technologies including remote sensing and archaeoastronomy, through academic activities in teaching, research and outreach.

The objectives of the Master's Program in Astronomy and Astrophysics of the CASAO are:

1. To develop, through international cooperation, a regional master's program for the establishment of astronomy and astrophysics as academic fields at the Central American universities.

2. To contribute to the training of qualified personnel responsible for introducing Central America to the fields of scientific research and knowledge, use and application of astrophysical instrumentation and of space technology.

3. To contribute to the creation of a basic infrastructure and to maintain functioning a Central American Astronomical Observatory where academic and research activities can be carried out.

In order to define the curricular profile, we first identified the regional needs in Central America. There are not enough local professional astronomers in Central America for us to be able to develop astronomy and astrophysics as academic fields. The few astronomers who live in Central America do not have the impact they wish to have on education, communications and on the enterprises in which knowledge and technology of basic space science could be of great application.

Graduate students who finish their master's program will be able to work at universities, astronomical centers, public and private institutions and enterprises with projects on science and technology or to continue higher education toward a doctorate.

Students graduating from this Master's Program will have acquired a broad knowledge of astronomy and astrophysics through 20 courses and a research project, covering seven fields. Fundamental Astronomy in courses Nos. 1 and 2, Physics in courses Nos. 3-7, Astronomy and Astrophysics in courses Nos. 8-13, Instrumentation and Astrophysical Techniques in courses Nos. 14-17, History of Astronomy in course No. 18, Education in Astronomy in course No. 19. and research through all the above mentioned courses plus the seminar and a research project leading to a thesis (see Table 5).

Astronomers graduating from this Master's Program will be able to teach at a higher level; to handle and to understand the functioning of several instruments, processing and analysis of data techniques, computer systems and international data bases; to design and to develop original and modern projects of research; and to use the abilities and techniques learned in other fields of research and applied technologies.

Astronomers graduating from this Program will have an attitude of continual training in new theories and techniques as they are developed; they will frequently examine current publications and make contact with active researchers; they will maintain a competitive level of training in order to develop satisfactorily their professional career.

For the Master's Program there have been approved academic rules in which methodology and procedures for evaluation of staff and students are included. As part of the syllabus, a synthetic program for each course is included, however these programs could be enriched for the benefit of the students for each professor at the time they have to teach each course.

As strategies for developing the Master Program we never have more than two courses at the same time; because classes for each course last from two to twelve weeks, methodology is intensive; however evaluation is over extended periods covering up to half a year and pass grade is 80%. Historical records for each student are kept and maintained updated.

Table 5.   Curriculum

No.	Subject Matter
1	Introduction to Astronomy and Astrophysics
2	Celestial Mechanics and Classical Astronomy
3	Spectroscopy
4	Nuclear and Particle Physics
5	Radiative Processes and Radiation Transfer
6	Physics of Fluids
7	Physics of Plasma
8	Solar System
9	Solar Physics
10	Stellar Structure and Evolution
11	Galactic Physics and Stellar Systems
12	Physics of Interstellar Medium
13	Extragalactic Physics and Cosmology
14	Astrophysical Instrumentation
15	Data Processing Techniques
16	High Resolution Techniques
17	Remote Sensing
18	Archaeoastronomy
19	Didactics of Astronomy and Astrophysics
20	Seminar- open
21	Research Project (with thesis)

For logistic support of the Master's Program, at CASAO, we have a library with several hundreds of journals and books on Astronomy and Astrophysics; a computer center equipped with astronomical software (i.e.: IRAF operating over LINUX) and Internet links and access to e-mail; a conference room equipped with slides and overhead projectors; a permanent exhibition of more than 100 pictures of the universe; offices for staff and students; and observational facilities such as a 42-cm telescope equipped with a CCD camera and Johnson filters, a photometer and other accessories.

2.4. Research

Research is a rather slow process that needs, on the one hand, well trained researchers and students with a solid background, who are clear about what they want to do and what they are looking for; and, on the other hand, good facilities and optimum conditions to pursue their goals, i.e. computers, software, journals, books, Internet, and so forth.

In Honduras, a few years ago, conditions in astronomy and astrophysics were very poor. Even though one may have been a well-trained researcher, one had nobody to talk to or work with. Now, at CASAO, with the Master's Program in Astronomy and Astrophysics, after having taught several courses on astronomy and astrophysics, after having interacted with several research institutions via almost two dozens of active researchers, things have changed and research has begun to grow.

At CASAO we have been working on Education Projects: "An Astronomical Observatory for Central America" is the main project, " "The Master's Program in Astronomy and Astrophysics" is one of its components. "Teaching of Astronomy" at different levels: elementary, secondary and university is another project in which schoolteachers also participate (Pineda de Carias 1993).

Also, because of the size of the telescope ($\sim$ 0.4-m) we have been involved in "Near-Earth Objects" projects. We were lucky to have a meteorite falling in Honduras. We were thus able to organize research groups with international collaboration. A paper was published in *Meteoroids* last year.

Now, with NASA and the Central American Commission of Environment, we are collaborating in a project that uses satellite observations for understanding environmental change in Central America. This is part of a broader project for monitoring the whole Earth.

With the collaboration of Visiting Professors from Argentina and Brazil, graduate students have begun to work on: "Kinematics of stars toward Collinder 121 open cluster", "Dynamical Structures in the outer region of Saturn" and "The effect of new degrees of freedom in the Cosmological Model". It is important to emphasize that the students chose these topics and fields of work themselves, with the collaboration of our staff. This is a new stage of development for our astronomical center.

2.5. Academic Organization

The Master's Program is part of CASAO, which is an academic unit of the National Autonomous University of Honduras. At the Astronomical Observatory we have the Secretary, Administration, Services, Computer Center, the Library and the dome, which houses the telescope and the observational equipment. At present, the Director of CASAO is also President of the Central American Assembly of Astronomers (CAAA) which, each year, organizes a CURso Centraoamericano de Astronomía y Astrofísica (CURCAA) –a course conducted in Spanish for the whole region. (See Figure 1.)

There are three academic divisions at the Central American Astronomical Observatory.

> **Astronomy and Astrophysics** which includes the Master's Program, a general course of Introduction to Astronomy for all students, Astronomy via Internet, Education Projects and visits to the Observatory and the *Astronomical Ephemeris*.

> **Archaeoastronomy**, in which we develop seminars and a project on Mayan Archaeoastronomy, taking advantage of the Mayan site of Copán.

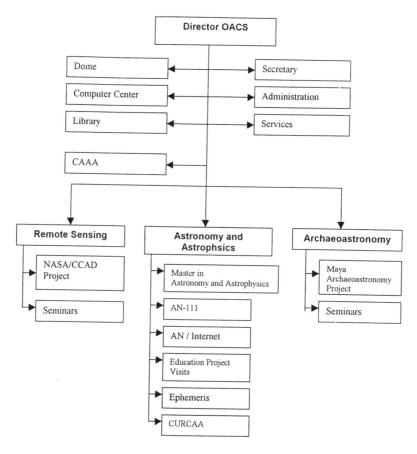

Figure 1.   Central American Observatory of Suyapa: Organigram

Remote Sensing, in which we develop some seminars on remote sensing and digital image processing, and in which we also have the NASA/CCAD Project.

At the National Autonomous University of Honduras, graduate studies are responsible to a Graduate Studies Direction which in turn is responsible to the President of the University. An Academic Committee, integrated by two staff and two student members, is also functioning. See Figure 2.

Because every month we receive a visiting professor, we can say that on average we have a proportion of 2:5 staff versus administrative members. However, these five administrative members are the Secretary, Administrator, Janitor and two Instructors, so we feel we need to raise the number of technicians.

The proportion of staff to students on average is 2:6, which is too low. Therefore, we need to raise the number of permanent professors, because for the year 2001 the number of students will increase.

The environment at the Astronomical Observatory is pleasant and everybody feels comfortable doing his or her own work.

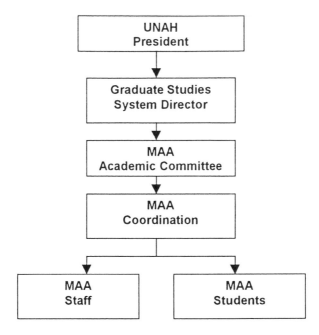

Figure 2.   Astronomy and Astrophysics Master's Program: Organigram

2.6. Collaboration

For the development of the Master's Program, the main collaboration has come from Visiting Professors who teach the courses. They have come mainly from the Observatory of Córdoba (Argentina), Observatory of La Plata (Argentina), Space Physics and Astronomy Institute (Argentina), Theoretical Physics Institute of São Paulo University (Brazil), Department of Astrophysics of Madrid Complutense University (Spain), Canary Island Astrophysics Institute (Spain) and Astrophysics Laboratory of Costa Rica University.

External financial support has come mainly from the Organization of American States (OAS – see Proyecto OEA-040/98, 1999). They first funded the Project "Inter-Institutional and Academic Strengthening of CASAO". For next year we have presented the project "Strengthening and Integration of the Centers for the Development of Astronomy and Astrophysics in Central American Countries" which is an international project jointly presented by Honduras, Costa Rica and Nicaragua. Next year, Costa Rica will host the VI Central American Course in Astronomy and Astrophysics (VI-CURCAA). Nicaragua, after having the first astronomer graduated from the Master's Program of Honduras, will be working toward developing astronomy as an academic activity. In Honduras, the Central American Astronomical Observatory will continue its activities and projects.

In Central America, we want to work together within the region. The underlying philosophy of CASAO is that each country should have a core of astronomers, with their own facilities in their own country, all working as one regional observatory.

At CASAO, we have also received collaboration from the Teaching of Astronomy Development Commission of the International Astronomical Union (TAD/IAU) and from the United Nations and European Space Agency (UN/ESA) Workshops on Basic Space Science Group.

2.7. Budget

More than 50% of the budget of CASAO comes from the National Autonomous University of Honduras. This covers salaries for staff and administrative personnel, educational material, travel and equipment.

About 40% has come from the Organization of American States (OAS) to fund: scholarships for students, trips for Visiting Professors, books and equipment for the Central American Astronomical Observatory.

Two other sources have provided 2% of the budget for the Astronomical Observatory. The European Space Agency and the United Nations donated some computers. From the TAD/IAU we have received funds for regional observational campaigns, some equipment and funds for repairs, after Hurricane Mitch damages.

3. Some Results

Astronomy and astrophysics are academic fields now established at the National Autonomous University of Honduras for the benefit of Honduras, Central America, and the international community.

As a graduate program, the Master in Astronomy and Astrophysics is a model of a well organized academic unit, with international recognition and support from prestigious international universities and organizations.

The success of the Master in Astronomy and Astrophysics program has depended on the local capacity to manage it (staff and students) and on the opportune assistance from the international community. This program is now a real professional alternative for the youth of the area and a good model to be followed by developing countries.

4. Recomendations

The international community of astronomers may help:

by creating links between developed and developing countries;

by providing visiting professors willing to come to developing countries to collaborate as they may be needed;

with funds for scholarships for graduate students to finish their degrees up to the highest level, while guaranteeing and stimulating them to remain in their own countries.

Acknowledgments. I want to thank the International Astronomical Union and the Local Organizing Committee of the 24th General Assembly for allowing me, the first Central American Astronomer to attend a IAU-General Assembly, to present and discuss a paper. I know that after this benchmark or "candle", colleagues and students of mine will follow the path, and in future meetings you will be hearing more about the Central American astronomers and their work.

References

Documents and Archives 2000, Dcouments and Archives at the Central American Observatory of Suyapa. National Autonomous University of Honduras.

Pineda de Carias, M.C. 1993, *Proyecto Un Observatorio Astronómico para Centroamerica*. UNAH. Madrid. España.

Pineda de Carias 1997 in *Report on the 7th UN/ESA Workshop on Basic Space Science: Small Astronomical Telescopes and Satellites in Education and Research.*, Tegucigalpa.

Proyecto OEA-040/98 1999, *Fortalecimiento Interinstitucional y Academico del Observatorio Astronómico de Suyapa* Organization of American States.

Discussion

Fierro offered congratulations on the fine work done in Central America but thought more effort should be put into sending students to developed countries to work on their doctorates. She pointed out that it has been found difficult to persuade visiting astronmers to spend long enough in Honduras to give doctoral-level courses. She also recommended that students should be taught English. Pineda de Carias emphasized that the Master's Program was intended to prepare students for several different career paths. One possibility is indeed for successful students to go on to doctoral studies and the universities from which visiting staff have come are willing to help. She admitted that the Program had experienced some diffciulties, as all new projects do at first, but she believed that these problems were being successfully dealt with. There is an English-language requirement for admission to the Program and many students voluntarily take courses beyond this requirement. Rosenzweig suggested that the Central Americans should contact Venezuelan astronomers, who have degree programs at the mater's and doctoral levels. (*Editor's Note*: see the paper by Rosenzweig, pp. 205-209.)

Astronomy for Developing Countries
IAU Special Session at the 24th General Assembly, 2001
Alan H. Batten, ed.

Astronomy Education in Universities of China

Cheng Fang and Yuhua Tang

Department of Astronomy, Nanjing University, Nanjing, China. e-mail: fangc@nju.edu.cn

Abstract. In China, more than twenty universities have astronomy education and research work. In four key universities, a complete series of educational programs of undergraduate, master's, doctoral and post-doctoral levels has been formed. After four-year study at the undergraduate stage, students can be enrolled in master degree specialties. Three years later, some of them begin their three-year Ph.D. education. Only a few students enter into post-doctoral programs. Master's and doctoral education systems are also established in the Chinese Academy of Sciences. In this paper, we give a description of the astronomy education of universities in China. After introducing the overall situation, we describe teaching materials, graduate-degree courses and facilities. We also discuss some problems and prospects for the new century.

1. Introduction

After a rapid development in the 1990s, more than twenty universities in China have established astronomy education and research work. In these universities, there are courses of astronomy for the students in astronomy departments and/or over the whole university. Some groups of student amateur astronomers have been organized. Besides, since the end of 1970s, astronomy education at M.S. and Ph.D levels has also been developed in the Chinese Academy of Sciences (CAS). Unlike the situation in some other countries, students can get their M.S. and Ph.D degrees in CAS. Nowadays, more than one third of M.S. and Ph.D students come from CAS. However, the education system is more or less similar to that in universities. Thus, here we will concentrate on astronomy education in universities in mainland China.

Among the twenty universities, there are four universities that are key ones and have relatively long histories. After a description of the astronomy departments in these four key universities in Section 2, the astronomical education system is described in detail in Section 3. The astronomical activities of students are illustrated in Section 4, followed by a brief discussion of the problems and prospects of the education system in Section 5.

2. Astronomy Departments in Four Key Universities

There are four key universities, namely, Nanjing University (NU), Beijing Normal University (BNU), Beijing University (BU) and University of Science and

Technology of China (USTC), in which astronomy departments (AD) or an astronomy specialty (AS) or Center for Astrophysics (CFA) have been established for a shorter or longer time. Table 1 gives some information about them.

Table 1.   Astronomy Departments in Four Key Universities

University	Date of Set Up	No. of Prof.	No. of Underg.	No. of M.S.	No. of Ph.D.
AD of NU	1952	18	80	20	10
AD of BNU	1960	5	70	15	10
CFA of USTC*	1978	7	20	10	9
AS of BU**	1960	3	30	14	4
Total		**33**	**200**	**59**	**33**

* Since 1999, an astronomy department has been established.
** In 2000, an astronomy department has been established.

The main facilities for student training in these universities are as follows:

1. 60-cm solar tower telescope (NU);
2. 65-cm reflector (NU);
3. 40-cm Schmidt-Cassegrain telescopes (NU and BNU);
4. 40-cm reflector (BNU);
5. 28-cm refractor (NU);
6. Radio telescopes (3cm in NU, 2cm in BNU);
7. 15-cm refractors (NU and BNU);
8. PC and Sun workstations (NU, BNU, USTC and BU).

As an example, Figure 1 shows the picture of the solar-tower telescope in Nanjing University. It is the only one of its kind in China. Figure 2 shows the picture of the 40 cm reflector in BNU. Generally, students use these facilities to do some practical training. However, once time in the four-year study, students also go to astronomical observatories and spend about 2-3 weeks workimg under the direction of researchers. This is very useful for the students to know the recent research in progress.

A complete series of educational programs of undergraduate, master's, doctoral and post-doctoral levels has been formed since the 1950s, and highly improved in recent years.

2.1.   Undergraduate education

Based on a national examination, students will be enrolled in universities according to their scores. There is a strong competition for enrollment in the universities. Each year, about 50 students are admitted to astronomy departments. After entry to university, four-year undergraduate education begins.

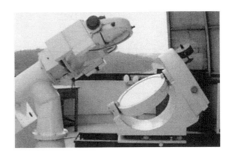

Figure 1.   The 60-cm solar-tower telescope of Nanjing University (left) and its coelostats (right). This is the only instrument of its kind in China

Figure 2.   The 40-cm reflector of Beijing Normal University

The universities have been perfecting the credit system, and have introduced the major-auxiliary course program and the dual bachelor's degree system.

During four years, students have to obtain 150 credits, among them 60% should be for the important necessary courses. In astronomy departments, all students should mainly study mathematics and physics during the first two years. The main astronomy courses during the next two years are as follows:

1. General astronomy;
2. Theoretical astrophysics;
3. Practical astrophysics;
4. Celestial mechanics;
5. Astrometry.

2.2. Master's-degree specialties

Students can be enrolled in the master's-degree specialty through recommendation, which is only for few excellent students, or by passing successfully the university examination. In recent years, there are about 30-35 students over the whole of China being admitted to astronomy master's-degree specialties of the universities. They have to spend 2.5-3 years to study courses of the master's-degree program. The main courses are as follows:

1. Radiation mechanisms*;
2. Stellar interier structure and evolution*;
3. High energy astrophysics;
4. Plasma dynamics;
5. Radio astrophysics;
6. Galactic structure and evolution;
7. Physics of cosmology;
8. Physics of solar active regions;
9. Numerical simulation in astrophysics;
10. Molecular astrophysics.

The courses with * can be chosen by students.

The students should obtain certain credits (30-35) and publish at least one scientific paper during the three-year period. Having passed successfully the defense of their theses, students can get their M.S. degrees.

2.3. Ph.D program

Based on university examination and oral test, M.S. degree students can be enrolled in Ph.D programs. In recent years, there are about 15-20 students being admitted each year in the astronomy Ph.D programs of the universities. They have to spend 2.5-3 years studying a few special courses, such as:

Physics of γ-ray bursts;
Pulsar physics;
Supernovae and supernovae remnants;
Physics of compact stars;
Observational cosmology;

Figure 3.   Observation of a partial solar eclipse organized by the student astronomical society of Beijing University

Physics of active galactic nuclei etc.

However, the main requirement in the Ph.D program is to do research work. Students have to publish between one and three scientific papers in the main astronomical journals. By the end of study, a student must write a comprehensive thesis, which has to be evaluated by about ten scientists, and well pass the defense of the thesis, in order to obtain the Ph.D degree.

2.4.   Post-doctoral program

Based on application and competition, only a few Ph.D students can be accepted in the post-doctoral program. Generally, it takes 1.5-2 years and universities provide special funds and accommodations for them. The people in the post-doctoral program can do any research work they like, but they must provide a report by the end of the period of study.

3.   Astronomical Activities

Some astronomical societies for young students have been organized in universities. They attract not only students in the astronomy departments, but also many amateur-astronomer students. A variety of astronomical activities, such as lectures, observations of celestial phenomena and popularization of astronomy etc., have been launched. As an example, Figure 3 shows the observation of a partial solar eclipse organized by the young student astronomical society of Beijing University.

Students in the astronomy departments participate frequently in society activities. They contribute to the popularization of astronomy, including organization of astronomical lectures, responses to inquiries about problems in astronomy raised by people etc. Figure 4 shows one such society activity.

4.   Problems and Prospects

In recent years astronomy education in China has been greatly improved. It should be mentioned that, besides the universities, there is also an education

Figure 4.   Responding to inquiries of people in a square in Nanjing

system of graduate students in the Chinese Academy of Sciences (CAS). Every year there are about 20-30 graduate students obtaining astronomical M.S. degrees in CAS, while about 10 students obtain astronomical Ph.D degrees in CAS. Thus, we have sources of well-educated students for astronomy. However, one problem is the loss of quality young people. After getting M.S. degrees, some excellent young students go abroad, especially to the United States, to enrol in Ph.D programs. Unfortunately, After obtaining their Ph.D degrees, many of them have no chance to get astronomical jobs. This is probably a common problem in developing countries. It is hoped that this situation will be improved in future.

Now we place great emphasis on basic scientific research and teaching and consolidate the student's theoretical foundation and astronomical surveying skills. Special training is offered to those students of greater creativity in an attempt to turn them into quality personnel. While continually reforming the graduate-degree courses, we have adopted the modes of running consecutive undergraduate and master's, master's and doctoral, or undergraduate, master's and doctoral programs. Interdepartmental, inter-institutional, and international joint training programs have been offered to students. Especially joint programs between universities and the CAS have been recently strengthened. Two astrophysical centers have been established: one is in Beijing University as a joint program between BU and the CAS and the other is in Nanjing as a joint organization between NU, USTC, the Purple Mountain Observatory and the Shanghai Observatory.

In summary, a complete series of educational programs of undergraduate, master's, doctoral and post-doctoral levels has been much improved in China. It provides a continuous good source for Chinese astronomers engaging in teaching, research and popularization of astronomy. We would like very much to learn the experience of foreign countries, especially of developing countries and to develop further the education system in China.

Acknowledgments. We would like much to thank Professors Qiao Guo-jun, Ma Wen-zhang and Zhang Yang for their help in providing good data. This work was supported by a fund from the Doctoral Program of the Ministry of Education of China.

Discussion

Orchiston and Raharto asked if Chinese graduates in astronomy could find jobs in astronomy. Fang replied that there is no difficulty for those with doctorates. For those with only master's degrees, employment depended on their own choice and ability. Kochhar asked if Chinese students returning from the West experienced any difficulties in re-adjusting to their home culture. Fang thought not, except for some who stayed abroad for a long time. He believed that most young Chinerse scientists valued their home culture and could make any necessary adjustments themselves. Martinez observed that the fact that some excellent scientists migrated abroad was not entirely bad. Some of them will eventually return, bringing with them experience of areas of astronomy not represented in the home country. They will also bring their own network of contacts, thus expanding the influence of the home country's astronomy. Leeuw suggested that countries like China, India and South Africa (naming only those who had reported here) could do more to employ their young citizens with astronomy degrees from abroad, who could prove to be valuable new blood. Fang agreed with both the foregoing comments, stating that the Chinese suthorities are not opposed to students going abroad. They did hope some would return and measures are now being taken to encourage them to do so, including creating opportunities for temporary returns (upto some months).

Section 3: Initiatives in Astronomy Education

Astronomy for Developing Countries
IAU Special Session at the 24th General Assembly, 2001
Alan H. Batten, ed.

Hands-On Astrophysics: Variable Stars for Astronomy Education and Development

Janet A. Mattei

American Association of Variable Star Observers (AAVSO), 25 Birch Street, Cambridge, MA 02138-1205, USA; email: jmattei@aavso.org

John R. Percy

Erindale Campus, University of Toronto, Mississauga ON, Canada L5L 1C6; email: jpercy@erin.utoronto.ca

Abstract. Amateur astronomers, and students, can contribute to astronomical research by measuring the brightness of variable stars. *Hands-On Astrophysics* (HOA) is a project which uses the unique methods and the International Database of the American Association of Variable Star Observers (AAVSO) to develop and integrate a wide range of science, mathematical and computer skills, through the measurement and analysis of variable stars. It is very flexible and can be used at many levels, in many contexts — for classroom use from high school to university level, or for individual projects. In this paper we describe HOA, and how it can be used to promote international astronomy education and development, through research-based science education.

1. Introduction

In any part of the world, *practical (or laboratory) activities* can be an important part of science education at the school and university level, especially when they include a strong element of inquiry and discussion. Carefully-chosen *research projects* can also be used; they are almost obligatory at the graduate level, highly desirable at the undergraduate level, and quite possible at the senior high-school level — both because they contribute to effective learning of science processes and skills, and because they motivate students through the excitement of doing real science with real data. Research projects enable students to contribute to science; this is especially desirable in the developing countries, where the supply of trained research personnel is low.

All over the world, thousands of amateur astronomers and students are contributing to astronomical research by measuring the changing brightness of *variable stars*. The basic techniques are simple; the equipment required is inexpensive; and the demand for these measurements from the professional astronomical community has grown by a factor of 25 in the last three decades.

Variable stars are those which change in brightness. If measured sufficiently carefully, almost every star turns out to be variable. The variation may be due to geometrical processes, such as the eclipse of one star by a companion star, or the rotation of a spotted star; or it may be due to physical processes such as pulsation, eruption, or explosion. Variable stars provide astronomers with important information about the properties, processes, nature, and evolution of the stars.

Many variable stars display changes in brightness that are large enough to be detected with the eye, using a small telescope, binoculars or (in the case of a few dozen bright variable stars) no optical aid at all. Amateur astronomers and students can make a useful contribution to astronomy by measuring these variable stars in a systematic way. The American Association of Variable Star Observers (AAVSO) was established in 1911. Its purpose is to co-ordinate variable-star observations made largely by dedicated "backyard" astronomers, evaluate the accuracy of these observations, compile, process, and publish them, and make them available to researchers and educators. The AAVSO now receives over 350,000 measurements a year, from 550 observers worldwide. The measurements are entered in the AAVSO electronic database, which contains close to 10 million measurements of several thousand stars. The demand for these measurements, from researchers and educators, has grown rapidly in the last three decades — partly as a result of major collaborations in space astronomy. To contact the AAVSO, write to: AAVSO, 25 Birch Street, Cambridge MA 02138, USA; e-mail: aavso@aavso.org; web site: **www.aavso.org**.

It occurred to us, several years ago, that the measurement and analysis of variable stars could help students to develop basic science and mathematics skills. Variable-star measurement and analysis is inherently simple; it can be done by any high-school or university student. The analysis and interpretation of the data involves a wide range of scientific and mathematical skills, some of which would be understood and appreciated by a beginning high-school student, and some of which would tax an expert in the field.

Both of us have a long-standing interest and involvement in variable stars and education. Together and with the help of many teachers (particularly Donna Young), the AAVSO Headquarters staff, and funds from the US National Science Foundation (NSF), we developed *Hands-On Astrophysics (HOA)*. It brings the excitement of astrophysical research and discovery to science, mathematics and computer classes through a flexible set of activities and projects on variable stars. It also helps to develop an understanding of basic astronomy concepts, to provide interdisciplinary connections, and to take students through the whole scientific process — while they have fun with real data!

HOA supports US National Standards for Science and Math Education — which are representative of standards in many other countries — by directly involving students in the scientific process. It uses 600,000 measurements of variable stars from the AAVSO International Database, and the techniques and materials that the AAVSO has developed and refined over nine decades. It includes an extensive teacher/student manual; computer software to analyze data and to create new data files; 45 star charts to make more observations; prints and slides for indoor activites; and a video-cassette with three short segments: the inspirational *Backyard Astronomy*, the informational *Variable Stars*, and the instructional *How to Observe Variable Stars*. It has an extensive website : **http://hoa.aavso.org** to provide ongoing support.

Students begin by using slides or prints of variable stars indoors, and by exploring the data through the software for graphing and analysis. They can then progress to making measurements of variable stars in the real sky or analyzing on-line archival data on variable stars. Some students may be able to access remote robotic telescopes to carry out their measuments. Students can take their own 35-mm slides of the night sky, for further indoor measurement or instruction. The *HOA* materials cover five northern constellations, but the *HOA* philosophy and process can be applied to other parts of the sky.

Students — especially those with a special interest in mathematics and computer science — can also enjoy variable star analysis indoors on the computer. *HOA* includes three PC software programs – a computer based tutorial HOAFUN, a data entry program HOAENTER, and a powerful data-analysis program VSTAR. VSTAR gives students a hands-on introduction to time series analysis — a technique which is now used in almost every branch of science, engineering and commerce.

There are many variable stars that are bright enough to be observed with small telescopes, or binoculars, or even with the unaided eye. Projects can involve assigning a star to each student, where they feel an "ownership" of the star and/or there can be a communal project where each student makes a few measurements of brightness each month or each several months. *HOA* can be especially successful when carried out by a group of students or amateur astronomers. The measurements can be combined for analysis and interpretation. These types of activities and projects develop and integrate a wide variety of skills, starting from background reading and planning; research judgement, strategy and problem solving; continuing with pattern recognition, interpolation and measurement; recognizing and understanding random and sytematic errors; computer programming and data management; processing and graphing of data; construction, analysis and interpretation of graphs; concepts of regularity and prediction; curve fitting and other statistical and numerical procedures; all the way to preparing oral and written reports.

HOA can be used in science, mathematics, and computer classes or for independent projects. It can be adapted to various levels, and can be used as a complete course of study or in parts. It can be used by any individual at any age to learn the art and science of variable-star astronomy. *HOA* materials are suitable for amateur astronomers who wish to learn more about the fascinating nature of variable stars. There is a wealth of information which can be utilized for science projects, for astronomy-club activities, and family learning. It is self-contained; no previous knowledge of astronomy, or variable stars is assumed. It is open-ended, and can lead to sophisticated research-based projects. It actively involves the students in the scientific process, and motivates them by enabling them to do real science with real data. Students can observe variable stars in the real sky, add their brightness measurements to the existing database in *HOA*, and analyze the data using the software provided. In addition, they can send their measurements to the AAVSO to be added to the International Database and to be used for further research projects by astronomers and educators worldwide.

2. Amateur Astronomers

An *amateur astronomer* is someone who loves astronomy, and cultivates it as a pastime or hobby. A more stringent definition might be that an amateur astronomer is someone who does astronomy with a high degree of skill, but not for pay (Williams 1988). Amateur astronomers make important contributions to astronomical research (Dunlop & Gerbaldi 1988) and education (Percy 1998a).

Given the fact that many science students in the astronomically-developing countries will go on to careers in engineering or business, amateur astronomy gives them a way to continue their interest in astronomy, and *HOA* gives them a way to continue to contribute to science. The Internet can also be a powerful tool for those who are "connected", since it provides access to large and sophisticated databases (such as the HIPPARCOS catalogue and epoch photometry –i.e. the individual photometric results contained in the catalogue) and instructions for analyzing and interpreting the data.

Astronomical research and education done by amateur astronomers is "democratic" in the sense that it enables anyone to do astronomy, whether they have formal credentials or institutional affiliations or not. It is "science for the citizens". That is certainly true of variable-star astronomy because, in the words of the song, "the stars belong to everyone".

3. Applications to Astronomically-Developing Countries

3.1. A Prelude to Research

Variable-star observing can be a prelude to more advanced astronomical activity. Wentzel (private communication, also 2001), for instance, has found *Hands-On Astrophysics* very useful in workshops with physics teachers in Vietnam, as part of the IAU's "Teaching for Astronomical Development" program. This is because students (and teachers) in many countries are not accustomed to working with real data — only with knowledge which appears in the textbook. Several developing countries have acquired (or are planning to acquire) small telescopes which will eventually be equipped with photoelectric photometers or CCD cameras. Is there a way to start doing real science, even before the telescope arrives and is operational? Yes! The solution is to begin doing serious visual measurements of variable stars, using binoculars or a very small telescope if one is available. The AAVSO, either through *Hands-On Astrophysics*, or by mail, by e-mail, or by its web site, can provide assistance in setting up an observing program. The measurements, so obtained, can then be contributed to the AAVSO International Database, to be used by researchers and educators (Mattei 1999).

The next step is photoelectric or CCD photometry with the newly-acquired telescope. These measurements are more precise than visual measurements, but the general principles of analysis are the same. The AAVSO has both a photoelectric program, and a CCD photometry program. Other international collaborative photometry programs are listed in Percy (1998b), who pointed out the value of beginning research as part of an international collaboration.

3.2. A Stepping Stone to other Databases

Databases from space astronomy missions are increasingly available on CD-ROM and/or the Internet, and they provide a practical way for astronomically-developing countries to begin research at very little cost — a PC with a CD-ROM drive and/or Internet connection. One example is the HIPPARCOS catalogue of astrometry and epoch photometry (Turon 1997, Perryman 1999). AAVSO observers provided crucial support for this mission (Turon 1997). In turn, the mission has provided dozens of new variables to be studied by photoelectric or CCD observers (Perryman 1999). It has also provided millions of photometric measurements of "unsolved" stars which require detailed analysis.

The HIPPARCOS mission has excellent research and education web pages (astro.estec.esa.nl/Hipparcos/hipparcos.html), with information on variable stars, interactive tutorials on variable star analysis, as well as data. Additional information on variable stars can be found on the AAVSO web site (**www.aavso.org**), along with user-friendly software for analyzing them (the TS11.ZIP program for time-series analysis, and the WWZ11.ZIP program for wavelet analysis), can be downloaded from **www.aavso.org/software.stm**. *Period98*, a powerful period-analysis package, can be downloaded from the web site **dsn.astro.univie.ac.at/period98** at the Institute of Astronomy, University of Vienna, along with an instruction manual.

For a continuation of this discussion, see the paper on "Simple Science, Quality Science", elsewhere in this volume (Percy 2001).

References

Dunlop, S., & Gerbaldi, M. 1988, *Stargazers: The Contributions of Amateurs to Astronomy*, Springer-Verlag.

Harmanec, P. 1998, A&A, 335, 173.

Mattei, J.A. 1999, ApSpSci, in preparation.

Percy, J.R. 1998a, in *New Trends in Astronomy Teaching*, ed. L. Gouguenheim et al., Cambridge University Press, 205.

Percy, J.R. 1998b, ApSpSci, 258, 357.

Percy, J.R. 2001, this volume, pp. 250-254.

Perryman, M.A.C. 1999, *Sky & Telescope*, 97, No. 6, 40.

Turon, C. 1997, *Sky & Telescope*, 94, No. 1, 28.

Wentzel, D.G. 2001, this volume, p. 46-48.

Williams, T.R. 1988, in *Stargazers: The Contributions of Amateurs to Astronomy*, ed. S. Dunlop & M. Gerbaldi, Springer-Verlag, 24.

Discussion

Melek enquired about the credit needed for this program. Mattei replied that it was designed either for one semester or two. Aguilar said that her experience was that research on variable stars, in particular observations of them, does help to stimulate interest in science and astronomy. She felt that this sort of work with a small telescope provided the opportunity for people in developing countries to have access to scientific culture. Kozai asked for information on what was needed to participate in the project: what types of telescopes, photometers, star catalogues, etc. are necessary? The answer is that *Hands-On Astrophysics* includes star charts that can be used for visual observing with any available binoculars, small telescope, or the unaided eye. It also includes slides and prints for indoor activities, so it is not essential to observe the "real" sky in order to profit from the program.

Astronomy for Developing Countries
IAU Special Session at the 24th General Assembly, 2001
Alan H. Batten, ed.

Exchange of Astronomy Teaching Experiences

Rosa. M. Ros

Department of Applied Mathematics IV, Technological University of Catalonia, Jordi Girona 1-3, Modul C3, 08034 Barcelona, Spain.
e-mail: ros@mat.upc.es

Abstract. The Working Group of the European Association for Astronomy Education responsible for Teacher Training organises an annual Summer School for teachers under expert guidance. For a week the teachers participating can exchange experiences, increase their knowledge and discuss different ideas and perspectives.

In general, the instructors are professional astronomers, professors and teachers from different countries. The papers presented offer very practical activities, paying special attention to didactic aspects, and take the form of general lectures to all 40 participants and workshops to reduced groups of 20 participants. There are also day and night observations, without expensive equipment or complicated procedures, that are easy to set up and based on topics that it is possible to use in the classroom.

The Summer Schools promote a scientific astronomical education at all levels of astronomy teaching, reinforce the link between professional astronomers and teachers with experience of teaching astronomy, allow debates among the participants on their pedagogical activities already carried out in their own classroom and help them to organise activities outside it.

Astronomy teachers need special training, access to specific research, to new educational materials and methods and the opportunity to exchange experiences. All these things are provided by the Summer School.

1. Presentation of EAA

The European Association for Astronomy Education (EAAE) is an association which brings together European teachers and lecturers interested in astronomy. The EAAE was born under the auspices of ESO in 1994 in Garching, but EAAE started its work during the Founding General Assembly in Athens (November 1995). The Web page of EAAE is:

http://www.algonet.se/~sirius/eaae/workgrps.htm.

The aims of the EAAE are to improve and promote scientific astronomical education in Europe in schools of all levels and other institutions involved in teaching Astronomy, to reinforce the link between professional astronomers and school teachers with experience of teaching astronomy at various levels, to allow debates between participants on the pedagogical activities already carried out in their own classroom, and to organise activities outside it.

The EAAE has a set of working groups in various fields (Table 1) and has 20 member countries (Table 2).

Table 1.   The Working Groups of the EAAE

- No. 1. Astronomical Concepts
Chairperson: Rainer Gaitzsch
e-mail: gaitzsch@ikra.med.uni-muenchen.de
- No. 2. Didactic Materials
Chairperson: Lidia Nuvoli
e-mail: palici@aerre.it
- No. 3. Training Teachers
Chairperson: Rosa M. Ros
e-mail: ros@mat.upc.es
- No. 4. European Student Projects
Chairperson: Mogens Winther
e-mail: mw@posthu.amtsgym-sdbg.dk
- No. 5. Public Education
Chairperson: Thomas W. Kraupe
e-mail: TWK@artofsky.com
- No. 6. Projects for EAAE
Chairperson: Bernard Pellequer
e-mail: bernard.pellequer@cnusc.fr

Teachers need to receive in-service education in order to teach astronomy effectively. They need special training, access to specific research, to new educational materials and methods and the opportunity to exchange experiences. This happens during the summer schools which the EAAE organises.

This is the real spirit of the EAAE-WG3 working group for teacher training. The principal objective of this group of professors and teachers, established in Athens in November 1995, is the organisation of the EAAE International Summer Schools. Good teacher education leads to better student education.

2.   Past European Summer Schools

The working group for teachers' training (EAAE-WG3), organises a summer school each year for primary-schol and secondary school teachers under the guidance of experts. For a week the teachers participating can exchange experiences, increase their knowledge and discuss different ideas and aspects in a friendly atmosphere. A general view of this training and a variety of pedagogical activities can be seen in the poster entitled "Summer Schools for European teachers to

Table 2.   The National Representatives of the EAAE

- Austria: G. Rath
e-mail: rath@borg-6.borg-graz.ac.at
- Belgium: E. Perotte
e-mail: eddy.pirrotte@skynet.be

- Denmark: Bent Klarmark
e-mail: klamark@post4.tele.dk

- Estonia: L. Leedjrv
e-mail: leed@aai.ee

- Finland: M. Sarimaa
e-mail: markku.sarimaa@ursa.fi
- France: B. Pellequer
e-mail: bernard.pellequer@cnusc.fr
- Georgia (Republic of): T. Borchkadze
e-mail: tenat@dtapha.kheta.ge
- Germany: Roland Szotak
fax: +49-(0)251-833669
- Greece: M. Metaxa
e-mail: mmetaxa@compulink.gr

- Italy: Cristina Palici di Suni
e-mail: palici@aerre.it

- Latvia: Ilgonis Vilks
e-mail; vilks@latnet.lv
- Luxembourg: F. Wagner
e-mail:
fernand.wagner@ci.educ.lu
- The Netherlands:
Gert Schooten
e-mail: Gert.Schooten@quant.nl
- Norway: F. Pettersen
e-mail:
franck.pettersen@unikom.uit.no
- Portugal: M.F.S. Martins
e-mail: felisbela.martins@ip.pt
- Russia: Michael G. Gavrilov
e-mail: gavrilov@issp.ac.ru
- Spain: E. Zabala
e-mail: ezabala@sinix.net
- Sweden: B. Lingons
e-mail: bjlin@avc.edu.stockholm.se
- Switzerland: M. Reichen
e-mail:
michael.reichen@obs.unige.ch
- U.K. M. Cohen
e-mail: amc@rmplc.co.uk

increase the quality and presence of Astronomy in European Curricula" (see p. 166).

The first Summer School was organised in La Seu d'Urgell (Spain, 1997), a small town in the Pyrenees, near the border with France, in a good observational zone. The second EAAE International Summer School was held in Fregene (Italy, 1998). This town, in the Mediterranean area, is near Rome and this location was used to offer the participants an interesting walking tour of Rome from an astronomical perspective. The third Summer School took place in Briey (France, 1999), a small village in the central area of the solar eclipse of August 11th. The participants were able to observe the eclipse and take part in different activities related to it. The fourth Summer School (Portugal, 2000) took place in Tavira, a small village in the Algarve on the Atlantic coast, near Sagres, birthplace of Portuguese maritime exploration. The theme was "Astronomy and Navigation", and the participants had the chance to practice using some old astronomical tools, on a boat trip. In all cases, there was a group of around 50 participants from several European countries which appear in Table 3.

In general, the participants were European teachers from secondary schools and to a lesser extent, there were some teachers from primary schools and planetarium employees. The instructors were professional astronomers, professors and teachers from different countries. The papers that were presented covered very practical activities, paying special attention to didactic aspects, and took the

Table 3.   Countries at the EAAE Summer Schools

Country	First School	Second School	Third School	Fourth School
Austria	0	1	1	1
Belgium	1	1	2	2
Denmark	0	0	1	0
Finland	2	3	2	4
France	14	11	13	3
Georgia	0	2	0	1
Germany	3	1	2	3
Greece	0	1	0	1
Italy	5	15	5	3
Latvia	1	0	1	3
Luxembourg	0	0	1	0
Netherlands	0	0	3	1
Portugal	1	2	4	13
Russia	0	0	0	1
Spain	20	10	14	9
Sweden	0	0	2	0
Brazil	0	0	0	1
Chile	0	0	1	0
Total	47	47	52	46

form of general lectures to all participants and working groups, and workshops to reduced groups of 15-20 participants. There were day and night observations based on topics possible to introduce in the classroom.

At the end of the Summer School a questionnaire is answered by the participants to evaluate the activities carried out. The results are used for planning subsequent Summer Schools. This kind of work leads to the Summer School being modified in some aspects every year, in order to adapt its contents to the general interest of the teachers who attended the conference.

At the first Summer School the activities were of a more theoretical nature than at the last one, which included as many workshops as possible about practical subjects and models for building. The time for encouraging general discussion was also increased. Our participants greatly enjoy explaining the situation in their countries and comparing it with others. Despite the initial opinions of those attending, the differences between the European countries are minimal and the general situation is common. In the majority of European countries, astronomy appears as topics included in different subjects. The largest number of topics can be found in physics, but some of them appear in geography, mathematics, geology and other scientific subjects. This interdisciplinary aspect of astronomy currently causes its dispersion among many subjects with the consequent problems. However, this situation can be used to introduce astronomy from all the teaching fields if the teachers are adequately trained. This is our objective in

Figure 1.   Teachers engaged in various Summer-School activities

the Summer School: to give teachers a lot of ideas, resources and facilities in order to explain more and more astronomy from all the different fields.

The importance of the poster session has gradually increased at every Summer School. This session offers participants the opportunity to present their works in the classroom environment with their students. On the first or the second day of the conference, each participant who wants can present his poster to the rest of the members of the Summer School. Throughout the week, the poster can be visited by the rest of the participants. This promotes the exchange of information between all the teachers, lecturers, and astronomers who attend the Summer School.

At the last event organised, a new point of interest was introduced: the Bazaar. This is a marketplace area which can be visited during the Summer School where the participants place astronomical materials to exchange, sell and buy (booklets, photocopied material, dossiers, models, tools, instruments, computer programs, CD-ROMs). The main aim of the organisers at all times is to facilitate contacts between all the participants. A very friendly general atmosphere promotes this situation, and is maintained despite the intensities of a week living together.

The topics presented at the summer schools appear in the poster mentioned previously.

The Summer School has three different languages to facilitate teachers' participation. The languages are the language of the host country, as well as

English, and one other, which can be used by the majority of participants. Although the lingua franca of the Summer School is English, a printed document (Ros 1997, 1998, 1999, 2000) with the contents of all the papers in two languages (English and one of the other official languages) is given to participants and instructors so that they can follow the meeting more easily. This document promotes understanding between all the participants.

3.   Future Summer Schools

The EAAE-WG3 is organising the 5th EAAE International Summer School in Bad Honnef (Germany, 2001). The Summer School will be held from 2nd to 7th July and the topic will be Astronomy at the Dawn of the Third Millenium. The place selected, Bad Honnef, is a small village near Bonn. The participants will be able to visit the Radio-Telescope at Effelsberg.

The WG3 on teacher training is also preparing the next Summer Schools. We are able to announce that the 6th EAAE International Summer School will be held in Pallas, in the Lapland area of Finland, to observe the Midnight Sun, during the second week of July 2002.

References

Ros, R.M. (ed.) 1997, *Proceedings of 1st EAAE International Summer School* Institut de Cincies de l'Educaci. Universitat Politcnica de Catalunya. Barcelona, 1997. ISBN: 84-89190-21-6.

Ros, R.M. (ed.) 1998, *Proceedings of 2nd EAAE International Summer School* Institut de Cincies de l'Educaci. Universitat Politcnica de Catalunya. Barcelona, 1998. ISBN: 84-89190-24-0.

Ros, R.M. (ed.) 1999, *Proceedings of 3rd EAAE International Summer School* Unitat de Formaci de Formadors. Universitat Politcnica de Catalunya. Barcelona, 1999. ISBN: 84-89190-25-9.

Ros, R.M. (ed.) 2000, *Proceedings of 4th EAAE International Summer School* Unitat de Formaci de Formadors. Universitat Politcnica de Catalunya. Barcelona, 2000. ISBN: 84-89190-26-7.

Discussion

Hingley suggested that one problem of astronomy texts is variable and erratic selection of topics. Could the EAAE compile lists of topics at various levels which would constitute a "core curriculum"? Ros replied that this was indeed a topic of interest for the EAAE and it would be discussed at the fifth school. (*Editor's note*: see also the abstract of Ros's poster on p. 166.)

Public Education in Developing Countries on the Occasions of Eclipses

Jay M. Pasachoff

Hopkins Observatory, Williams College, Williamstown, MA 01267, U.S.A. e-mail: jay.m.pasachoff@williams.edu http://www.williams.edu/astronomy/eclipses

Abstract. Total solar eclipses will cross southern Africa on June 21, 2001, and on December 4, 2002. Most of Africa will see partial phases. The total phase of the 2001 eclipse will be visible from parts of Angola, Zambia, Zimbabwe, Mozambique and Madagascar. The total phase of the 2002 eclipse will be visible from parts of Angola, Botswana, Zimbabwe, South Africa and Mozambique. Public education must be undertaken to tell the people how to look at the eclipse safely. We can take advantage of having the attention of the people and of news media to teach about not only eclipses but also the rest of astronomy. I am Chair of a "Public Education at Eclipses" subcommission of IAU Commission 46 on the Teaching of Astronomy, and we are able to advise educators and others about materials, procedures and information releases.

Though science does not usually make it to the front pages of newspapers, total solar eclipses bring widespread attention. People are eager to find out about the Sun being visibly extinguished in the midst of the day, and are often frightened both about the sun itself and about the possibility of their being injured. Much confusion abounds, in the developed and the developing countries alike, about how hazardous solar eclipses are and about how to view them. The Working Group on Eclipses of the solar commissions of the International Astronomical Union runs a Web site,

http://www.williams.edu/astronomy/IAU_eclipses,

at which we have many links to information about observing eclipses safely in addition to information about eclipse expeditions and maps. Commission 46 on Education and Development in Astronomy has a Program Group on the educational opportunities given to astronomy educators and scientists when a country's attention is drawn to science through the occasion of an eclipse. I have the honor of being Chair of each.

For those in the path of totality, people usually don't understand the difference between totality and a partial phase. I liken the difference to (1) going on the Underground in London to the Covent Garden tube stop, and then returning home immediately, and (2) buying a ticket and going inside to hear La Traviata. In both cases, you can say "I have been to the opera," but only in the second case have you had the full benefit. At the 1998 eclipse in Aruba, I was able to appear on television in advance to try to explain when and why people should look at totality, but nobody really believed in advance how dramatic the onset of totality would be and therefore how easy it would be to tell when you could look safely without filters.

Without a clear understanding of the difference between totality and partiality, people living near the band of totality do not understand why they should bother to travel to see the total eclipse. Further, many people mistakenly think that there are extra rays during an eclipse, and therefore that it is hazardous to watch. Actually, there is less of everything during each part of an eclipse, and it is only that you are more tempted to watch the partial phases than you are to stare at the Sun today that could lead to any eye damage.

During the total phase of an eclipse (Figure 1), which occurs somewhere on Earth about every 18 months, the Moon entirely covers the surface of the Sun that is visible every day. Then the bright sunlight doesn't hit the Earth's atmosphere to make the sky blue, and the Sun is aloft in a dark sky in the middle of the day. Without the blue sky, we can then see the faint corona of light that always surrounds the Sun but that is normally overwhelmed.

People don't realize in advance how dramatic the overall change in the atmosphere around you is. For that reason, watching an eclipse on television misses most of the effect, which includes the rapid darkening of the sky in the last few minutes before totality, the drop in temperature, the eerie sharpening of shadows, and the change in the color balance of the incident light. The actual phenomenon of the Moon going in front of the Sun is almost less spectacular than these other effects.

The band of totality is only 100 km or 200 km wide, and the people within that band usually see about 90 minutes of partial phases before totality and a similar length of partial phases afterwards. So we must discuss both partial phases and totality.

Further, the partial phases are seen in a very wide area to the sides of the band of totality. Whole continents or equivalently large areas are included (Figure 2). The August 11, 1999, total solar eclipse crossed from Western Europe over Hungary, Romania, and Bulgaria, which benefitted from the arrival of many tourists for the event, as did Turkey, further along the path.

The next total solar eclipses, on June 21, 2001, and on December 4, 2002, both cross southern Africa. For the 2001 eclipse, most of Brazil sees a partial eclipse as does all of Africa from the Sahara to the south (Figure 3). For the 2002 eclipse, most of Africa again sees a partial eclipse. In this latter case, the partial phases are also visible from the western two-thirds of Australia, with totality itself visible at sunset from a narrow band that comes inland in the middle of Australia's southern coast (Figure 4).

It is thus important to mount education programs throughout Africa, Brazil, and Australia on what an eclipse is, what the difference is between partial and total phases, and how to observe an eclipse safely. I discuss how to observe eclipses and when they occur in my observing guide (Pasachoff, 2000).

It is important to educate the people correctly about the benefits and hazards of observing solar eclipses. All too often, the people become scared by overzealous warnings about eclipse hazards. When they later hear that they

Figure 1.   The Corona on the day of the 1999 Eclipse (Williams College photograph).

were not given correct information by groups of ophthalmologists or by the government, for example, they may be less inclined to accept even the correct warnings that may be given later on how to prevent the transmission of AIDS, for example (Pasachoff, 1996).

The easiest way to observe an eclipse safely, while still being outdoors to benefit from the overall experience, is to make and use a pinhole camera. To do so merely requires punching a hole a few millimeters across in a piece of paper, cardboard, or foil. Then you hold the material with the hole overhead and look away from the Sun, at the image projected on another surface, usually a piece of paper. Since you are looking away from the Sun, it is completely safe to do so.

To view the partial phases many people manage to obtain solar filters from one of many suppliers, often for a cost of only about $1. I am sorry to see the recent trend of making solar filters in the shapes of eyeglasses, since then people tend to keep them on and to stare at the Sun for a long time. Though if the eyeglasses are of proper optical density and are unscratched and untorn, it is safe to look through them for long periods, I always only glance at the partial phases for a second or two and then look away. In any case, the partial phases

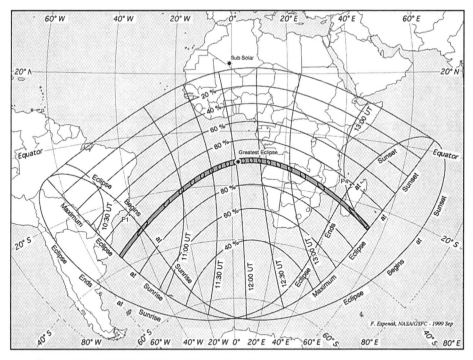

Figure 2.   Path of the eclipse of 2001, June 21.

change slowly, so there is no need to stare. I often cut the two halves of the eyeglasses apart, making two separate filters for two different people. Welders' glass in dark shades, No. 13, No. 14, or No. 15, are also satisfactory.

Old-fashioned methods are less encouraged. Smoked glass can be of uneven quality and the smoking can be wiped off, which would make it hazardous. Fogged and developed black-and-white film can make good filters, but some newer black-and-white films don't incorporate silver and so are unsafe. Color films, no matter how dark they look, are never safe, since they don't incorporate silver and so don't absorb enough in the infrared. Photographic neutral-density filters, such as the Kodak Wratten filters, are unsafe for the same reason. Some CDs or CD-ROMs have enough reflective coating of the correct type on them to make filters, but not all. And we have heard of people looking up through the hole instead of through the shiny reflective part of the CD. Similarly, though X-ray photographic film, when fogged in light and then developed to full density, can make a satisfactory filter, we have heard of people looking up through the almost-transparent part of someone's X-ray where the bones reduce the photographic density, which defeats the point of using the filter.

Still, the glory of the eclipse is worth seeing, and we should be encouraging people to watch even partial phases, with proper filters, and totality whenever possible, for which no filter is necessary or useful. And we should take advantage of interviews on radio and television or in the newspapers to demonstrate to the public how science works. My own team's scientific results are discussed in part at http://www.williams.edu/astronomy/eclipse. It is interesting that we can

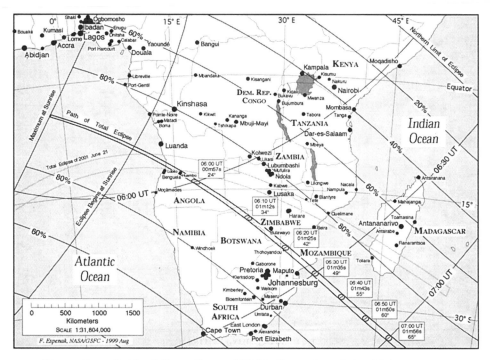

Figure 3.   Path of the eclipse of 2002, December 4, in Africa.

predict eclipses and their locations for thousands of years in advance, and that we can use our observations to understand how the Sun shines. People all over the world benefit from the demonstrations of the methods of science, and at the times of eclipses, we have a special platform to demonstrate astronomy.

Acknowledgments.   My most recent expeditions were funded in large part by grants from NASA in the Guest Investigator Program for SOHO (EIT): NRA-98-03-SEC-051; from the National Science Foundation's Atmospheric Sciences Division: ATM-9812408 and ATM-0000545; and by the Committee for Research and Exploration of the National Geographic Society: 6449-99, as well as by support from the Science Center of Williams College and the Brandi and Milham Funds.

References

Pasachoff, Jay M., 1996, *Public Education and Solar Eclipses*, in L. Gouguen-heim, D. McNally, and J. R. Percy, eds., *New Trends in Astronomy Teaching*, IAU Colloquium 162 (London), published 1998, pp. 202-204.

Pasachoff, Jay M., 2000, *Field Guide to the Stars and Planets*, 4th edition (Houghton Mifflin).

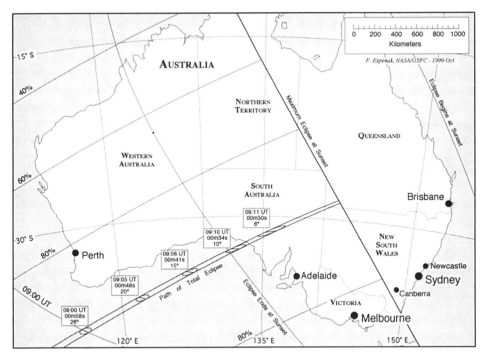

Figure 4.   The eclipse of 2002, December 4, in Australia

Discussion

Podmore referred to his poster (see p. 364-5) and added that a new book had just appeared, edited by Aisling Irwin, entitled *Africa and Madagascar: Total Eclipse 2001 and 2002* (published by Bradt Publishing ISBN 184162 015 7) and that the Astronomical Society of Southern Africa will hold a three-day symposium in Harare, just before the eclipse of 21 June, 2001.

Much discussion centred on bad advice about eclipse watching. Ratnatunga had been in Curaçao in 1998 and the tourist authorities were very proud that they had advised the public to watch the eclipse on TV. No-one seemed to be aware that the total phase could be quite safely looked at without protection – and so people missed the glory of a total eclipse. Dworetsky recounted similar stories from the U.K. in 1999, although the Royal College of Ophthalmic Surgeons found not one case of eye damage from proper use of filters.

Torres-Peimbert said that the 1991 eclipse in México provided much public exposure for astronomers on TV and in the newspapers. Even during totality, TV commentators discouraged people from watching directly. Rosenzweig said that in 1998 in Venezuela cooperation of government officials was needed to go to remote sites. The Universidad de Los Andes developed educational materials especially for distribution in schools. Astronomers also visited schools personally. This strongly motivated people to watch the eclipse. The materials are offered to Zambia for the forthcoming eclipse.

Using Television for Astronomy Teaching

Julieta Fierro

Instituto de Astronomía, UNAM, Apt 70 264, México DF 04510,
México. e-mail: fierroju@servidor.unam.mx

Abstract. The full potential of television for education has not been used in developing nations. It is relatively inexpensive to produce astronomy programs that can be broadcast taking advantage of satellite transmissions.

We suggest that these programs should have the following elements in order to be efficient:

1 Be in the local language.

2 Be short enough so that the teacher has a chance to comment on them during a one-hour lecture.

3 Show experiments specially if they are meant for schools that do not have laboratory facilities.

4 Be produced for several educational levels, including programs aimed for teacher training.

Inexpensive books should be edited in the local language in order to serve as an educational complement to the television series.

1. Introduction

Developing nations have a great need to enhance their science teaching. For instance, in México, middle school has only recently become compulsory and it has been necessary to train teachers and to produce new materials for a student population of two million. As many studies have shown, in order to teach science efficiently, one must adapt to one's particular audience. So it is important to produce teaching materials that are meaningful to local students. In particular Spanish, as spoken in México, should be used in order to satisfy the needs of our students. Similarly, examples that have to do with students' everyday lives should be included. Thousands of students live in rural communities where a single teacher must deal with every subject for the three-year secondary school. Recorded programs on specialized topics such as astronomy provide, therefore, a good teaching aid.

2.   Production

Two sets of programs have been produced. The second one drew experience from the preceding one. The program structure was the following: Introduction; several series of explanations; humor breaks and a conclusion. The explanations included experiments. In some cases the experiments were complemented. During the breaks a professional actor presented a small sketch and said something funny.

This structure was chosen bexause if the astronomer explains the same topic in several different ways, with animation and adding a touch of humor, the audience would have a better chance of grasping the idea. This procedure made the 15-minute programs agreeable to a wide audience.

The topics that were recorded are: Galileo, celestial sphere, telescopes, solar system, binary stars (determination of diameters and masses), stellar evolution, planetary nebulae, black holes, astronomical distances (to Venus, parallax, spectroscopic parallax, cepheids, supernovae, cosmological expansion), galaxies, search for extraterrestrial life and cosmology.

3.   Teacher Training

One of the reasons for producing these videos was to offer an aid for teacher training. In México elementary-school teachers receive very little formal science education. We would like them to learn how to learn, to teach, to adapt and to change. So it is important for them to have good books, workshops and videos as a complement. We also feel teachers need to have more time available to spend on their education. When a teacher uses a video he or she should interrupt its viewing in order to ask questions and discuss the topic with students to make sure they understand.

4.   General Considerations

Video production is a very time-consuming activity. It not only requires serious planning but one must work hard on the script and production. Nevertheless, once the videos are ready they can be distributed widely, even to remote localities, and can be transmitted by the educational television channels which reach tens of millions of students. In other words, one must not underestimate the media for educational purposes. If someone wants to begin video production my suggestion is to start with a very simple project and build new materials from it.

5.   Complementary Materials

A special book for teachers was produced to accompany the television series. It includes the main topics we feel a teacher must handle in order to have a basic astronomical understanding and to be able to answer some of the most frequent questions their pupils have. Small books for the general public are in the process of being edited, since we believe it is important to have written materials along with the videos, the topics are the following: a brief history of astronomy, so that teachers can understand how scientific knowledge was developed; a little ancient prehispanic astronomy related to the calendar is included and planetary nebulae, which is one of the topics to which Mexican astronomers have contributed to worldwide knowledge. The rest of the materials tend to cover classical topics: solar system, stellar evolution, astronomical instrumentation, search for extraterrestrial life and cosmology.

Discussion

In reply to Tancredi, who asked if the videos and educational materials were available to other Latin American countries, Fierro replied with a definite "yes"! Send enquiries to her at: Directora General de Divulgación de la Ciencia, Universum-Cu., C.P. 04510, D.F. México, or by e-mail at the address given at the head of the paper.

Astronomy for Developing Countries
IAU Special Session at the 24th General Assembly, 2001
Alan H. Batten, ed.

The Vatican Observatory Summer Schools in Observational Astronomy and Astrophysics

Christopher J. Corbally

Vatican Observatory, University of Arizona, Tucson AZ 85721, U.S.A.
e-mail: corbally@as.arizona.edu

Abstract. Two seemingly incongruous components have come together about every two years: the serene terraces of the Pope's summer residence at Castel Gandolfo, and the noisy exuberance of 25 beginning-level graduate students. Add in a small faculty of first-rate professors and a resourceful local support team, and one has the ingredients for the month-long Vatican Observatory Summer Schools. The eighth School takes place in the summer of 2001, and its goals are the same as when the series started in 1986: to encourage and motivate a mix of young people from industrialized and developing countries who are at critical moments of their research careers, and to make a small, but significant contribution to the progress of developing countries by exposing some of their most talented young citizens to people involved in high quality research in astrophysics. This account outlines the nature of the Schools, their follow-up, and something of how the spirit of sharing of personal and institutional resources is achieved.

1. Introduction

It certainly can seem an odd combination to the students who apply: the Vatican hosting a Summer School in astronomy. Will they be expected to spend as much of their days at the School in church as in the classroom? My sense is that the students do find our Vatican Observatory Summer Schools unique, but not in the way they might have thought initially.

The uniqueness stems from the way in which the idea was conceived. Martin McCarthy, now retired from the Vatican Observatory, was pacing on the upper terrace of Castel Gandolfo during a night in 1984 before one of his departures back to the United States. He was musing about our fruitful exchanges with colleagues, both those that happened at observatories or at formal meetings, sometimes run by the Vatican. But what *more* could we do to help make the wonders of the heavens better understood and appreciated? Then came the inspiration: share our facilities at Castel Gandolfo with talented younger scientists, from both industrialized and developing nations. From this start the Vatican Observatory Summer Schools (VOSS, for short) took initial shape in his mind.

The shape was refined when McCarthy discussed the idea with George Coyne, the director, on the way to Fiumicino airport the following morning.

As Coyne would emphasize to the Vatican authorities, the Vatican Observatory was not aspiring to become a teaching institute but, by running a Summer School about every other year, wanted to encourage and motivate a mix of young people from industrialized and developing countries who were at critical moments of their research careers. In addition, the School's hope was to make a small, but significant contribution to the progress of developing countries by exposing some of their most talented young citizens to people involved in high quality research in astrophysics.

So, this was to be a School in which a mutual sharing would benefit astronomy in all nations, even while significantly impacting developing nations. If this mix was to happen, then about two-thirds of the students at the School would have to be substantially financed. The Vatican authorities supported the idea and preparations for the first School, held in 1986, started.

2. Short Description

From our goals it follows that for each School we try to choose 25 students, coming from all over the world, who are at an early graduate level, and who show ability and real passion for astrophysical research (a broad term to include all the branches of astronomy, theoretical and observational). Our website (http://clavius.as.arizona.edu/vo/voss.html) links to the current announcement and application details. You will see there that we put a great deal of weight on the brief statement given by each applicant on why they wish to attend a School. We also much rely on the two letters of reference to distinguish applicants who are highly motivated towards research from those who are merely brilliant in their studies.

The academic structure of the VOSS is quite simple. The topic is chosen by the dean and the director. These two, to complement our Vatican Observatory's staff, seek a small faculty from people who are actively engaged in astrophysical research in that topic, who are renowned in the international community of scholars, and who are good communicators of the passion they have for their work (Table 1). (Note that the topic and faculty are limited. In this way the subject is treated at depth and the relationships between students and faculty can become relaxed.) The Schools last one month, ending before the Pope's official summer visit to Castel Gandolfo in mid-July. The daily pattern has evolved into two substantial lectures on the School's topic being given in the morning, Monday to Friday, and presentations by the students or guest lecturers following these immediately before lunch. Guest lectures broaden the scope of the School. One such lecture focussed on working with media, a vital way to harmonize the seemingly opposing needs in a developing country of fostering national pride and of accepting foreign collaborations (a problem raised by R. K. Kochhar in the panel discussion that ended this Special Session –see pp. 367-370).

Our intent for the Schools is serious, to help young scientists develop research careers in astrophysics, but the secret of achieving this is that, above all, we should all have fun at the Schools. The learning is fun, even in class. We are learning purely for the sake of learning and not for any examination results. A certificate of successful attendance is all that the students take home. That

Table 1.   VOSS Faculty and Themes

Year	Dean(s)	Faculty [1]	Theme(s)
1986	M.McCarthy	D.Latham, V.Rubin	Stars and Galaxies
1988	C.Corbally & C.Lada	J.Goad/Keppel, F.Shu	Star Formation and Galaxies
1990	C.Corbally	R.Garrison, J.Stocke, J.Van Gorkam	Stellar Spectroscopy and Galaxies
1993	R.Boyle	P.Biermann, G.Rieke, M.Rieke	Activity in Galaxies
1995	W.Stoeger	C.Frenk, C.Impey	Cosmology
1997	G.Consolmagno	M.A'Hearn, H.Campins, F.Vilas	Comets, Asteroids and Meteorites
1999	C.Corbally	R.Gray, M.Richards, R.Kudritzki	Single Stars and Close Binaries
2001	G.Coyne	G.Schmidt, R.Narayan, W.Stoeger, F.D'Antona	Stellar Remnants

[1]The regular faculty are often assisted by other senior scientists in residence. Notable among these have been Emmanuel Carreira, S.J., and Vilppu Piirola.

makes for a totally different classroom situation from what is dictated back in home institutions. It is one where "egos are left at Fiumicino airport upon arrival," in the words of one of the faculty. No question is stupid, and no question is asked to show the superior knowledge of the questioner or to put the professor on the spot. This is a unique opportunity to probe foundations.. "I was asked questions for which I did not know I had the answers," was the typical experience of another of our faculty. Yes, we all (re-)learn that teaching and learning can be fun.

Meeting people coming from different countries and situations is obviously fun too, once initial apprehension or shyness is overcome. Castel Gandolfo's situation in Italy gives us a rich cultural context in which to have those encounters take place. There is a "free lunch" for all on the terrace overlooking Lake Albano after the morning's lectures. The afternoons are free to siesta, to study, to play soccer, to swim in the lake, to visit Rome, etc. Most evenings are free also, though the students might arrange a party featuring a national cuisine, a formal discussion, or some stargazing through our rooftop telescopes -no longer really used for research observing, but Mattei and Percy (2001) suggest a way of doing useful science with them. Encounters with each other and with different cultures are the themes of outings at the weekend. These are generally organized for the Saturday only, but the central weekend is devoted to an overnight trip for everyone, funded by the School, to a place of astronomical and cultural interest such as Florence.

Amid all this fun is there serious study too? From what I have seen of individuals and groups around the Papal Palace hard at work on projects, the answer is a definite, "yes." We encourage the students to work in groups for several reasons: it extends our resources such as computers and library, it promotes interactions, and it helps the English-language skills of the students. Overcoming the English-language barrier, as Mahoney (2001) well points out, is a major problem for some aspiring astronomers in developing countries. With this in mind, our Schools are run in English, but with help, and cheerful misunderstandings, available in other languages. Each student has to give at least one short talk to the whole School. For some this is a major undertaking, but it is a talk given to a supportive audience (who know what it feels like themselves) and so it can be a great help in breaking through the language barrier.

For those interested in further details, Joson & Aguirre (2000) provide a recent description and pictures of the Schools.

3.   Follow-up

The students feel that the Schools are all too short. Castel Gandolfo has become a second home to them. The end comes into sight just as they are really enjoying the science and the new friendships and are becoming more at ease with English. (The faculty have a slightly different perspective since day after day lectures, with many questions and interactions in and out of lecture time, are quite exhausting.) It is in that feeling of wanting to continue that one of the keys to a successful School is developed: follow-up.

The follow-up is first started among the students themselves as current addresses are exchanged. E-mail has obviously made a great difference, since the Schools began in 1986, to the ease of follow-up contacts (and to organizing the Schools in the first place), but it is heartening to find how much ordinary mail has been used among our alumni(ae) when necessary. Something of the "Third-World networking", which Narlikar (2001) discusses as being so necessary, starts as the School itself ends.

Moreover, the students have also made friends with the invited faculty and with the staff of the Vatican Observatory, and contacts at *all* levels continue. Such contacts are vital for placing some of our alumni in graduate schools that will suit their talents and enthusiasm for astronomy. We have had many former VOSS students come to the University of Arizona, partly due to a scholarship from the Jesuit Community of the Vatican Observatory, now called the Martin F. McCarthy scholarship in honor of VOSS's inaugural dean. The Harvard-Smithsonian Center, the Max Planck Institute, Bonn, and the University of Toronto are prominent in showing a similar welcome. All the faculty and deans, and especially Coyne as director, have written many effective letters of recommendation to help place alumni elsewhere. For, though there are no exams in the Schools, one can get to know a student's ability and potential very thoroughly over the course of a month through questions asked, general conversations made, and projects completed.

I am always glad to hear of and to experience times when VOSS alumni meet again at conferences. There the bond of friendship is renewed. So it was wonderful that, with the hard work of Chris Impey (faculty in 1995) and others, alumni from the first six Schools returned to Castel Gandolfo in June 1998 for a week-long International Symposium on Astrophysics Research and Science Education. Impey (1999) points out in the introduction to the meeting's proceedings that it soon acquired the appropriate nickname: Super-VOSS. It was unfashionably broad in its sweep of topics, reflecting the many research interests of the alumni, and refreshingly inspirational, both on the progress of astrophysics and on the challenge of educating in science, particularly in developing countries. It also accelerated Third-World networking. For once the alumni of each year had met and exchanged news among themselves (horizontal networking), I noticed how the groupings then became more those of continents, thus linking all the different VOSS years (vertical networking). In this way resources became pooled, at least potentially, to tackle common problems, whether scientific or educational.

There is one further thought concerning follow-ups to a School. Hearnshaw's (2001) "critical factors" for success of research in developing countries ring true in pointing to the contrast facing young researchers between the abundance they may have experienced in their graduate days and the relative isolation and lack they have awaiting them at their home institutions. With this in mind, the Vatican Observatory Foundation is raising funds to endow two post-doctoral fellowships at the University of Arizona. Through these fellowships we hope to help those from developing countries, such as our alumni, complete their training in astronomy in a stimulating environment and one where they can forge suitable, lasting collaborations for when they return to their own countries. We also want to take steps to ensure that their access to prime observing facilities and to stimulating meetings continues even when back home. Post-doctoral programs such as outlined would partly address Snowden's (2001) point on the lack of motivation among scientists in developing countries. The contrast is a problem, but every bit of collaboration helps to overcome it, as the success story of astronomy in Venezuela witnesses (Rosenzweig 2001).

4.   Impact

Impact can be hard to assess adequately. The numbers are that in seven Vatican Observatory Summer Schools we have had 173 students from 48 different nations, with 57% of the students from developing nations. Table 2 gives the breakdown by continent and nations. (The cover devised by Chris Impey and Dana Irvin for the Super-VOSS proceedings – Impey 1999 – more dramatically shows the Schools' world coverage.) All but 10 alumni(ae) are either following professional careers in astrophysics or preparing to do so. Half of those from developing countries have spent at least two years with graduate programs or in research at leading institutes in industrialized nations. So the numbers tell of success in helping young scientists, including a good number from developing nations, to develop research careers.

Somehow though, on reading the accounts of faculty and alumni in *The First Ten Years* (Maggio 1997), it becomes clear that the impact of the Summer Schools is more to be measured by each person's growth into a more complete, more productive, more integrated researcher. Our alumni have come to realize that the friendships forged between all participants, students and faculty, are not an optional extra to success in research and teaching; they are the means to make research and teaching more productive because more human. These relationships are not meant to be exclusive; other colleagues are included as the years go by. It is such open-ended, professional, cross-cultural friendships

Table 2.   VOSS Distribution up to 1999 by Continents and Nations

Continent	Nation	Total
AFRICA:	Nigeria 4, South Africa 5	9
NORTH AMERICA:	Canada 8, Cuba 1, Mexico 4, United States 16	29
SOUTH AMERICA:	Argentina 14, Brazil 9, Chile 2, Colombia 1, Peru 1, Uruguay 1, Venezuela 6	34
ASIA:	China 5, India 9, Indonesia 1, Japan 1, Korea 3, Sri Lanka 4, Taiwan 2, Vietnam 1	26
EAST EUROPE:	Armenia 1, Bulgaria 5, Croatia 4, Estonia 2, Hungary 1, Iran 1, Lithuania 1, Poland 5, Romania 1, Russia 3, Slovenia 1, Ukraine 5	30
WEST EUROPE:	Austria 2, Belgium 2, Denmark 2, Finland 4, Germany 4, Greece 7, Italy 7, Netherlands 1, Portugal 1, Spain 2, Sweden 1, Switzerland 1, Turkey 4, United Kingdom 1	39
OCEANA:	New Zealand 6	6

that provide the real hope for astronomy in developing countries – and indeed everywhere.

Acknowledgments.   His Holiness John Paul II is thanked for his support of the Schools, and all the VOSS faculty, staff and students are thanked for contributing their inspiration to this account.

References

Hearnshaw, J. B. 2001, this volume, p. 15-28.

Impey, C. 1999, in *International Symposium on Astrophysics Research and Science Education*, ed. C. Impey (Vatican Observatory Foundation), xi.

Joson, I. B., & Aguirre, E. L. 2000, S&T, 99, No.5, 82.

Maggio, E. J. 1997, *The Vatican Observatory Summer Schools in Observational Astronomy and Astrophysics: The First Ten Years* (Vatican Observatory Foundation).

Mahoney, T. J. 2001, this volume, p. 333-339.

Mattei, J. A., and Percy, J.R. 2001, this volume, p. 89-94.

Narlikar, J. V. 2001, this volume, p. 324-328.

Snowden, M. 2001, this volume, p. 266-275.

Rosenzweig, P. 2001, this volume, p. 205-209.

Discussion

Hearnshaw commented that six graduate students from New Zealand had gone to VOSS and all had returned absolutely enthusiastic about the experience. He believed VOSS to be one of the best programs in the world for young astronomers. He wanted to know the topic for the 2001 school. Corbally thanked Hearnshaw for the graduate students he had sent, who had all brought much to the schools they attended. The 2001 school is to be on the theme "stellar remnants" – details on the website.

Kochhar suggested that another series of schools might be initiated dealing with topics such as those covered in this Special Session. Corbally said he would be delighted to consider it but the Observatory's manpower is limited. One year they ran a summer school for bishops (called a workshop) and had to skip a year in the sequence of student schools.

Snowden was puzzled that motivation was mentioned as one of the purposes of VOSS. First-year graduate students are surely already highly motivated. He also wondered if there were a clash of aims between the astronomers and the Vatican authorities. Corbally replied that, given the difficulties in the paths of many graduate students on their way to becoming full-tim researchers, they can well use a boost to their motivation at the VOSS and the continuing support that comes from the friendships made there. The Vatican authorities support astronomy as a value in itself. There is no clash between the Observatory staff and the Vatican over the schools. While this point of view may seem strange to some, its truth is contained in the Christian doctrine of the Incarnation.

Wentzel enquired how VOSS encouraged really useful letters of recommendation. Corbally said that this was done by getting the word out at meetings such as this and by responding honestly to queries why someone's brilliant student had not been selected. Alumnae of the schools, whether student or faculty, write particularly helpful letters. They know the kind of help needed in the difficult task of selection.

Initiatives in Astronomy Education in South Africa

Case L. Rijsdijk

South African Astronomical Observatory, PO Box 9, Observatory, 7935, South Africa. e-mail: case@saao.ac.za

Abstract. A brief review of the issues affecting the current status of science education in general, and astronomy education in particular, is given. The paper looks at the present situation at primary, secondary and tertiary levels. South Africa has unique educational problems and the initiatives by local observatories and universities at school level are described. The problems encountered by the South African Astronomical Observatory (SAAO) Science Education Initiative (SEI) are typical, as is the SEI approach to addressing some of these. The experience of the SEI is described, as are some of the resources developed by them for primary and secondary schools. Finally a brief look is taken at future developments, in particular, ways in which the Southern African Large Telescope (SALT) can contribute to astronomy and science education.

1. Background

The history leading up to the present crisis in South African education in general, and science education in particular, is well documented (Blankley 1994, Medupe and Kaunda 1997). Since 1994 several people (Grayson 1996, 1997, Rutherford 1997) at the tertiary level have suggested and tried models to address some of these problems.

2. Problems

For learners at South African schools the problems in science education start with under-qualified and unqualified science teachers: only 16% of science teachers in the country have one or more years of a science subject at university (Arnott and Kubheka 1996). Of the remaining 84% many are teaching learners[1] at a level that they themselves have not achieved! It is not unusual for a teacher who left school with a grade-10 certificate now to be teaching grade-12

---

[1]Because many students in South Africa are older than their peers, especially in the lower grades, and because many people are entering formal education as adults, pupils and students are often referred to as learners.

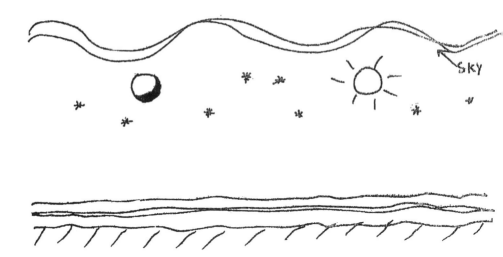

Figure 1.   Flat Earth...

learners. In many cases, especially among the black teachers, the science subject studied is biology and it is these teachers who are expected to teach physics and chemistry as well. Many have had little or no exposure to physics or chemistry other than that which they picked up at school or in their studies of biology.

Before 1996 astronomy was taught only as a small section of the geography syllabus as an introduction to the Earth's geography. It started by looking at a very simplistic portrayal of the "Big Bang", as a rationale for the existence of matter out of which the solar nebula formed, followed by the formation of the solar system and ended by looking at the Earth's place in the solar system/universe. Textbooks were frequently written in the northern hemisphere and so the Moon's phases were wrong: just full, new and quarter were shown to try to offer some sort of explanation for the tides! Eclipses were sometimes covered in physics as an example of "rectilinear propagation of light": an explanation was seldom offered for the causes of the Moon's phases or why there was no eclipse every month. In fact this section was usually omitted for two reasons: it was NEP (not for examination purposes) and, more commonly, the teacher knew very little, if any astronomy: frequently misconceptions were perpetuated, (see below).

Many of our black children do not live in an environment that is conducive to learning or science. Their parents had minimal if any schooling, are illiterate and as a result there are no books, magazines or newspapers in the home. Overcrowding within the home is also a problem in that children have little or no space in which to try to do homework and many need to get jobs to help the family survive.

At school many of these children are not much better off, resources at schools are poor and very limited: many do not have electricity let alone a phone line. In addition classrooms are overcrowded, teachers often have to cope with classes that have 60+ learners in them.

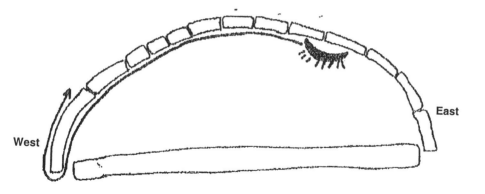

Can account for the occurrence of day, night, stars.

Figure 2. Geocentric Flat Earth. Conflict between untutored views and scientific views!

Added to the above is the language problem: South Africa has 11 official languages and many dialects. These languages were oral, only formalized and given a linguistic structure relatively recently and use a Western alphabet. Since these written languages are new, their vocabulary has not yet developed properly and often one word is used to describe several phenomena. This becomes particularly critical when scientific words with precise meanings are used. In one local language for example the word "amandla" can be used to describe "energy", "force" and "power": learners understandably become confused.

At high-school and university levels, instruction is often given in a language other than the learners' or students' mother tongue. [2] Ongoing research (Clerk, 2000) with first-year university students seems to indicate that students benefit from conversation with instructors, both on and off the topic: something that didn't happen at school.

3. Prior Knowledge and Misconceptions

It is well known that there are many misconceptions about even the simplest astronomical and scientific phenomena. Many people do not really understand how the phases of the Moon occur, or why seasons happen (Schneps and Sadler 1988). Many teachers in South Africa have these and other misconceptions. Research carried out at Potchefstroom University (Wesi, Lemmer and Smit 2000) on a group of local teachers, who were asked to make a sketch of their view of the universe, produced some interesting representations :

Many showed other interesting features, and some common ideas were observed, amongst which were:

- a flat Earth with the sky beyond the Sun, Moon and stars,

---

[2] I will continue to use the term "students" for those in tertiary institutions.

Figure 3.   Random Universe: spherical and bounded universe.

- living things such as trees, animals and people on a flat Earth,

- often the crescent Moon and Sun were in such a position that the dark side was illuminated.

When one talks to students after a workshop or slide presentation it becomes clear that many still see the Sun as a small bright object moving across and above the plane of a flat, stationary Earth and that the phases of the Moon are caused by the Earth's shadow.

Because of the impoverished background of many learners, a lot of concepts, examples and material are "learnt" but often not understood: learners are unable to see how and where to apply the knowledge gained. Assessment might show that students "know" their physics yet struggle to apply it to solve problems: their misconceptions are often complex and deep-rooted (Enderstein and Spargo 1998).

Teachers frequently stifle questioning for fear of having their own lack of knowledge exposed or to save themselves from embarrassment when learners ask questions of a speaker the teacher thinks are stupid. On one occasion a learner was reprimanded for asking what his teacher thought to be a stupid question: she intervened by saying ". . . everyone knows those dark patches on the Moon are shadows caused by clouds!" This after just having painstakingly explained that there was no air (atmosphere) or free water on the Moon!

4.   Addressing the Problems

The efforts by the South African Astronomical Observatory (SAAO) are fairly typical of the approach adopted by the astronomical community in South Africa. The Science Education Initiative (SEI) at SAAO realized that there was little point in teaching astronomy to a population that had little or no science. It therefore adopted the approach to use astronomy as a vehicle for science education, (Rijsdijk, 2000): at least until such time that the levels of science education had reached a stage where some proper astronomy could be profitably taught. There are schools where the levels of education are on a par with, or exceed, their First-World counterparts, and here the SEI was able to support teachers with some proper astronomy education using modern resources such as computers, the Internet and small telescopes.

The SEI identified at an early stage that the target group for its limited resources, both human and financial, were teachers rather than learners. The main reason being that there were just too many learners to interact with profitably: one teacher would be able to share her/his gains with many learners for the rest of his/her teaching career. Later this was narrowed down to those teaching at the upper end of the primary-education levels: grades 5-9, (Rijsdijk, 2000). The main reason being that grade 10-12 teachers were too involved with preparing their learners for the matriculation exams to want to get involved with "new stuff that is not even in the syllabus". This did not mean however that learner groups were excluded: often learner groups provided an ideal testing ground for newly developed materials and resources.

The SEI then focused on two things: developing resources to help teachers and run workshops where they could come to grips with the material and take it into the classroom with confidence. However for this to be practical, the first task was to get astronomy re-introduced into the curriculum. This was achieved in 1996, and since then a wide variety of resources and teacher support material has been developed by the SEI.

The South-African government declared 1998 to be the Year of Science and Technology (YEAST): an ambitious project to promote science and technology throughout the country, with a focus month allocated to each province. In addition the Science Councils were also asked to focus on an area of their speciality: the National Research Foundation (then the Foundation for Research Development) chose Astronomy and Space Science. As a result of this the SAAO developed the "Friends with the Universe" project, (Rijsdijk, 1998). This proved most successful and has been retained as the SAAO outreach programme, more or less replacing the SEI.

4.1. Resources

It became clear, as has been mentioned on previous occasions, (Rijsdijk, 1998, 2000), that whatever resources were developed would need to be:

- cheap and simple,

- use readily available materials, i.e. scrap cardboard tubing, paper, wire, marbles etc.,

- easily reproducible,

- of a modular format,

- needed to be relevant/applicable to the local environment,

- focus on group work and skills development,

- supported with activities and assessment material,

so that after the teachers had themselves worked through the materials in a workshop environment, these resources could be taken straight into the class-room. Teachers were always given some sort of "resource pack" to take away with them: usually this took the form of a set of "how to" notes of the work-shop accompanied by exemplars of the materials used that could be photocopied. This approach appears to be in line with thinking elsewhere, (Percy, 1999).

During the last few years, "Friends with the Universe" has developed a large range of workshop modules: some of which are unique to the "Friends with the Universe", whilst others are the evolution of other materials from around the world. Some of these can be downloaded straight into common word-processing packages from the SAAO homepage (http://www.saao.ac.za/education) for those schools or teachers that have access to the Internet. As better techniques are developed for transferring images, further modules will be added to the SAAO homepage. To date over 20 different modules have been tried and proved suc-cessful. They vary in difficulty and can be used in different grades, see Table 1. Some modules have not been assigned a grade: teachers are encouraged to decide for themselves which ones are appropriate. Learners from different environments have different needs.

Teachers are also given workshop demonstrations on a wide range of astron-omy related topics:

- Why do Balloons Float?,

- Weightlessness,

- Magnetic Braking,

- How Rockets Work

- Modeling Orbits

being typical examples.

In addition to these practical, hands-on exercises, teachers can also borrow posters, books, videos and small telescopes (typically 11.5-cm Newtonian reflec-tors) to enrich their teaching. In order to borrow a telescope a teacher would normally come to the observatory's Science Education Resources Centre and learn how to set up and use the telescope, after which it can be borrowed for about a week.

As a part of YEAST, "Friends with the Universe" developed a set of ten full colour posters that were to be distributed to the bulk of schools in South Africa. They were a multipurpose set in that they were both decorative and educational. A smaller pamphlet on the solar system was also produced. Both are still available to teachers and learners.

The "Friends with the Universe" has also worked closely with, and advised, several NGEO's in developing resources for Curriculum 2005 and in the prepara-tion of local textbooks, newspaper posters and a wide variety of other hand-outs: newspapers especially are increasingly being used as resources by teachers.

Finally, "Friends with the Universe" uses many local, indigenous and uni-versally available resources to help illustrate scientific principles and concepts:

Table 1. Some of the Modules developed by "Friends with the Universe"

No.	Title	Description	Grades
1	Where is Up?	Why people don't fall fall off the Earth	1-3
2	Making Shadows	What causes shadows and how to make them using simple well-known objects	2-4
3	Looking at the Moon	Simple recording observing of the Moon's phases	6-9
4	Where is the Moon?	Using sunrise/sunset tables learners predict future full/new Moons and make a model to find the Moon.	6-12
5	Making Comets	A teacher demonstration using dry ice to make a comet	teachers
6	The Seascope	A device to enable learners to look under water in rock pools and rivers	6-12
7	Measuring	Using different techniques to measure the area of irregular shapes	7-9
8	Large Numbers	Making sense of large numbers by analogy and by taking a trip to α Cen.	6-12
9	Quadrants	Using scrap material a simple quadrant is made to measure heights	5-10
10	Inverse Square Law	Using a simple grease-spot photometer this important law is verified	10-12
11	The Power of the Sun	Using 10 above, the power of the Sun is calculated	10-12
12	How far is that star?	Using 10 and 11 above the distance to α Cen. is calculated	10-12
13	Observing the Sun	Safe methods of looking at the Sun and explaining what sunspots are	6-12
14	Making telescopes	Using cardboard tubes and plastic lenses learners make a telescope for about $1	8-12

Table 1 *cont.*: Some of the Modules developed by "Friends with the Universe"

No.	Title	Description	Grades
15	Alexandrian Astronomy[3]	Using a lunar eclipse and the size of the Earth to find the size and distance to the Moon.	12+
16	How big is the Earth?	Using the method of Eratosthenes to find the size of the Earth	10-12
17	Eclipses	What causes them and how can they be observed	all ages
18	Seasons	By making two simple models learners discover what really causes the seasons, if there are any, and what is the meaning of the solstices and equinoxes	8-12
19	Patterns in the sky	What are the constellations and where to find them by making a planisphere	6-12+
20	Making a spectroscope	Using a small piece of diffraction grating and some cardboard, learners make a spectroscope to measure to within 5 nm, cost U.S. $0.20.	12+
21	Using a spectroscope	Demonstrates the use of 20 above, including the Doppler effect	12+
22	Making Craters	What causes craters and how are they formed	6-10
23	Planning trips	A skills development exercise - using assorted timetables learners are expected to plan a trip	8-12
24	Talking to Mars	An exercise demonstrating how time delays occur and what effect they have	6-9
25	Timelines	Learners use the launching of satellites as basis for a timeline in the classroom. This runs in parallel with another showing birthdays, news items of the year	4-8

[3]See Rijsdijk, 1999.

local star-lore and in particular the two forthcoming eclipses are seen as invaluable resources to educate many people using newspapers, TV and radio. In the past both solar and lunar eclipses have been used. The forthcoming eclipses in the sub-continent are particularly important and appropriate resources are under development. The logistical problems of tackling large, sparsely populated areas is being researched and addressed.

4.2. Workshops

Typically these cater for around 25 learners/teachers at a time, although teacher groups are usually smaller. Workshops for teachers last for about 2 hours, take place after school hours and are frequently interspersed with associated activities such as slide/video shows, a look at the Sun/Moon or other related demonstrations. Workshops for learners are shorter, usually occur during school time or vacation time and are planned as a package accommodating the needs of the teacher.

No special facilities are required: at the SAAO an old store-room has been converted to accommodate these workshops. This was deliberately done to show teachers what can be achieved with minimum of facilities. Everything required for the workshop is supplied: scissors, glue, paper, writing equipment, staplers etc. The reason being that then everyone starts off with the same and the correct stuff!

During the workshop participants are encouraged to help each other for a variety of reasons: the main one being in that it gives those "helping" a chance to build confidence, and frequently language problems are overcome this way. The facilitator often does not have a good enough (if any!) knowledge of the languages used by participants.

One thing that is not tolerated is for workshop facilitators to become surrogate teachers. Frequently teachers who bring classes to the Science Education Resources Centre are under the impression that these workshops are designed to give them a break from teaching! Facilitators should involve them from the outset, and when arranging these workshops for learners this should be made clear.

4.3. Starbus

One of the stated objectives of "Friends with the Universe" during YEAST was to reach the rural youth and teachers, (Rijsdijk, 1997). It was obvious that the distances and costs involved made it impossible for these people to travel to major population centres. It was decided instead to take the resources, workshops, demonstrations and talks to the people.

To achieve this a mini-bus was bought, decorated and equipped with all the same equipment and resources as the Science Education Resources Centre and taken into the rural communities. This proved most successful and became known as the Starbus, (Rijsdijk, 1998, 2000). Usually the Starbus would visit a school and use that as a "base": with teachers from surrounding schools coming there to attend the workshops. The "base school" would be carefully selected to maximize the impact of the Starbus visit.

5. New Strategies

In order to find solutions to these problems, the government and a wide range of other organizations have developed and have begun to implement many new strategies.

5.1. Curriculum 2005

In 1997, the new curriculum was finalized. Called Curriculum 2005: (Department of Education, 1997) it is an Outcomes Based Education (OBE) system, loosely based on that of New Zealand, Holland and Canada. It has done away with traditional subjects and replaced them with eight Learning Areas. One of which, the Natural Sciences, is broken into four themes or strands: Life and Living, Matter and Materials, Energy and Change and Earth and Beyond. Astronomy now features as a substantial part of the latter.

Implementation started in 1998 but the new curriculum was soon found to be too cumbersome and too far removed from the traditional system. The language was difficult, the structure too theoretical and it was generally beyond the scope of many teachers. Training courses took time and at the end teachers were often confused and had little knowledge of what actually to do in the classroom. It has been reviewed (Chisholm, 2000), and many of the complexities have been removed: implementation is expected to proceed more smoothly after June 2001 when the revised C2005 will be introduced.

5.2. Tirisano

A word meaning "working together". The plan was implemented by the South African government in 1999 (Department of Education, 1999) with the aim of building an education and training system for the 21st century. Its main goals are:

- breaking the back of illiteracy within five years,

- developing schools as centres of the community,

- developing the professional quality of teachers,

- developing a life-long learning process, and

- combating HIV/AIDS through education.

This is a new programme and there is as yet no evaluation of its success and it will be some time before any statistics become available. There is also an active advisory panel to advise the minister of education on new strategies by consulting with a broad spectrum of science, technology and education experts, giving the minister a clear indication of the country's needs, (Kahn, 2000).

5.3. Other Outreach Programmes

From an astronomical point of view, all other facilities involved with astronomy run some sort of outreach, education and popularization programmes. The national radio astronomy facility at Hartebeesthoek (HartRAO) in Gauteng is very active and runs many programmes similar to those of the SAAO. When the occasion arises, such as at the National Science Festival, SASOL SciFest, SAAO and HartRAO run joint workshops.

Boyden Observatory, near Bloemfontein, in the Free State is becoming active again and is run by Free State University. They run an education programme using some SAAO resources and also have many open nights. The planetaria in Cape Town and Johannesburg are kept busy and they too are active in educational outreach. The Cape Town planetarium offers an introductory course on descriptive astronomy twice a year and consists of five, two hour lectures.

Over the last few years a few large science centres have started up and there are also many smaller ones scattered around the country. They have formed an umbrella organization known as the South African Association of Science and Technology Centres, (SAASTEC). One on the north of Kwazulu-Natal is particularly active and runs practical science classes for over 6000 final year grade 12 learners a year!

6. Universities

The university with the most accessible course in astronomy is the University of South Africa (UNISA). This is a distance-learning university and due to a rationalization process that is being implemented in the country, it is now the only one to offer a full undergraduate course.

The University of Cape Town has close links with SAAO and post graduate students studying for M.Sc. and Ph.D degrees would be jointly supervised by staff from both institutions. Witwatersrand, Natal (Durban and Pietermaritzburg) and Pothchefstroom all offer some tuition in astronomy, whilst Rhodes University specializes in radio-astronomy.

As is apparently the case in other parts of the world, the number of students studying physics is decreasing and several universities are considering combining physics with astronomy and space science. At present the council for higher education is reviewing faculties at universities and technikons throughout the country, and at this time there is no clarity yet: it may well be that there will be some rationalization and that only certain universities will be offering astronomy as an undergraduate and post-graduate course.

7. The Future

When looking at the needs of South Africa in the future there is no doubt that research will have to be relevant and that it should address the social, economic, educational and political needs of the country, (Medupe and Kaunda, 1997). The government has however realized that there is also a need for the people within country and the rest of the world not see South Africa as just another Third-World nation (DACST 1996):

> Scientific endeavour is not purely utilitarian in its objectives and has important associated cultural and social values. It is also important to maintain a basic competence in "flagship" sciences such as physics and astronomy for cultural reasons. Not to offer them would be to

take a negative view of our future - the view that we are a second-class nation, chained forever to the treadmill of feeding and clothing ourselves.

There is a need to establish some self-esteem, especially amongst the country's youth. In the past learners were taught a range of myths and legends about astronomy. They were taught how science changed the perception of the universe in which we lived by describing theories of the universe of the ancient Greek, Roman, Babylonian, Aztec, Mayan and many other cultures, but never any of their own, the theories of the San, Batswanas, Zulus and Xhosas were ignored: science was a white man's thing, (Medupe and Kaunda, 1997).

There is a rich heritage of ethno-astronomy in Southern Africa: it is an oral tradition, but this makes it no less relevant to the people of the region. In the past, as in a large number of other cultures, many aspects of life were controlled by the stars. The time to plant, initiation ceremonies and other festivals were determined by the rising and setting of identifiable single stars, groups of stars or constellations. What in the West are called the Pleiades or the Seven Sisters, are known as "isiLemela" to many people in South Africa: they are the Digging Stars, their rising early in the morning meant it was the time to start preparing the soil for planting. Similarly, there are indigenous names for many other stars and constellations, (Snedegar, 1995).

This rich heritage should be used to make South-African education more relevant to its own people: intellectual and cultural enslavement is as bad as economic, physical or political enslavement. Indigenous knowledge is slowly coming to the fore: there is an indigenous culture of science and as Walter Massey, former director of the U.S. National Science Foundation said, "Science provides a set of common experiences, a kind of cultural glue to hold together multicultural societies."

The Southern African Large Telescope (SALT) should, and will, become an icon for South African youth: physical evidence that science is not something that happens only in Europe, Japan and the United States of America; it happens in South Africa as well! A workshop is planned for 2001 in Cape Town where the educational potential of all the large telescopes in the world, optical and radio, will be discussed to see if there is a possible synergy. For the country's youth to see that South Africa is a part of international big science is as important as, *no*, more imoprtant than, that same youth seeing the national football side taking part in the world cup.

It is the Southern, not South-African Large Telescope, this subtle difference clearly indicates that it will provide an access to the skies for the scientific and astronomical community, not only of South Africa, but of the sub-continent as a whole. Together with the telescope is a Collateral Benefits Plan which clearly sets out the role that SALT will play in the larger community, it will:

A. support industrial empowerment,

B. provide educational empowerment,

C. enhance Public Outreach and provide direct educational benefits,

D. develop a visitor centre,

E. be an African facility.

"Friends with the Universe" has already started on aspects of B and C, and SAAO is at present working on developing D: two such centres are envisaged, one in Cape Town and one in Sutherland in the Karoo where the SAAO observing site is located. To be known as Stargates the centres will focus on spectroscopy with an exhibit called "Fingerprinting Light". With construction on SALT starting in 2001, the Sutherland Stargate is now a priority.

Acknowledgments. I would like to thank Drs Bob Stobie, Patricia Whitelock, and Peter Martinez for their continued support for the educational programme at SAAO and Frank Andrews and Tony Fisher of Carter Observatory, Wellington, New Zealand for encouragement and sharing the wealth of experience in producing material for OBE education. In addition I would like to thank the IAU and the SAAO for funding to attend the 2000 General Assembly.

References

Arnott, A. and Kubheka, Z. 1996, *Access, enrolment and pass rates in mathematics and science at secondary schools.* NARSET report: Issues Relating to Access and Retention in Science, Engineering and Technology in Higher Education. FRD, Pretoria.

Blankley, W. 1994, *The abyss in African school education in South Africa*, S. Afr. J.Sci. 90, 54.

Chisholm, L. 2000, *Report of the Review Committee of Curriculum 2005.*

Clerk, D. 2000, *—Just what is so difficult about the language encountered by first year university students in science courses. An ongoing investigation* Physics Conference, Rand Afrikaans University (unpublished).

DACST (Department of Arts Science and Technology) 1996, White Paper on Science and Technology, *Preparing for the 21st Century.*

Department of Education. 1997, *Curriculum 2005 - a discussion document.*

Department of Education. 1999, *Tirisano: Working Together to Build a South African Education and Training System for the 21st Century.* Policy document.

Enderstein, L. G. and Spargo, P. E. 1998. *The effect of context, culture and learning on the selection of alternative options in similar situations by South African pupils.* Int. J. Sci. Educ. 20, 711 - 736.

Grayson, D. J. 1996, *A holistic approach to preparing disadvantaged students to succeed in tertiary science studies. Part I. Design of the Science Foundation Programme.* Inter. J. Sci. Educ. 18, 998-1013.

Grayson, D.J. 1997, *A holistic approach to preparing disadvantaged students to succeed in tertiary science studies. Part II. Outcomes of the Science Foundation Programme.* Inter J. Sci Educ. 19, 107-123.

Kahn, M. 2000, *Toward a Strategy for the Development of Science, Mathematics and Technology Education.* Working document presented at a Strategic Planning Workshop to advise the Minister of Education, NRF Pretoria. Unpublished.

Medupe, R. T. and Kaunda, L. 1997, *The Problems of Science in Africa*, Mercury 26, no. 6 Nov./Dec 1997, 16 - 18.

Percy, J. R. 1999, *Effective Learning and Teaching of Astronomy. Teaching of Astronomy* in Asia-Pacific Region Bulletin No. 15, 41 - 44.

Rijsdijk, C. L. 1997, *Friends with the Universe.* Business plan submitted to the Foundation for Research Development, internal document.

Rijsdijk, C. L., 1998. *Taking Science to the People.* Proceedings of the 2nd National Conference on the Public Understanding of Science and Technology in South Africa: Science and Society, Pretoria.

Rijsdijk, C. L. 1999, Quantum Magazine, NSTA.

Rijsdijk, C. L. 2000. *Using Astronomy as a vehicle for Science Education.* Combined conference of the Astronomical Society of Australia and the Royal Astronomical Society of New Zealand, Sydney, July, 2000. PASA www.atnf.csiro.au/pasa.

Rutherford, M. 1997, *Opening access to quality education.* S. Afr. J. Sci. 7, 211-215.

Schneps M. and Sadler, P. M. 1988. *A Private Universe* (video), Pyramid Films, Santa Monica. CA

Snedegar, K. V. 1995, *Vistas in Astronomy*

Wesi, R. P., Lemmer, M. and Smit, J. J. A., 2000. *Science teachers perceptions of the Universe.* Paper delivered at the South African Institute of Physics Conference, Rand Afrikaans University, July, 2000, unpublished.

Discussion

Fierro asked what kind of follow-up material would be produced for teachers. Rijsdijk replied that very little was produced at present, but he hoped to make use of some of the assessment and follow-up material used by the Carter Observatory in New Zealand. Many teachers return of their own volition and some come back often, bringing *new* teachers with them.

Astronomy for Developing Countries
IAU Special Session at the 24th General Assembly, 2001
Alan H. Batten, ed.

Distance Education and Self-Study

Barrie W. Jones

The Open University, Milton Keynes MK7 6AA, United Kingdom,
B.W.Jones@open.ac.uk

Abstract. Distance education and self-study are defined and described and their possible application to developing countries is discussed.

1. Introduction

It is *not* my purpose to specify how distance education and self-study should be used to teach astronomy in developing countries — my teaching experience in such countries, though not negligible, is rather too limited for that. I do, however, have extensive experience of distance education in the U.K., and therefore my purpose in this article is to display the characteristics of distance education, and of self-study, in order to promote debate about how they can best be used in astronomy education in the developing world. Distance education has the potential to overcome

- a shortage of astronomy teachers

- a shortage of non-human resourses

- difficulties that students face in attending a campus.

Additionally, distance education materials can be used to increase self-study *on* campus, and hence overcome the first two of these problems.

2. What is Distance Learning(and Self-Study)?

The main defining characteristics of distance education are that

- it allows students to study at home or in their work place

- it allows students to study part-time or full-time

- course materials are delivered to students mainly by mail, or electronically

- face-to-face contact with a teacher is limited, even absent.

Face-to-face contact with other students can be less limited, particularly in urban areas, if several students of a course live or work near each other.

Self-study is that aspect of *any* learning process where students study on their own, i.e. without real-time human contact. In distance education, self-study accounts for a far larger proportion of the student's study time than it does in conventional education. However, it is vital to realise that if distance education is to be effective then the student *must* be supported — sending out materials and letting the student get on with it is a recipe for failure. Forms of support include

- preliminary advice to the student about which courses best match the student's needs, and in what order the courses should be taken

- a clear specification to the student of the knowledge and skills that the student should bring to each course

- course materials (printed or electronic) that are suitable for self-study

- a clear study guide for each course, with a study timetable, including assignments to be completed at specified times throughout the course

- a personal tutor, in telephone, e-mail, mail or (limited) face-to-face contact, who also marks the student's assignments, but as well as grading them, provides copious teaching comments on them

- student self-help groups, with telephone, e-mail or face-to-face contact

- residential schools in mid-course, with a focus on practical work in science (just a few days a year is of enormous benefit).

Even with all of these forms of support, distance education is not easy for teachers or for students. The students need to be motivated and determined, and the tutor needs to acquire the special skills needed to teach in this way. But distance education provides the possibility of education where otherwise there would be none.

At its best, distance education can provide an education as good as any obtained by conventional means. Experience in the U.K. and elsewhere is that any scepticism about whether this is an effective way to teach science in general and astronomy in particular, is *completely* unfounded.

3.   Distance Education in Science around the World

Distance education is rapidly spreading, in developing and developed countries. It is almost entirely targeted at students at least 18 years of age, and therefore distance-teaching is mainly through universities and colleges, rather than schools. But the level of courses and qualifications need not be at degree level, and indeed many are not.

Increasing numbers of conventional institutions are offering the option of distance education, though this is often a rather small part of their activities. In addition there are institutions *dedicated* to distance education, and many of these are national bodies. The first large national institution was the U.K.

Open University (UKOU), which admitted its first students in 1971. It now has about 170,000 students. The majority are studying for undergraduate degrees, but some are aiming for masters' degrees, and others are not aiming to get degrees at all. The countries with the largest national institutions include (in alphabetical order)

China	France
India	Indonesia
Korea	South Africa
Spain	Thailand
Turkey	U.K.

Among the national institutions worldwide, few teach science of any sort, and even fewer teach astronomy. Therefore, there is a great opportunity for expansion, though it will be necessary to convince the institutions and potential students of the value of an astronomy education, not just the cultural value, but particularly the economic value arising from the wide range of scientific and other skills that are well taught through astronomy. Astronomy need not be taught on its own. It can be included within science courses where it acts as an attractor of students, as a motivator, and as a rich source of fascinating examples. This builds the case for astronomy to be taught in places where, on its own, it might be seen as lacking relevance to the needs of a developing country.

Another argument for distance education in developing countries is the experience of many institutions that the total cost of producing a graduate by distance teaching is roughly half the cost of producing a graduate by conventional means. This reduction in costs is achieved when at least several hundred students enrol for a particular course.

An important type of student is the school teacher; distance education has enormous potential to increase the number of teachers in a developing country who are willing and able to teach astronomy in their schools. From such grassroots teaching, much can grow.

4. The Production of Effective Distance-Learning Courses

I address here the preparation of a *course* rather than a complete degree programme. In the UKOU a course can typically account for anything from about 3% of a degree to about 18%; these are 10-point and 60-point courses respectively. Students are free to take just a single course, for which they are awarded a certificate of completion.

4.1. The course team

At the heart of course preparation is the course team, and experience worldwide has shown that this is the single most important way of achieving quality and effectiveness in a course. The success of the course team approach rests in the discussions that define the course structure and content, and on the comments that an author receives from several others on the first and subsequent drafts of any material (s)he produces.

For all but the shortest courses at the UKOU, a typical course team has the following membership

- four to six academics, one of whom chairs the course team; they are responsible for the academic design and structure of the course, and they either produce all drafts of all the academic material (printed and any electronic material), or produce the final draft of material provided by consultants

- a course manager, who is responsible for the day-to-day running of the team, and who liaises between the team and the various design and production agencies for the course material

- an editor, who puts the final gloss on printed materials, e.g. by checking for consistency and by advising on literary style

- various other people depending on the course components, e.g. a TV producer if there is any TV or video, a software expert if there are any CD-ROMs or internet usage.

The course team might use consultants to prepare first drafts of certain materials, and there should always be one or more external subject experts and education experts to comment on the drafts.

It is possible for an astronomy course, or a science course with an astronomy content, to be prepared or *adapted* by a course team with rather few members from a developing country. This is important given the shortage of astronomy educators in some countries. But at the very least there should be one academic from the developing country, otherwise the course might not be shaped to serve the students' or the country's particular needs.

4.2. Course components

A common misconception about distance learning is that electronic media are the chief conveyors of course content. This is unlikely to be the case even with the emerging e-university approach (see below), and it is certainly not the case at the moment. Printed texts typically carry most of the course material.

Electronic media, which include videos, CD-ROMs, and broadcast TV, are very good for certain teaching functions, but unless they already exist they can be expensive to produce, particularly CD-ROMs. These media also require the student to have access to the appropriate equipment.

In science, and notably in astronomy, a course component of great value is a kit of very simple items for practical work. Indeed, the 'kit' here need be nothing more than a list of items that a student can readily obtain. There must, of course, always be a clear description of how the practical work is to be carried out. At the UKOU we have found that high-order skills can be developed in this way, e.g. recording relevant details, tabulating data, graph plotting, analysis of results including estimating uncertainties, report writing. Successful practical projects for which we have sent out *no* equipment have included measuring the length of the sidereal day, a quantitative study of light pollution, measuring the luminosity of the Sun, and many more.

An essential feature is some sort of interaction with a tutor — more on this in the next section. In addition, a residential school of up to a week, with a focus on practical work, is of enormous benefit, but can be expensive to set up, and it might be unfeasible for students to travel to it. However, if it is a requirement of course accreditation that there is supervised practical work then a residential school will be essential, and it will almost certainly be more feasible to have a single, week-long school than, say, three week-end schools.

4.3. Active learning

It almost goes without saying that the course materials must build on prerequisite skills and knowledge that are clear to the author and to the student. But even when a student comes to a course adequately prepared, it is *essential* that the course encourages the student to engage in active learning.

One feature of active learning is in one's approach to reading a text — we attempt to teach active learning skills such as note-taking, highlighting, and summarising. Another feature of active learning is to test whether what has been studied has been understood. This must not only involve recall, but the application of knowledge and skills in situations that are new.

UKOU science materials typically include three types of active-learning self-test. First, there is the 'stop-and-think' device. This is a short question, clearly delineated, and placed at any point in a text (print or electronic). It tests whether the student has understood the immediately preceding material. It can do this by requiring recall, or a simple application, or that the student continues the 'story'. The correct response immediately follows the question.

Second, there are questions placed at the ends of sections. These test understanding of a greater span of material. A full, acceptable response is always given (usually at the back of a text), along with comments that can be extensive. Third, there is a longer piece of work, often called an 'activity'. There are fewer of these than of the questions. Examples include summarising an extended piece of material, solving an extended problem, carrying out a practical investigation, building a conceptual model. Acceptable responses plus comments are again normally included at the end of the text.

We design science courses so that if an average student does all the self-tests (as they are strongly encouraged to do), then about 25% of the total study time will have been spent on them. Time spent on assignments that are graded by the tutor is additional to this.

Most textbooks are rather weak on active learning. End-of-chapter questions are often about as far as it goes, and it is common to get just a single number as an answer to a numerical question, and no answer at all to other questions. Some UKOU science courses do however use textbooks at the higher levels of study, along with a specially prepared study guide that, among other things, provides an opportunity for active learning.

5. The Effective Delivery of Distance-Learning Courses

As much care needs to go into the delivery of a distance learning course as into its production, otherwise many students will not complete the course who otherwise would have done. The academic aspects of delivery are in the hands of a presentation course team, much smaller than the production course team.

The first requirement (as mentioned above) is that prerequisite skills and knowledge must be specified to the student beforehand, and the course must be built *only* on these. Among the skills are those for the self-study that is at the heart of distance learning; these study-skills need either to be specified for entry, or they need to be taught within a course. They include

- time planning

- active reading (notetaking, highlighting, summarising)

- dealing with difficult material (including how not to get unduly 'hung up')

- written communication (including essay writing, but other forms too)

- problem solving, mathematical and otherwise.

To assist students with time planning there must be assignments throughout the course. These will be prepared by the course team, new ones for each presentation of the course. The student has to send the completed assignments to the tutor by clear deadlines. In the UKOU a course lasting nine months will have at least four such deadlines, roughly equally spaced. There will then be an end-of-course examination. The UKOU assignments are graded by a tutor, and the overall grade, along with the examination mark determines the student's course grade. This gives the student a strong incentive to meet the assignment deadlines!

5.1.   Tutors

Tutors have a crucial role to play in raising the quality of course delivery. In the UKOU a tutor of a course will have about 20 students, though this number could be increased.

The tutor, by writing clear and copious comments on the student's assignments, targeted at the student's individual needs, thereby carries out an important teaching function. The UKOU physics and astronomy courses have been much praised in the recent round of external review of university teaching for this aspect of tuition. The tutor also has an important teaching function through other forms of contact. There might be a few hours of face-to-face tutorials, but more extensive contact is usually by telephone, and increasingly by e-mail.

Tutors must clearly be trained to help deliver a distance-education course, not only in their general role, but also sometimes in the subject they are tutoring. In the UKOU, tutoring a single course, even a 60 point course, is a part-time job, so most of our tutors have full-time jobs in universities or in schools. We have found that for the more introductory science courses, teachers of senior-school science make admirable tutors.

Some of the tutors of our introductory astronomy course initially had very little knowledge of astronomy - many had backgrounds in physics or in geology. Therefore, in developing countries, the shortage of astronomers to act as tutors for an introductory astronomy course might not be a major problem, provided there are adequate numbers of willing senior school science teachers.

5.2. Completing the feedback loop

To achieve high quality it is essential that there are systems that encourage feedback from tutors and students to the presentation course team, and it is equally essential that the course team takes swift and appropriate action. Errata in course materials, problems with assignments and with the examination, can then be put right for the current cohort of students. It might only be possible to correct other problems, such as a course being overloaded or out of date, in time for the following cohort, but without feedback some problems are unlikely to be corrected at all.

About every four years a course should be thoroughly reviewed by subject experts, to see if any updating is required. The course team should also look out for opportunities to use new media, notably electronic media.

6. The e-University

An e-university is one that relies on the Internet for a substantial proportion of its teaching operations. An e-university need not necessarily be a distance teaching institution, but many distance teaching institutions are moving in the direction of becoming e-universities. Quite at what point the UKOU could be called an e-university is a matter for a not very relevant discussion; some say that it is already an e-university.

An e-university offers great promise because it could enable students in any country to study the distance-learning courses of an institution in another country. The drawback is that the student needs ready access to the Internet.

It is sometimes claimed that all aspects of study should ultimately be handled over the Internet. This is not a desirable goal; there are things for which the Internet is not the most appropriate medium. These include

- anything that could as well be on a CD-ROM

- long texts

- the end-of-course examination.

Let us look at each of these in turn.

Unlike the Internet, a CD-ROM is not subject to slow telephone lines and to loss of connection. Regarding long texts, there are problems with reading from a screen for long periods, and only to be able to access most of a course via the Internet is quite a restriction on flexibility of study. It is not proper to suggest that the student overcomes this problem by printing out the text - this would load huge costs for colour printing on to the student.

An end-of-course examination needs to be such that the resulting student award is credible. Therefore, some form of invigilation is required, and so a credited invigilator would have to be present throughout the examination. It would be impractical for an invigilator to travel to a single student's home or place of work, and so, to use the Internet, there would have to be a sizeable group of students in a room equipped with many terminals, and the invigilator would have to be able to ensure that each student was not getting inappropriate help from the Internet. This could well be impractical and so a conventional examination would then be required.

Appropriate uses of the Internet include

- information on a course

- registration for a course

- delivery of assignments by the course team to the student; delivery of completed assignments by the student to the tutor; the return by the tutor to the student of the assignment grade and teaching comments

- e-mail between student and tutor and between students, including 'notice-board' discussions

- delivery by the course team to the student of errata, stop-presses, updates, and other news

- feedback between the course team, the students, and the tutors

- guided use of databases

- use of robotic telescopes

- use of on-line journals.

The issue of the credibility of an award does not arise in relation to the assignments provided that the award system requires threshold performances in the assignments and separately in the end-of-course examination.

Clearly an e-university could bring astronomy education to a developing country, provided that the problem of Internet access can be solved, and provided that *conventional* invigilation for examinations can be provided.

7.   Self-Study of Astromony on Campus

Course materials designed for distance education can be used to great effect on campus to alleviate a shortage of astronomy teachers and a shortage of non-human resources. This is because self-study reduces the number of contact hours required for a learning outcome, and because certain distance education materials can reduce the need for high cost resources, e.g. laboratory space. It is however essential for the student to be acculturated to the use of self-study for acquiring core content.

Such contact hours as are available can be used for higher quality forms of contact, such as

- inspirational lectures (rather than lectures to deliver core content)

- problem-solving classes

- small-group tutorials.

All in all, it is my view that distance education has considerable potential for extending the teaching of astronomy in developing countries on campus as well as off it.

Discussion
Dworetsky asked if the costs quoted were the full costs or only the fee charged. Jones replied that the figure he gave (approximately 50% of the cost of producing a graudate by conventional means) included *all* costs. Gerbaldi, speaking from her own experience of distance-education commented that costs should not be reduced too far since distance-learning is much more effective when there is an on-line tutor (working through e-mail). She also felt that a tutor was needed if students were using distance-education material on a conventional campus. Finally, she pointed out that methods of learning differ from country to country for both cultural and political reasons. There is need for caution in extending the use of distance-learning to some countries. In reply, Jones reiterated that his 50% figure included provision for a tutor. He agreed that students using distance-education materials on campus needed some tutorial help, but it is also important to acculturate students to self-study. He agreed that cultural and political problems could arise in the use of distance education in some countries. In fact, many of the U.K. Open-University materials had been used successfully in a variety of countries. More experience is needed, especially in astronomy, and might lead to the production of several versions of a course, to take account of local conditions.

Orchiston mentioned that the Swinburne University of Technology in Melbourne and the University of Western Sydney now offer courses in astronomy. It is possible to take adult-education courses that lead to both bachelor's and master's degrees in astronomy. Jones replied that he had not discussed adult-education courses *on campus*, although they are certainly an important means of offering astronomy to a wider public. The University of Central Queensland used to offer *distance* education, but he was not sure if it still did so.

Pasachoff commented that, in the U.S.A., several venture capitalists are funding major on-line courses. He is most familiar with GEN, the Global Educational Network, which is now putting together some dozens of courses in a wide variety of fields, which will be delivered over streaming video (this apparently saves some copyright expenses compared with providing images on CD-ROM or the Web, since legally no copies are being made. They hope to sell these courses for $300-$600 apiece, not mainly to traditional 18-22 year-old University students but, rather, to advanced-placement high-school students or alumni some years or decades past graduation.

He also commented on Jones's remark that on-line courses free academics for higher-level work. Pasachoff felt that that requires good faith on the part of the institution. It is all too possible that once a course is on tape, or on the Web, staff could be made redundant or not rehired, or that the steady-state of faculty and staff could be lower than it otherwise was. He knew of at least one faculty union was wary of the faculty tape-recording courses for a remote campus, because of the possibility that the faculty member would not be rehired.

He also asked how much a single course cost at the Open University? Jones replied that a full degree, three or four years' worth of courses, is about £3900 pounds. The fee to a student for a single course £300, which is probably about half the total cost to the Open University. The figures are roughly comparable to those quoted by Pasachoff.

Astronomy for Developing Countries
IAU Special Session at the 24th General Assembly, 2001
Alan H. Batten, ed.

Principles for Tertiary-Level Astronomy Courses

Derek McNally

Department of Physical Sciences, University of Hertfordshire, College Lane, Hatfield, Herts., AL10 9AB, U.K. e-mail: dmn@star.herts.ac.uk

Abstract. Any worthwhile tertiary-level course of study should, as its highest priority, reflect the discipline it represents as it is contemporaneously practiced. Were it not to do so, students would be intellectually underprovided. This paper sets out general principles which a first degree-level course in astronomy should aim to provide for its students. No specific syllabi will be attempted but, rather, the paper will outline ranges of topics and their level of treatment. While all students taking such courses should have as professional experience as possible, it must be recognised that most students taking tertiary-level astronomy courses may not become professional astronomers and that such courses will necessarily have to have flexibility to meet local circumstances.

1. Introduction

On a global basis, the objectives of tertiary education (education beyond school, in particular university education) are legion. Local, national needs and requirements, educational tradition, expectation of tertiary education and availability of resources will all, in varying proportions, shape the content of tertiary-education courses. Consequently little point will be served in setting out detailed syllabi for a universal astronomy/astrophysics degree course. However, astronomy is an eclectic discipline – it is one of the purest of sciences, being concerned with the study of matters having apparently little connection with day-to-day life. On the other hand astronomy is very much an applied science drawing together other sciences - chemistry, physics, biology, while drawing heavily on engineering skills and the whole being heavily influenced by mathematics. Astronomy is therefore a marvelous vehicle to extend horizons and to generate awareness of the interconnectedness of scientific endeavour. Astronomy is also pushing at the boundaries of scientific knowledge – the frisson of the possibility for fundamental discovery is always present. And astronomy – however eclectic – is a useful science of great value in the affairs of mankind. Sadly, astronomers have, in the past century, been reluctant to push the educational value of their science and have abdicated science teaching very largely to others. The final decades of the twentieth century have seen renewed growth in astronomy teaching – often in the form of add-on and optional courses in other science-degree courses. The value of astronomy as science education, because of its applied nature, is something that awaits exploitation and, together with other similar sciences, has the potential to reinvigorate interest in the physical sciences.

There are three major pathways in which astronomy is currently used in tertiary education:

i science for those not studying a physical science discipline;

ii courses with another degree such as physics or mathematics;

iii degree courses specialising in astronomy.

The first pathway has been exploited in many countries - in particular the United States - and has had the benefit of extensive course material and textbook development. A criticism of the pathway is that it does not develop a feeling for the disciplines of the physical/mathematical sciences for the very good reason that the undergraduates taking such courses lack an adequate physical-sciences/mathematics background. This pathway, while it spreads awareness and limited understanding of astronomy, maybe even generating a lifetime interest, has allowed a lacuna to develop between the first and third pathways which the second does not fulfil - namely courses in physical science for those adequately qualified but who may not find a specialist course attractive and turn, for their tertiary education, to other disciplines. This is a serious neglect since the twenty-first century will need a large cadre of people with a broad scientific education, appreciating the inter-relatedness of the sciences and having sufficient depth of knowledge to understand the processes, the value and the limitation of scientific endeavor. The second pathway does not contribute to bridging that gap since the astronomy is often restricted to one course in one year of study and is not a sustained study over the full degree course.

Pathway (iii) - the specialist degree course on which I will concentrate - can take a variety of forms. It can concentrate on astrophysics, combine astrometry and astrophysics, be directed towards detector systems with the mix weighted to taste. To give a local example, the original astronomy degree at University College London had no component of cosmology as the then professor of astronomy considered cosmology equated with speculation. He attached overwhelming importance to evidence, obtaining it, critically evaluating it and, where considered sound, to using it. It is such that gives variation and colour to different degree courses. There are those who consider that astronomy degrees should be astrophysics degrees and should not contain anything on the astronomy of position, determination of time or celestial mechanics. Clearly vast advances have been made in astrophysics which enhance its attractiveness but, post-HIPPARCOS, such omissions may ignore astrophysically valuable techniques.

There is an interesting, very small, but persistent, group of undergraduates who are passionately devoted to celestial mechanics and the astronomy of position who find astrophysics tedious and can feel let down even by an "astronomy" degree course. Consequently in designing an astronomy degree one must consider carefully the motivation of prospective undergraduates and ensure that the range of such degrees is as wide as resources can sustain.

In this paper I shall look at the basic entry knowledge an undergraduate should possess on entry to an astronomy degree in section 2; the fundamental structure of astronomy will be addressed in section 3 and the development of that structure will be sketched in section 4. I will look at the role of the practical and the project in degree-level work in section 5 before finally looking again in section 6 at some of the themes for the future of astronomy degrees.

2. Pre-Entry Knowledge

A person contemplating a degree in astronomy must recognise that it is a degree which demands a sound understanding of mathematics and physics. From this there can be no escape. The prospective undergraduate must therefore prepare by attempting to reach the highest levels of competence in both mathematics and physics that are offered by the secondary education system. There should be competence in the use of algebra, coordinate geometry and calculus. There is considerable value in experience of pure mathematics as well as applied. The student should have an understanding of mechanics, light and optics, electricity and magnetism and some introduction to basic atomic and nuclear physics. They should also have had considerable experience in laboratory physics. There has been a trend away from laboratory experience at all levels of education. Laboratories are expensive to establish and maintain; laboratory work is also expensive in time available in crowded curricula. This is probably one of the major weaknesses in science teaching since it is only by investigation that a student can develop any appreciation of why some scientific concepts at first seem to lack contact with the real world. Experiment also introduces concepts of error of measurement and brings in some element of computation. Laboratory work may indicate that theoretical studies may be a more productive route through tertiary education!

Clearly the basic background possessed by a student on entering tertiary education must be reinforced, developed and extended at the tertiary level. No degree course in astronomy can omit the development of understanding of physics and mathematics in its undergraduates. Indeed studies of physics and mathematics may take up to 50% of the course loading, particularly in the earlier years at university.

3. Basic Elements of Astronomy

Astronomy depends on the reception and interpretation of electromagnetic signals for the bulk of its information on the universe. Other forms of information carrier exist such as study of meteoritic bodies which impact the Earth, direct investigation of other solar-system bodies, detection of neutrinos and oof particles from cosmic rays. These last are important sources of information but are highly specialised in specific areas. For the purpose of this section, I will address only information that comes via electromagnetic radiation.

Gravity is a dominant force throughout astronomy from the orbits of planets and their satellites, through the structure and evolution of stars to the large scale structure of the universe. Accordingly therefore, undergraduate teaching must contain at least basic Newtonian dynamics and for advanced studies an introduction to the Special and General Theories of Relativity.

An understanding of electromagnetism and gravitation are essential for a study of astrophysics but for a full appreciation of astronomy there must also be the additional element of the astronomy of position and the measurement of time. This area of astronomy has become seriously undervalued with the result that there is a declining number of astronomers with a good understanding of what positional astronomy can do in an astrophysical context. This gives cause for deep concern, given the valuable information in the HIPPARCOS catalogues and the prospects in the foreseeable future for microarcsecond astrometry. It can be argued that positional astronomy is unnecessary for the successful prosecution of astrophysical research - and there is an element of validity in such arguments given the very wide canvas of astrophysics. However, positional astronomy and measures of time should remain an important part of astronomy degrees - perhaps defining the essential difference between a degree in astronomy and a degree in astrophysics. Celestial mechanics is an interesting case in point in this connection. Celestial mechanics has been recalled from obscurity by the demands of space science. It has become extremely sophisticated. It does not have wide appeal and it is by no means unequivocally clear to me whether it should be presented as the highly mathematical end of astronomy teaching or form a part of applied mathematics. That it should remain available is essential even though it is never going to appeal to more than a handful of dedicated (often awesomely so) undergraduates.

3.1.   Electromagnetic radiation

Any student of astronomy has to be aware of the emission, transmission and detection of electromagnetic radiation. Fundamental physics of reflection, refraction and polarisation should be understood at the level of application of Maxwell's Equations. Emission and absorption of radiation by atoms andappropriate atomic/quantum physics. Photoelectric phenomena should be addressed before developing the theory of emission, transmission and detection of electromagnetic radiation.

Emission of electromagnetic radiation   An important aspect of emission of electromagnetic radiation, now often neglected, is a proper treatment of black-body radiation. Black bodies are best approximated under astronomical conditions where also departures from thermal equilibrium are clearly apparent. The black body relates emitted radiation to other properties of matter, e.g. temperature. The physics of black-body radiation is also an interesting study in thermodynamics and quantum statistics. An appreciation of black body radiation can often give a zero order approximation to a real situation and allow understanding of the essence of a situation which otherwise demands very difficult analysis - as with most problems encountered in the study of stellar atmospheres.

Emission of electromagnetic radiation also extends into line emission by atoms, ions and molecules and the relevant branches of atomic quantum physics. In the radio astronomy of today, the fine and hyperfine structure of energy levels of atoms and molecules are of considerable importance in the detection of interstellar species, for example.

Transmission of electromagnetic radiation   The transmission of electromagnetic radiation through gases whether stellar, interstellar or intergalactic is of crucial astrophysical importance. In passing through matter, electromagnetic radiation can be absorbed, re-emitted, scattered and polarised. The consequences of that absorption, re-emission, scattering and polarisation can be detected and quantified with important consequences for understanding the physical state of the matter that produced those effects.

Again the opacity of stellar material is determined by the absorption and scattering properties of the constituents of stars. Opacity plays the dominating role in determining the structure of stars and that opacity is determined by the quantum properties of the atoms, ions and molecules which comprise the material of stars. It is crucial that any degree course in astronomy has (at least) a basic course on radiative transfer. Such a course will be the foundation for a wide variety of highly disparate advanced topics. No astronomer/astrophysicist can be without an appreciation of radiative transfer.

Detection of electromagnetic radiation   Unless the radiation from cosmic objects can be detected and measured, there is little knowledge available from these objects. Again all astronomy undergraduates should be aware of how electromagnetic radiation can be acquired (basic telescope design), significant signal detected and how such signals are calibrated. This is not the same as a course on instrumentation which is highly dependent on frequencies observed and specific detector technologies. All undergraduates should be aware of how photons are collected, aware of such matters as resolution and signal strength. They should know that signals can be differentiated by frequency (e.g. spectroscopy) or that a number of frequencies can be integrated to enhance detector response and they should know about the available and projected detector systems of their generation. The detection of electromagnetic radiation lends itself to both the taught course and practical/observational/laboratory work. It is in this context that undergraduates should learn the all-important consequence of signal-to-noise ratio and the trade-offs that they will have to make in acquiring useful observations for themselves. Signal-to-noise ratio leads directly to error - a topic that is rarely touched on within the context of school science. Observational error is not a criticism but a fact of life. Adroitly used course work and practical work based on the detection of electromagnetic radiation can be a singularly useful way to engender good scientific practice.

The type of physics involved here is at some remove from mainstream physics teaching and is one reason why there should be specialist degrees in astronomy and astrophysics.

3.2.   Gravity

Gravity is an associating force. On the scale of astrophysical enquiry it is the force which keeps planets in orbit around the Sun, maintains the Sun as a stable star and plays a key role in the dynamics of galaxies. Basic gravitational concepts appear in both school and university physics. One can therefore expect undergraduates to understand gravitation up to the extent of central-force elliptical orbits and Kepler's laws of planetary motion.

However, this is not the stage at which gravitation becomes astronomically useful. Orbits have to be related to path on the sky. Parabolic and hyperbolic orbits need introduction. The concept that six elements are required to define an orbit needs to be introduced together with elementary concepts on how such elements are to be determined (i.e. the interface to celestial mechanics) and the information that can be determined from objects such as binary stars when it is not always possible to determine all six orbital elements. The value of measuring radial velocities in place of spatial dimensions also needs to be introduced.

The concept of gravitational potentials needs to be developed in order to examine the equilibrium of stars. This leads to an important integrating study – stellar structure. Like a course on radiative transfer, so too should any degree course in astronomy have at a minimum, an introductory course on stellar structure. Stellar structure brings together many physical strands - gravitational forces, dynamics, energy transfer by radiation, energy generation by nuclear processes, thermodynamics, quantum physics. While modern stellar-structure theory is sophisticated and dependent on high-powered computers and computational methods, the fundamental aspects of stellar structure are readily demonstrable in a simple way. A mass-luminosity-radius relation can be derived in a straightforward way that demonstrates unequivocally the dependence of structure in stars on atomic physics and the evolution of stars driven by nuclear physics. Again the basis of fundamental techniques used in constructing stellar models can also be made clear in a simple way. A basic stellar-structure course can therefore establish criteria on which to come to grips with the sophisticated stellar models that are now being investigated and builds up basic understanding of the subject.

Astronomy and astrophysics degree courses have a basis of understanding of gravitation on which to build. It is important, however, to extend orbital concepts so that direct connection can be made to astrophysical technique actually in use in deriving astrophysical data from observations via gravitation. Stellar structure is one area of astrophysics where gravity plays an important direct role, in interaction with other physical process, to produce a highly sophisticated understanding of the internal structure of stars and their evolution which can be related, in turn, to relatively straightforward observations of the outer layers of stars.

3.3.   Positional astronomy and the measurement of time

Sadly, positional astronomy and the measurement of time have become the poor relations in astronomical education. Because positional astronomy reached a peak of development towards the end of the nineteenthth century and because it required very painstaking work to produce a small progressive step, the blossoming of astrophysics in the early twentieth century displaced positional astronomy from its prime position as an astronomical activity. The success of the HIPPARCOS project should help to bring positional astronomy back towards the cutting edge but to be successful, there will have to be greater emphasis on positional astronomy courses within astronomy degree structures. It is here that a possible distinction might be drawn between astronomy and astrophysics degrees in that the latter need not pursue positional astronomy beyond that needed to locate objects. However, if positional astronomy is to be fully exploited as an astrophysical tool, then there must be good understanding of the nature of that tool.

Positional astronomy, post-HIPPARCOS, now offers a quality of data on stellar distances, possible stellar multiplicity etc. that has hitherto been unavailable. That information is beginning to make an impact on the refinement of certain astrophysical parameters. A proper appreciation of these data may have, as yet unforeseen, revolutionary consequences. It is therefore important that facility in the use of the concepts of positional astronomy should be widely developed so that the full potential of the HIPPARCOS data-set can be exploited and appropriate astrophysical input generated to guide the planning of microarcsecond positional astronomy projects now being considered.

It is an even sadder reflection of the state of physical science in our time that the basis of time-measurement is so little understood within the physical sciences. In mitigation it could be argued that the procedure of timekeeping has become so refined, that accurate time services can be taken for granted - the measurement of time is a Cinderella subject area because of its exceptional success. Yet as a global society we are critically dependent on the provision of accurate time services in every aspect of daily life. If there is one place where reference to the science of time-measurement should find a prime place, it is in the astronomy degree. Time-measurement may seem obscure and difficult in the search for precise definition that can be implemented practically. This has led to leap years, leap seconds, ephemeris time and all manner of pitfalls for the unwary. At present, there is a strong lobby in the world of satellite-based communication and navigation-service providers to get rid of leap seconds. There are few people with the necessary expertise and knowledge to argue that question or indeed critically assess the long term benefits/losses that might flow from a change of present timekeeping and management practices. Revolutionary changes are likely to occur in the near future on the formulation of celestial reference-frames and time dissemination. The astronomical community, more than any other, should acknowledge its responsibilities and ensure that astronomy degrees maintain adequate teaching on the measurement of time. It is acknowledged that this may not be the most electrifying experience for the average undergraduate but it does not have to be the dull grind it often is.

There is still a powerful necessity to maintain positional astronomy and measurement of time as a key part of basic astronomy degree level teaching - for the benefit of astrophysics as much as for retaining a body of people knowledgeable in this area for the purpose of maintaining proper public policy in accurate timekeeping.

4. Gilding the Lily

In the previous section I have looked at three key areas in tertiary astronomical education. Once these fundamental issues have been addressed the range of topics to be offered are legion. The world is indeed your oyster. Degrees can be developed around range of topics related to the specialist knowledge of the staff available to teach. In my view degree courses should be built on the premise that it is better to inculcate understanding in depth rather than see superficial comprehensive coverage.

Clearly any degree course is going to start with some form of introductory course describing in simple, straightforward terms, the entire range of the astronomical canvas from the Earth to the latest in cosmological thinking. However, it is unlikely that within the limitation of staff expertise and sheer limitation of time and resources available, that a satisfying comprehensive degree course in astronomy could be developed giving adequate depth to all components. Each degree will have to decide its principal focus, the range of departure from that focus, the type of student it wants to attract and the scientific value and capability of the graduate it will ultimately deliver to the world of work. Not all of these aspects pull in the same direction by any means and significant compromise will be needed to achieve scientific integrity and coherence within the framework of staff abilities and interests, university regulations and availability of resources. However, the ultimate criteria against which an astronomy degree should be judged will be its scientific integrity and coherence.

A degree should be further judged against its choice of scientific focus. For example a degree in astronomy might be focussed on planetary studies. But planetary studies cannot be properly pursued without good knowledge of the structure of the Earth and atmospheric physics. Nor can planetary studies be focussed without good knowledge of the structure and evolution of the Sun (and by implication the nature of stars other than the Sun). Such a degree might not necessarily offer an advanced course on the structure and evolution of the universe!

Some astronomy degrees should retain a place for celestial mechanics. The basis for celestial mechanics is very much focussed in advanced applied mathematics and numerical methods. However, celestial mechanics does have practical significance in the development of space navigation and should, as such, find a place in degree level astronomy teaching. The effect is twofold - the first to encourage interested students to consider celestial mechanics as a possible path towards a career but also to ensure that the expertise of celestial mechanics for space flows back into astronomical developments.

Celestial mechanics is a special case but there are many other special cases and every astronomy degree should offer at least one really advanced course which is offered in few other places worldwide - even if only relatively few students actually take it. Rarity disciplines are endangered species but they must not be allowed to become extinct. Sadly, the exigencies of modern economics may ensure such extinctions.

5. Practical Astronomy

It is often taken as axiomatic that practical work is such a good thing, that there is little need to justify its inclusion in a degree course and even in association with non-degree taught courses. After all says the mantra, access to the sky is freely available to all. That is so, but it is free, at best, only as far as the sixth visual magnitude. Thereafter the costs rise dramatically. Astronomy practical classes are costly to provide. Since a degree course should be aiming at professional standards and at inculcating a professional approach, then the equipment used in such practical courses must be at a professional standard, particularly for advanced classes. This means that an optical telescope should be sought at about 0.5-m aperture and radio telescopes at about 2-3 m. Since most telescopes for undergraduate teaching will be located in brightly lit and often radio noisy urban areas, there is not very much point seeking larger telescopes unless there are other reasons, e.g. a parallel research use, a specially favoured site. Any telescopes should have focal-plane arrangements which offer the strength and space to take relatively heavy detectors. The mountings for such telescopes should be stable and as free from locally generated vibration as is possible. Such provision will generate expense - particularly in respect of building works. Optical telescopes should be housed in domes (or equivalent structures) as a protection against wind buffeting. It should always be borne in mind that a carefully selected optical telescope, for example, mounted in an adequate dome can be written off on a period of 25 years, if not longer. An investment of £500,000 can therefore equate to an annual investment of £20,000 over a 25-year period.

Focal-plane instrumentation changes rapidly but, for example, good CCD cameras are highly affordable and can make excellent use of a 0.5-m aperture optical telescope. It is important to be able to provide imaging, photometric and spectroscopic capability, i.e. to maximise the range of types of observation that can be attempted.

Clearly the level of provision in the early years of a degree course need incur less expense but even here there should be considerable flexibility for focal-plane instruments. It is a good maxim to buy the best possible telescope that can be afforded - both in respect to the quality of the optics and in respect of the construction of the mounting and drive. Good engineering is never cheap.

Unfortunately the costs of maintaining undergraduate laboratories in the physical sciences is high – they demand floor space, services, specialised equipment and adequate levels of staffing – both technical and teaching. In astronomy, optical-wavelength practicals are done at night, so the staffing levels must be high in order to ensure student safety while working in dark conditions. It should be borne in mind that radio astronomy and optical solar astronomy can be carried out during standard university working hours.

The positive gain from laboratory work is that students are forced to assess their capabilities, the capabilities of experimental systems, to recognise the omnipresence of error and, particularly in astronomy, to learn the value of proper preparation – a unique observational opportunity may be lost because of poor preparation just as much as through the vagaries of the weather. Laboratory work allows informal contact with teaching and research staff and thereby an opportunity for students to assess, and reassess, their future career plans. Practical work is a highly valuable pedagogical tool and it should have a high priority – sadly this is not always the case in our financially hard-pressed universities.

Project work is an important part of degree work. If adroitly used, projects can contribute greatly to the scientific maturation of undergraduates. On the other hand, projects can be difficult to assess, in that the contribution of individual students to a project may be hard to separate clearly from the contribution of peers and academic supervisors. A well considered assessment scheme can normally cope with such problems. Projects can take many forms – from an individual driven project, an academic-supervisor driven project to a group project comprising several different undergraduates. Projects need not be observational (in temperate climates, weather can be a considerable inhibitor of optical/IR observational projects though the advent of well-sited robotic telescopes could remove much of this problem) and can be as varied as the resources of the teaching department can permit. A limiting factor can often be the range of experience within the academic supervisors.

I tend to favour individual undergraduate projects as this builds up individual creativity (but at the expense of loss of experience of group working) though it must be recognised that individual projects are demanding of supervisory capability. Providing individual projects for a year of 20 students is demanding because self-driven undergraduates are relatively rare (and as such require little supervision but may need steering to maintain realism) and the necessary supervision needs the backing of a substantial research group. Projects are clearly for advanced students – but small projects can be incorporated within the body of practical teaching in all years to habituate students to this form of study.

Finally it is well worth considering observational field trips. One of the difficulties with the academic calendar is that classes must be scheduled – observational classes being no exception. If the local climate is not of the best, cloudy nights can apparently show an unhealthy correlation with scheduled classes. Field trips to a climatically well-found observatory can allow a week devoted exclusively to observational work from sunset to sunrise. Such work is particularly valuable in turning creative undergraduates into professional astronomers – the change can often be dramatic. Field trips are not expensive but not easy to arrange – few well-found observatories welcome the advent of undergraduates taking up valuable telescope time. The actual experience of real observing at a high-quality site is the essence of such trips – this cannot be obtained by using a robotic telescope to the same degree. Undergraduate field trips should be developed more widely in view of the beneficial personal development that they engender.

6.   The Future

In this paper I have set out an outline of what a degree course in astronomy/astrophysics might contain. I have argued for certain basic elements in section 3 and advocated that a significant amount of course time should be devoted to practical/project work in section 5. I purposely did not set out a full syllabus in section 4 since it is clear that degrees can take many forms with different emphases meeting local needs, interests and expertise. To some, what I have argued for may seem unduly elitist and expensive. I do not believe it is elitist to provide a specialist degree and to treat that degree in a highly professional way. While many individual students taking astronomy degrees may never become professional astronomers, it would be utterly wrong to deny them, as undergraduates, a proper professional training. A professional approach is a highly portable skill, even if learnt within the confines of one particular science.

The criticism of expense is more justified. To set up and to maintain an astronomy degree is expensive – in terms of specialist staff, library provision and, in particular, in terms of observational provision. However, one can only operate within the income one has and that income should buy the best – in any area – that can be afforded. In many cases it may not be financially possible to offer a specialist astronomy/astrophysics degree.

As I have argued elsewhere (McNally, 1998), it may be that the heyday of the specialist degree in the physical sciences is coming to an end. It may even be that the specialist degree, so good at reproducing our own kind, may no longer be the method of choice in developing a scientifically literate population. Why then spend on specialist degrees? Simply, to maintain a cadre of well-educated creative specialised physical scientists without whom the physical sciences will not develop. Nevertheless, numbers of undergraduates in the physical sciences may well continue to fall. Specialist-degree courses will remain but the specialisms will have to find new ways to generate income. This, I believe, will have to come through the development of widely based physical-science degrees, incorporating several specialist subjects, designed to produce knowledgeable scientists though not necessarily specialised scientists. Such courses will need to have wide appeal to generate sufficient (in the sense of meeting national needs) numbers of graduates. Astronomy should be a major player in the promotion and development of such degrees aimed at lifting the level of understanding science from its present abysmal levels (as is only too plain from the treatment of science by the media). The twenty-first century will be even more crucially affected by scientific advances than was the twentieth – to cope with these advances as a global society, it is imperative that we have a substantial number of people within the community who are not only scientifically literate, but literate and knowledgeable in the physical sciences – an area where undergraduates are still voting with their feet – if only to attempt to resolve the environmental challenges of the twenty-first century in a rational manner. Such more widely based degrees in physical science will have to coexist with, and complement, the specialist degrees - and vice versa. The physical sciences are just too important to allow them to wither into a minority interest. Physical sciences such as astronomy will be of great importance in recruiting and retaining the interest of undergraduates studying selected groups of physical sciences. Astronomy should therefore assume a high profile in developing the teaching of the physical sciences in the twenty-first century – if we fail to produce proper physical science teaching then it is very likely that our response to current and future environmental challenges will be more likely to be disastrous than successful. That is the challenge which cannot be shirked.

References

McNally, D., 1998, *University Education in the Next Century, New Trends in Astronomy Teaching*, IAU Colloquium 162, Eds. Gouguenheim, L., McNally, D. and Percy, J.R., CUP, Cambridge, pp. 12-15.

Discussion

Bhatia asked if students who took pure astronomy were as employable as those who took some astronomy as part of physics. McNally replied that some of his students became astronomers, others found positions as mathemtaicians or computer scientists.

Astronomy for Developing Countries
IAU Special Session at the 24th General Assembly, 2001
Alan H. Batten, ed.

SALT/HET Cooperation in Education and Public Outreach

Mary Kay Hemenway and Sandra Preston

Department of Astronomy, University of Texas, Austin, Texas 78712-1083, U.S.A. e-mail: marykay@astro.as.utexas.edu

Abstract. The "Science with SALT" meeting in March 1998 opened avenues of cooperation between SAAO and the University of Texas at Austin in education and public outreach. This paper will review past interactions and future plans. SAAO personnel have visited the HET and McDonald Observatory and have taken part in planning meetings for the Texas Astronomy Education Center museum area and educational programming. Discussions concerning the extension of the daily radio show StarDate (English), Universo (Spanish) and Sternzeit (German) versions to a southern hemisphere version are underway. In addition, we are cooperatively planning a workshop to discuss an international collaborative for educational outreach for state-of-the-art telescopes for which a regional collaborative in southwestern U.S. (SCOPE) serves as a model. The towns of Sutherland and Fort Davis are discussing forming a "twin-town" relationship. Projects and plans that link cutting-edge astronomical research to classrooms and the public will be reviewed.

1. Introduction

With plans for the Southern African Large Telescope (SALT is currently under construction at Sutherland, South Africa) largely based on the design of the Hobby-Eberly Telescope (HET is located at McDonald Observatory in Texas, U.S.A.), there has been significant technology transfer between the observatories. Both observatories share another goal beyond exploring the secrets of the universe with their unique telescopes. Both are committed to providing a motivating education and public outreach program to visitors and to students. Astronomy offers unique opportunities to motivate students to learn science, mathematics, and modern technology. Their fascination with the cosmos can form the basis for a robust educational and public outreach program that reaches far beyond the physical sites of the observatories to include a much wider population.

The unique design of HET and SALT provides an opportunity to capitalize on the technology as well as the science as topics. HET has 91 interchangeable one-meter spherical mirrors arranged to form a 10x11 meter hexagon. The telescope is permanently tilted with azimuthal motion allowable only between observing periods. This design allows the cost to be significantly lower than for a fully-steerable optical telescope of similar size and still allows about 70% of the sky to be accessed. Celestial objects are tracked as Earth's rotation carries them across the field of view. The telescope is designed to be especially efficient for spectroscopy. Eventually, several spectrographs will be available at different resolutions. HET is mostly operated in queue-scheduled mode. Resident astronomers can change the computer schedule queue for real-time observations.

2. Education and Public Outreach Interactions

As engineers and scientists, and lawyers, planned the cooperative agreements that exchange observing time for technical plans and assistance, the scientists and staff especially interested in Education and Public Outreach (EPO) at these observatories began to exchange visits and information (without the lawyers). The first major meeting took place at the "Science with SALT" conference at the South Africa Astronomical Observatory (SAAO) in March 1998. A significant portion of the meeting was devoted to increasing science awareness and providing educational opportunities. Formal presentations and focus groups involving many of the participants provided for a diverse and broad exchange of information. (Hemenway 1998; Preston 1998; Rijsdijk 1998)

Following that meeting, Case Rijsdijk of SAAO was invited to participate in the national Advisory Board meeting for the "Decoding Starlight" exhibit plans in May 1998, in Fort Davis, Texas. This exhibit will form a principle part of the new Texas Astronomy Education Center (TAEC). The 11,000 square-foot building with its theater, exhibit hall, classroom, and visitor amenities will be the focal point for outreach at McDonald Observatory. An outdoor observing area and telescope park adjoin the TAEC.

In July 2000, Frans Hanekom, Mayor of Sutherland, Alleta van Sittert, Town Clerk of Sutherland, and Peter Martinez of SAAO came to Fort Davis to participate in the TAEC groundbreaking ceremonies. They reviewed plans, toured current visitor facilities, and made arrangements for a Twin Town agreement between Sutherland and Fort Davis. In September 2000, Preston and Judge Peggy Robertson of Fort Davis participated in the groundbreaking of SALT and the signing of the Twin Town agreement.

Plans for a major meeting of EPO personnel of the world's largest observatories are underway. The Large Telescope Educational Collaboration (LATEC) Workshop will take place at the SAAO in February 2001. LATEC may be modeled on SCOPE (the Southwestern Consortium of Observatories for Public Education; see: http://hyperion.as.utexas.edu/mcdonald/scope/). By forming a partnership of observatories, we will be able to create a cooperative spirit, limit duplication of effort, leverage funds, provide a support network in a career arena where there are only a handful of institutions, create opportunities for professional development, and share challenges and solutions.

3.   The Texas Astronomy Education Center

The TAEC is more than a building and outdoor observing facility. It is part of a coordinated program to bring astronomy research to the public and to teachers. The TAEC theater will offer presentations to the public and the classroom will be used by visiting school groups and for teacher workshops; the remoteness of the observatory does not restrict access to astronomy to only those who can travel to the TAEC. A visit of a few hours can do little to fill in what is often rather poor background in astronomy. The goal of the TAEC is to use the beauty and excitement of astronomy as the motivator to increase science literacy at all levels. The Internet is a powerful tool that can be used to deliver pre- and post-visit information and activities to teachers, and to serve as an informative and exciting resource for the public and educators. The activities will also be delivered through workshops at teachers' meetings and within school districts.

The exhibits and workshops have a special emphasis: spectroscopy. Preparing museum exhibits demands special care. Rowan (1992) noted "An idea may be confusing because it involves (a) difficult language, (b) difficult-to-picture structures or processes or (c) difficult-to-believe notions (e.g. Earth is weightless)." In surveys conducted with McDonald Observatory visitors, we found that most people are unfamiliar with spectroscopy and, in fact, they stumble over the word the first time they try to say it. While most visitors know astronomers use telescopes to study the universe, they believe telescopes are used to collect "information." Many people don't know telescopes are collecting light or that astronomers are analyzing light –light that has been traveling to us for billions of years– to learn about the Universe. Feher and Rice (1985) identified two important characteristics of museum exhibitions about light that appeared to greatly improve their ability to produce a full understanding of physical phenomena. These characteristics are (1) that an exhibit is usable by the visitor in different ways, offering more varied exploration of a topic, and (2) that different exhibits explore the different aspects of the same phenomenon, providing a "richer learning atmosphere." For these reasons, the exhibits begin with the key ideas that "sunlight is starlight" and the nature of light and spectra. All the exhibits have captions and keys in both English and Spanish. The variety of exhibits allows visitors to explore topics though different media - reading, observing, interacting, and experimenting - that are suitable for different levels of understanding. For example, some exhibits are aimed at rather young children and others allow those with prior knowledge of chemistry and physics to probe deeper their applications in learning about the composition, evolution, and motions of celestial objects.

In preparation for the opening of the TAEC, Hemenway and Armosky (2000) have developed and field-tested a new series of instructional units aimed at teachers. Standard ways of teaching about the nature of telescopes must be augmented to explore the innovative technology used in HET and SALT. "Telescope Technology for Teachers" allow students to explore the construction, pricing, and structural design of HET. Students use small inexpensive flat mirrors to build a human model of the HET; the students serve as the mirror actuators that tilt, tip or piston the individual segments into position and then they aim a flashlight beam at a target. Their success is measured (as it is for the actual telescope) by how small a target they can hit and how long they can hold it steady. Other activities illustrate the use of the tracker and fiber optic. Armosky and Hemenway are preparing a parallel set of "hands-on, minds-on" activities that explore spectroscopy. All materials are aligned with the National Science Education Standards (National Research Council 1996).

Just as SAAO has posted a series of their activities on the Internet (http://www.saao.ac.za/education/modules.html) to share with the world, these McDonald Observatory activities will be available on a special web page (http://stardate.utexas.edu/marykay/ttt/ttt.html). SALT and HET are already freely sharing their educational activities with each other.

Among the other resources produced at McDonald Observatory are the daily radio show StarDate (in English, with its companion programs in German "Sternzeit" and Spanish "Universo"). (Barnes, 1996) Six issues per year of Star-Date magazine provide another avenue for the public to learn about astronomy. Special publications such as a guide to the solar system (McDonald Observatory, 1998a,b) in English and Spanish and educational posters with classroom activities are available for teachers.

4. Twin Towns

The twin town agreement of 1 September 2000 unites the similarly-sized, remote, rural communities of Fort Davis and Sutherland (respective populations of 800 and 2000) to explore the ways that will enable them to serve the increasing numbers of visitors drawn by the existence of the major astronomical facilities nearby. Both towns have immediate infrastructure developmental problems that need addressing over an extended period. The agreement is meant to be the beginning of a process of sharing information, expertise, and friendship between the communities. As part of the agreement, communications between local schools and individual children will help both explore the cultural dynamics of their communities. The private sector members will explore business practices, retailing products, marketing strategies and small business entrepreneurship as they discuss the changes that occur as a rural entity becomes a tourist destination.

Among the immediate plans are to exchange exhibits, photos, maps, and respective flags. The schools will identify projects to share. Individual children will be linked by mail or electronic mail to other children. Individual businesses will be encouraged to link with each other, and eventually exchange of retail products may take place. Websites will be created for each town to list business/tourism opportunities and historical backgrounds.

Cooperation between SALT and HET has led to an opportunity to broaden the cultural and economic life of two widely separated communities.

5. Conclusion

By working together in these efforts, we will be able to create a cooperative spirit, limit duplication of effort, leverage funds, create opportunities for professional development, and share challenges and solutions. The unique features of our similar telescopes bring us close together. Although each country can claim particular features of its educational system, we share similar problems. We each deal with a bilingual or multi-lingual population. We each want to use astronomy to inspire and motivate students and young people into considering careers in science and/or engineering. And, we each wish to increase the general science literacy level of our citizens. SALT and HET have provided us with an arena in which to work together.

References

Barnes, S. P. 1996, UNIVERSO: *A Spanish-Language Astronomy Radio Program* in *Astronomy Education: Current Developments, Future Coordination*, ASP Conference Series vol. 89, J. R. Percy, ed., 165-166.

Feher, E. and Rice, K. 1985, *Development of Scientific Concepts through the Use of Interactive Exhibits in A Museum*. Curator, 28, 35-46.

Hemenway, M. K. 1998, *Building Public Support for Astronomy though School Based Education* in *Science with SALT: proceedings of the SALT/HET Workshop*. D.A.H. Buckley, ed. 157-163.

Hemenway, M. K. and Armosky, B. J. 2000, *Telescope Technology for Teachers*. Bulletin of the American Astronomical Society, 32, 1559.

McDonald Observatory, 1998a, StarDate Guide to the Solar System.

McDonald Observatory 1998b, Universo Guia Del Sistema Solar.

National Research Council 1996, National Science Education Standards (Washington, DC).

Preston, S. 1998, *Astronomy for the Masses: Making Research Relevant for Everyone* in *Science with SALT: proceedings of the SALT/HET Workshop*. D.A.H. Buckley, ed. 151-155.

Rijsdijk, C. 1998, *Friends with the Universe: Taking Astronomy to the People* in *Science with SALT: proceedings of the SALT/HET Workshop*. D.A.H. Buckley, ed. 165-171.

Rowan, K. E. 1992, *What Research Says: No. 19 About Explaining Difficult Ideas*. Association of Science and Technology Centers Newsletter, Nov./Dec. 1992, 7-8, 10.

Problems Facing Promotion of Astronomy in Arab Countries

Anas M. I. Osman

National Research Institute of Astronomy and Geophysics, Helwan, Egypt. e-mail: amiosman46@hotmail.com

Abstract. Promotion of astronomy in Arab countries is facing many scientific and technical problems. Teaching astronomy starts very late in schools, with very simple and limited courses. Many teachers lack a suitable astronomical background, which can lead to incorrect understanding by students of many astronomical ideas and phenomena. Teaching astronomy at higher levels is also very limited, for example: aomng the 16 universities in Egypt, astronomy is taught in only two faculties of science, just for two years. Graduate students find many difficulties in obtaining jobs related to astronomical activities and this is a serious limitation on the attraction of the study of astronomy. On the other hand, astronomical institutions are suffering from a serious lack of the new sophisticated equipment, while the budget allotted for maintenance is very small, and there is a serious shortage of technical staff. The training of astronomers and technicians is badly needed, since good research work depends on modern technological equipment and the complicated software packages used in controlling such equipment and in data analysis. Good libraries are needed for promotion of astronomy especially, the Internet facilities available for the staff is very limited. The effects of culture are very clear; many authorities in developing countries believe that astronomy is a luxury. Finally, most of astronomers are engaged with a lot of administration for all matters, so the free time left for science is very limited.

1. Introduction

1.1. Ancient astronomy

Astronomy in Arab countries is an old and important science, and has flourished throughout history. This is very clear from the time of the ancient Egyptians, who observed the motions of the Sun, Moon and planets and the heliacal rising of stars, especially Sirius, whuch were used to establish the solar and sidereal calendars. The Egyptian calendar begins with the heliacal rising of Sirius, which defined the beginning of the Nile's annual flooding. They used astronomical means to construct and orient pyramids and temples as well. The sun shines on February 22nd October 22nd, every year, on the statue of Ramses II. One of these days was his birthday and the other was the anniversary of his coronation. Later, the Alexandria school was founded in the city which was the capital of the country during the time of Ptolemies, and much was achieved by many scientists and astronomers. Among them were Aristyllus, who was interested in stellar catalogues, and Aristarchus, who measured the relative distance of the Sun and the Moon and estimated the real diameters of these objects and the size of the Earth relative to the Sun and Moon. Eratosthenes determined the Earth's radius using the inclination of the Sun at noon in two different cities, Alexandria and Syene near Aswan, knowing the distance between them. Hipparchus was a great astronomer; he observed the stars positions and completed a star catalogue and discovered the phenomenon of precession. He devised a sophisticated geocentric model of the universe capable of describing and predicting most observed motions of the celestial bodies in the sky. The Julian calendar had been introduced in 46 B.C., by Sosigenes at the time of Julius Caesar. In the second century A.D. Ptolemy wrote his famous book, *Almagest*, in which he summarized all the then-current knowledge of astronomy in thirteen volumes. He extended the star catalogue to 1022 entries, correcting older reported positions for precession. The best-known contribution of Ptolemy in his book was a new cosmological model for the planetary system in which he assumed that the Earth is near the center of the system. The method for predicting the positions of the Sun, Moon and planets was called the epicycle theory.

On the other hand, the ancient Babylonian civilization in Iraq played an important rule in the field of astronomy. As early as 3000 B.C., they had developed a solar calendar using the Sun and the Moon to keep time and mark the seasons. They observed the motions of the Sun, Moon and the five visible planets among the stars. They recorded these motions and discovered and recorded the annual apparent path of the Sun through the sky, which is called the ecliptic. Also, they carefully studied the Moon and its phases and found that, there are roughly twelve complete changes of the Moon in a year. By 600 B.C., they were able to predict the future positions of the moon and planets within the band of constellations, which they called the Zodiac.

1.2. Arab astronomy

The Arab contribution to astronomy was not less important than that of other civilizations. In 640 A.D., much of the Alexandrian knowledge was transferred to Arabs, who further developed mathematics and astronomy. They had compiled tables for determining the positions of planets, the first visibility of the lunar crescent, times of prayer for different geographic longitudes andfinding the direction of Mecca, and they constructed curves for astrolabes and quadrants. As an example of Arab advancement, we note the recorded measurement of the Earth's circumference made near Baghdad in the year 820. The result was only 4% too large. Similarly, the Arab astronomer, Al-Battani had only 4% error in his measurement of the eccentricity of the Earth's orbit. He wrote his famous book *Elzig*, that contains astronomical tables for the motion of the Sun after long-term observations. These tables were used widely in Europe after they had been translated into Latin. From his observations, Al-Battani could define the time at which the length of the day equalled that of the night, and the length the year, with very high accuracy. Accurate observations helped him to obtain new important information regarding the orbits of Earth, Sun and Moon. The work of Ibn Yunis and his *Hakimit Tables* were very important in the field of astronomy. He observed a solar eclipse from an observatory on Mokattum mountain near Cairo and calculated the period of the Earth's rotation from the time and height of the solar disk before and after the eclipse. Al Byruni, Al Faraghani, Ibn El Heitham and others were also important Arab astronomers. Around 1000 A.D., the Islamic Empire had spread to Spain and astronomical tables were published with the reference longitude in Córdoba. Many astronomical phenomena (sunrise, sunset, moonrise, moonset, twilight, lunation...etc.) are strongly related to Moslem traditions (prayer time, Ramadan Fasting, beginning of lunar months and Pilgrimage). In spite of this fact, and in spite of their previous achievments, Arab countries did not continue their success in the field of astronomy. Nowadays, promotion of astronomy in Arab countries is facing many scientific and technical problems, the solutions of which need all efforts, if we are to push astronomy forward in this present era of space exploration.

2. Problems Facing Promotion of Astronomy in Arab Countries

2.1. Astronomy education

Education in Astronomy in Arab countries is still limited despite its importance for public and Moslem traditions. On the level of elementary schools and popular education, the astronomical knowledge and information included in books and presented and explained by teachers is very limited. Many teachers do not have a suitable astronomical background to explain the different astronomical phenomena, since they graduated in different branches of science. This serious factor affects directly the accuracy of astronomical information transmitted to the students, and is the main reason for some incorrect and inaccurate understanding and explanations. On the level of preparatory and secondary schools, astronomy is taught as small parts of geography and physics, concentrated on the solar system and the different elements of each planet, beside short notes about stars, galaxies and the seasons. The educational aids and demonstration facilities are very limited, for example, planetaria are not available in most cities in Arab countries. On these levels, training courses for teachers, through official seminars and visits to observatories and open discussions with astronomers, are badly needed. These courses must be sponsored directly by ministries of education. Also, the education authorities in Arab countries are requested to support the introduction of astronomy into secondary schools, to attract amateur astronomers and interested students to study astronomy in universities and higher schools. At the level of universities or higher schools, a subsidiary elementary course is taught in some geography and surveying departments, while some Arab universities teach a general course in astronomy for the first-level students, but not as a compulsory course. On the other hand, the lack of astronomy departments in Arab universities is very clear. Physics departments offering astronomy and astrophysics courses for the B.Sc. degree in physics and astronomy are very few, in comparison to the large number of universities. For example, among 16 universities in Egypt including 21 faculties of science, astronomy is taught only in two departments. The Astronomy Department of Cairo University, Faculty of Science awards the B.Sc. and the higher degrees of M.Sc. and Ph.D. in astronomy, although basic astronomical and astrophysical courses are taught in the third and forth years of study. This restricts the number of the students, who want to study astronomy, since they have no background in that science as they do in other sciences which are taught in the first years. The second department in Egypt is the Department of Astronomy and Meteorology in Al Azhar University, where only B.Sc. in astronomy or meteorology can be awarded. This causes a serious shortage of graduate astronomers who can work in observatories or astronomical institutions. As an average, there are two or three graduate students per year, and some years none graduate. Due to the absence of graduate astronomers, the astronomical institutions assign jobs for graduates from physics or mathematics. Also, in Jeddah and King Saud Universities in Saudi Arabia, there are two departments, although there are no observatories in which to work. In Jordan, there is an Institute for Astronomy and Space Science which awards the M.Sc. degree, although there are no astronomy departments in Jordan's universities offering undergraduate studies in astronomy. Accordingly, the chances to pick a good astronomer to join in high-standard astronomical activities are limited. The limitations of astronomy education in Arab countries are the main reason for the serious shortage in the number of high-standard graduate astronomers to teach at different levels and to do good research work.

2.2. Astronomical culture

In spite of the relation between religion and astronomy in Arab countries, where many astronomical phenomena have been used in daily life (Prayer times, etc.), much of the public does not feel the relation and its benefits. Many people know nothing about stars and planets, or even the Sun and Moon, except their positions in the sky. Moreover, some educated persons do not know the difference between astronomy and meteorology or the difference between astronomy and astrology. This confusion is mainly due to the lack of astronomical background. The simplified popular astronomy books are very rare and are often written by non-astronomers, while publishers are not willing to publish astronomy books, or even translated ones, because of the problem of distribution. Although some planetaria are found nowadays in some big cities, their numbers are still not enough to play effective role in popularizing astronomy among the public and youth. Astronomical societies in Arab countries are very weak and very small in number. In most cases, they depend upon the membership of some amateurs and students, who cannot prepare or organize a public seminars or general lectures in the field of astronomy. Accordingly, contact between astronomers and the public is absent and astronomical information is not transferred to society. Finally, TV and broadcasting can play a serious role in enhancing the astronomical background of the public. They can present programs and films about the heavens and space exploration and explanations of some astronomical phenomena. Unfortunately, there is no interest in this type of film or program and people who can prepare and present these programs are very rare and face many obstacles in these jobs.

2.3. Limited jobs for graduate astronomers

Although the number of graduate astronomy students is very small, there are not enough jobs for them. The limited number of observatories and astronomical institutions in Arab countries decreases the chance for graduate students to find jobs. Each year, observatories and astronomy departments have about two jobs available for brilliant graduates. Because of the economical situation, the students prefer to study other branches of science rather than astronomy, to have the chance for work after graduation. This is one of the main factors restricting the spread and promotion of astronomy in Arab countries.

2.4. Training

Training is very important for astronomers and technical staff if they are to use and to run the new sophisticated instruments used nowadays in astronomy. The use of CCD cameras and other modern detectors for astronomical observations requires a high standard from astronomers and technicians. In addition, training on software packages is badly needed, especially those packages used in controlling the various detectors and in analyzing the observational data. Good research work cannot be done without modern equipment and good computing facilities. Although, Egypt for example, has a 2-m telescope and many Arab countries could buy other telescopes very easily, there is no guarantee that they would be run continuously and produce good research work. The budget allotted for maintenance is very small compared to the price of the instruments themselves, so a telescope may stop working at any time. This problem can be solved by inviting experts from outside and sending students and technicians to other observatories and institutions in developed countries, where they can receive a real training.

2.5. Libraries, books and periodicals

There is a serious shortage in astronomy books and periodicals in most Arab libraries, since most of them are published outside the Arab region and are quite expensive. Since reference works serve only a very limited number of astronomers and students, the budget allotted by the authorities is very small, correlated to these numbers in spite of the high prices. Foreign astronomy books translated into Arabic are very rare, while our libraries do not contain Arabic astronomical references. Because of this situation, new researches and discoveries in the field of astronomy are not available to Arab astronomers and students. The Internet service needs to be introduced to all Arab astronomical institutions if all these difficulties are to be overcome.

2.6. Geographical position and international contacts

Because of the geographical position of most Arab countries, they are isolated from the advanced Western countries with their famous observatories and institutions. It is difficult for Arabian astronomers or students to attend a meeting or conference in Europe, since this needs a lot of money for travel, while that is no problem for any European, who can travel through Europe by train and without entry visas. This situation prevents many useful contacts with the advanced institutions in the fields of training and joint research work on which the promotion of astronomy in Arab countries can take place.

3.   Suggestions

In order to overcome the many difficulties facing the promotion of astronomy in Arab countries we recommend:

- Introducing astronomy as a special course for students in secondary schools.

- Training of teachers at all levels of education in the field of astronomy.

- More astronomy departments must be founded for more astronomers.

- Constructing observatories in Arab countries to attract graduate students and amateurs.

- Training of young astronomers and technicians abroad to improve their standard.

- Delivering popular lectures and seminars for public.

- It is important to allot a good budget by authorities for different astronomical activities.

- Introducing astronomy as a general compulsory course for the students of the first level in the universities.

- Encouraging the translation of astronomy books into Arabic.

- Ordering books and periodicals in the field of astronomy and increasing the budget for libraries.

- Supporting the publishing of simplified books for the public at cheap prices.

- Intoducing Internet service into all Arabic institutions.

Abstracts of Poster Papers

Abstracts of poster papers relevant to this section of the Special Session are presented below.

Astronomy in Cuba: Practice and Trends. An Effort to Develop a Non-Formal Education Programme

Oscar A. Pomares, Institute of Geophysics and Astronomy, Havana, Cuba

In recent years, a daily increasing movement of non-professional astronomers has become the center of development for astronomy, a subject practically absent now from the national education system in Cuba. A key role in this movement has been played by the professional staff of the Department of Astronomy of the Institute of Geophysics and Astronomy. A direct outcome of this joint effort with amateurs is the research on meteors and comets, presented at two national and one international scientific meeetings.

The opening last year of the "Palacio de las Ciencias" in the main building of the country, "El Capitolio Nacional", the participation in of professional astronomers in lectures and workshops for the general public, and their participation in prime-time TV and radio programs open a way for the growth of astronomical knowledge among the Cuban people.

Two national meeetings gathering together professional and non- professional astronomers have been held already. Future work in the NEOs international campaign is foreseen. Practice and trends of astronomy in Cuba clarify views of our future in the oldest natural science.

Astronomy Educational Activity in Jordan

Jack Baggaley, Department of Physics and Astronomy, University of Canterbury, Christchurch, New Zealand

Throughout the last three millennia the area of Jordan, as part of the Arab World, has a rich history in the foundation and development of astronomy. Jordan today is pro-active in developing community programmes of education. I will describe the outreach programmes and community participation that I experienced during my attendence in Jordan at a conference timed to coincide with the Leonid Meteor Storm of November 1999.

A Jump-Start for Astronomy Education in Taiwan

Wen-Ping Chen, Institute of Physics and Astrnomy, National Central University, Chung Li, Taiwan 32054, China R.

Spurred by the leaping developments of research activities (SMA, TAOS, AMIBA), Taiwan is catching up in virtually all aspects of its education in astronomy. The first astronomy research institute was established by the Academia Sinica about 10 years ago, which catalyzed within two years the first graduate school of astronomy, as well as an elaborate astronomy museum. Since then astronomy education at all levels, from colleges to primary schools, has been booming. More than a dozen universities are offering astronomy courses, and two more graduate schools will soon be instituted. Textbooks get written, and books on popular science, either translated or composed by local authors, have mushroomed on the market. I will outline these ongoing activities along with plans on the horizon.

An Interactive Approach to Planetary Orbits at Secondary Level

V.B.Bhatia, Department of Physics and Astronomy, University of Delhi, New Delhi 110 007, India

I describe an interactive unit for teaching/learning planetary orbits at secondary level. The unit consists of a pre-test, activites related to the planetary orbits and a post-test. The pre-test is designed to modulate the activities to suit the needs of the students. The activities are (1) Trajectory of a particle in a gravitational field when the field is constant, and when it varies with distance (invoking a fictitious planet WonderX); (2) Plotting the orbit of a planet round the sun starting from select initial conditions and discovering Kepler's laws; (3) Plotting orbits with force laws containing terms like $1/r^3$ and $1/r^4$ and discovering precession. Students work through graded steps. To help them, simple numerical algorithms and computer programs have been developed as they are not yet comfortable with calculus. A post-test gauges the progress of the students for a possible revisit to the topic. Such units can be very effective in teaching basic astronomy in developing countries.

An Undergraduate Program for Astronomy in México

Hector Bravo-Alfaro et al., Departamento de Astronomía, Universidad de Guanajuato, Guanajato GTO 36000, México

Astronomy in México has an ancient tradition, reinforced during the twentieth century by groups working in theoretical and observational astronomy. During the 1990s, the Great Millimeter Telescope (a single 50-m antenna) has been approved, and a 6-m infrared telescope is under study. Graduate and undergraduate programs must be improved to prepare future Mexican and Latin

American astronomers to take advantage of these facilities. To meet the challenge, two traditional Mexican programs (Instituto de Astronomia-UNAM and Instituto Nacional de Astrofisica, Optica y Electronica-INAOE) are updating their graduate programs. Similarly, the Departamento de Astronomía de la Universidad de Guanajuato is joining physicists in the first undergraduate program in México in physics and engineering with an option in astrophysics. This will prepare students for industry, academia or national laboratories, either in physics or astronomy. Jobs in academia have been scarce; many students had to give up their goals after one or two postdoctoral positions. Graduate and undergraduate programs must adjust, by broadening the scope of present programs so that students are better prepared for other job opportunities. We present a B.Sc. program designed by astronomers and physicists to try to address some of these concerns and to prepare the students for either continuing with graduate studies or finding employment in an ever-changing job market. (Co-author is Victor Migenes, Guamajato, México.)

Astronomy in Romanian Universities

Mihail Barbosu, Faculty of Mathematics, Babes-Bolyai University, RO 3400 Cluj-Napoca, Romania

In this work we present characteristics of the Romanian higher education related to the study of Astronomy. In spite of Romanian economic problems, opportunities for Bachelor's degree, Master's degree (at "Babes-Bolyai" University of Cluj-Napoca) and Ph.D. degree are provided for students enrolled in the Faculties of Mathematics or Physics. General regulations, description of courses, research resources and job opportunities are also described and discussed in this paper.

Astronomy Education in Thailand

Busaba Hutawarakorn et al., National Electronics and Computer Technology Center, Thailand

Thailand is one of the developing countries which tries to advance its economy, technology and science. Education in astronomy is considered a supporting factor; astronomy is a basic science from which the young generation can learn to understand and to conserve mother nature and at the same time develop analytical thinking. The poster reports the present developments in astronomical education in Thailand which includes (1) current astronomy education in school and university; (2) educational activities outside school; (3) development of programs for teaching astronomy in school (including teacher training); (4) the access of educational resources via internet. Proposals for future development and collaborations will be presented and discussed. (Co-authors are B. Soonthornthum and T. Kirdkao.)

Astronomy Books in Spanish

Julieta Fierro, Instituto de Astronomía, UNAM, Apt 70 264, México DF 04510, México

Great cultures have created language. They have discovered its strength among other reasons for education. For a long time the Bible was one of the few books available in western culture, its influence is beyond any doubt. Many developing nations have no science books in their mother tongue. They might carry a few translations but these do not convey the local culture so it is harder for students to grasp the concepts and to build on what they know. Books, even if they are extremely simple, should be written in local languages because that will facilitate the conveying of knowledge and the creation of scientific culture. In the books examples that pertain to every day local life must be given, in particular examples that have to do with women. Women play a central role in developing nations by child bearing; if they become literate they will influence enormously the quality of their children's education, in particular their science comprehension. In México a collection that includes astronomy books has recently been edited by the National Council for Culture and Arts. The books are small and light, which encourages middle-school students to carry them around and read them while traveling in public transportation, such as the subway. Every other page is a new subject, that carries illustrations, abstracts and conclusions. The astronomy books are on search for extraterrestrial life, the stars and the universe. These books are distributed nation-wide and are inexpensive. They have been written by Mexican astronomers.

Teaching of Astronomy in India

Mandayam N. Anandaram, Physics Department Bangalore University, Bangalore 560 065, India

Here I will describe the inclusion of astronomy and astrophysics in College level courses of Bangalore University. I will describe the role of the Inter - University Center for Astronomy and Astrophysics (IUCAA) at Pune in making available instruments such as photometers and CCD cameras at low cost to aid teaching of astronomy as well as the running of a large number of training programmes for teachers and students. I will also describe some outstanding problems and suggested solutions.

Summer Schools for European Teachers

Rosa M. Ros, Department of Applied Mathematics, Technological University of Catalonia, Jordi Girona 1-3, Modul C3, 08034 Barcelona, Spain

The Summer Schools have been organised by the European Association for Astronomy Education (EAAE) for European teachers. The first was organised in La Seu d'Urgell, Spain, the second was organised in 1998 in Fregene, Italy and the third in 1999, during the week of the eclipse in Briey, France, on the line of total darkness. We had a cloudy eclipse, but fortunately we could observe it. The fourth one was held in July 2000 in Tavira, Portugal. About 50 participants are involved in each Summer School. In the last, participants came from 14 countries. The activities are organised in General Lectures, Working Groups and Workshops for reduced groups and day and night Observations. To increase communication, each Summer School has three official languages: the language of the host country, English and another well-known by the participants. The proceedings are published beforehand with all the contents to facilitate participation. Each paper appears in English and another language.

The Leonids Observation Project by High-School Students all over the World

B. Suzuki et al., National Astronomical Observatory, Japan

We organized the Leonids observing network that comprised of 276 observation teams consisting of about 3,000 senior high-school students in Japan on Nov. 17 1998. We counted the visible meteors with our naked eyes. It was a simple method, but the many data enabled us to discuss the structure of the dust trail. However, the base-line is so short that we could not discuss the structure in full detail. In 1999, we organized a worldwide network of senior high-school students. The network is comprised of 307 teams from 23 countries. The base-line of our new project is the size of the Earth. We succeeded in making 20 hours continuous observations with this network. One of the observational teams in Tanzania encountered the Leonid meteor storm. We report on the scientific results and the educational aspects of these projects. (Web site: http://www.leonids.net/) (Leonids '98-'99 staff – co-authors – are A. Miyashita, M. Okyudo, H. Agata, T. Mizuno, T. Hamane K. Watanabe of Japan, C. Pennypaker, A. Gould, K. McCarron, G. Reagan and K. Meredith, of the U.S.A.)

New Student Laboratory Work about Pulsational Phenomena in Astronomy

Salakhutdin N. Nuritdinov, National University of Uzbekistan, Tashkent, Uzbekistan

The pulsation phenomenon is inherent to most types of object and it plays a great role at certain stages in the evolution of objects in the universe. That is why students must study this phenomenon in the framework of laboratory hours. Often the study of these phenomena is reduced to an analysis of some differential equations with variable coefficients. A class of these equations is connected with the stability problem of the oscillations of self-gravitating systems (S.Nuritdinov, Sov. Astron., 1985, 29, 293) . In order to carry out this laboratory work every student is required to compose a computer program using the periodical solution stability method and the parameter resonance theory. The program will find the critical amplitude of the pulsation and some dependences between physical parameters.

Astronomy and Astronomical Education in the FSU

Nikolai G. Bochkarev, Sternberg Astronomical Institute/Euroasian Astronomical Society, Universitetskij Pr. 13, 119899 Moscow, Russia

The current situation in astronomy and astronomical education over the territory of the Former Soviet Union is traced. New facilities for radio-astronomy are being put into work – the most important of them being the two coupled 32-m dishes, VLBI network "Quasar"; a number of observatories are acquiring an international status (in the frame of CIS); The Internet is becoming available for an increasing number of astronomical institutions. Azerbaijan astronomers have overcome their isolation from the rest of the world and cooperate actively with the astronomical community.

All-Russia and international Olympics in astronomy for high school students are held and attract participants from increasing number of regions of Russia and other states.

The outcome of the 9th JENAM in Moscow and of the events attached to the Meeting is presented.

Conceptual Aproach to Astronomy and Basic Science Education

M. Melek, Cairo University, Cairo, Egypt

An approach is developed, in which the major dynamical and physical concepts of astronomy and basic space science are used, to build a scheme (prototype model) for education on the undergraduate level. A way to teach different theories and observational facts is shown, in which those concepts are built in or used; within the suggested educational scheme. The computational techniques which are needed in astronomy and basic space science are disscussed at which steps through the suggested educational scheme might be introduced.

Research Oriented Astrophysics Course for Physics Students

Tapan K. Chatterjee and A Pedroza-Mendelez, Universidad A.Puebla A.P. 1316, Puebla PUE 72000, México

Section 4: Current Status of Astronomy Research in Developing Countries

Astronomy for Developing Countries
IAU Special Session at the 24th General Assembly, 2001
Alan H. Batten, ed.

Astronomy in Algeria: Past and Present Developments

Abdenour Irbah, Toufik Abdelatif and Hamid Sadsaoud

Observatoire d'Alger - CRAAG, BP 63, Bouzaréah 16340, Alger,
Algeria. e-mail: irbah@pleiades.unice.fr

Abstract. Astronomy studies have been developed in Algeria since 1890 when the Algiers Observatory was built. Several instruments were soon installed on the site and have contributed to many scientific projects such as the international sky-map program. However, the observatory activities were suddenly interrupted by the departure of all French astronomers in 1962. Twenty years were needed before new astronomy programs were developed at Algiers observatory. They are principally based on imaging through atmospheric turbulence, solar physics and studies of pulsating variable stars. Only one observational program, however, has so far been developed. This consists of solar observations in the framework of an international program to survey the Sun's radius. The astronomers now form a relatively important team since more than twelve researchers have permanent status. This is a good start taking into account the fact that astronomy is not taught in Algerian universities. We will begin first by giving an overview of the history of Algiers Observatory, including its instrumentation. We will then present the existing Algerian team and all their current scientific work and proposed projects.

1. Introduction

Although astronomy greatly affects the daily life of Algerian people as in all Muslim countries (hours of prayer, religious celebrations etc.), the people do not know the basic explanations of current astronomical phenomena. Thus, there always are problems to define the Muslim celebration dates, which are all linked to the visibility of the lunar crescent. The celebration dates are often fixed before the conjunction between the Sun and the Moon. More astonishing was the last partial eclipse that occurred in August 11th 1999 in Algeria. In fact, the Algerian people dreaded this natural phenomenon and many of them stayed at home. These examples show clearly the state of astronomy in Algeria, a state principally due to the fact that astronomy is taught neither at school nor in the universities. Nevertheless, there are some structures in which astronomy is developed. In fact, there are several amateur clubs that cater for young people interested in this discipline. Although the clubs do not have many resources, they make considerable efforts that deserve to be mentioned. The most important institution in which astronomy projects are developed is, however, the Algiers Observatory. This professional establishment is more than one hundred years old but Algerian society did not benefit from its learning and long experience. This paper presents, then, astronomy in Algeria focusing, however, on what has been developed at the Algiers Observatory. Thus, after a brief recall of the history of the Algiers Observatory, the present research center will be described. We will then introduce the Astronomy and Astrophysics Department and all current projects. We will make some remarks about the existing international collaboration and about the problems encountered in the full development of astronomy.

2. History of the Algiers Observatory

After several observing campaigns near Algiers during 34 years, an observatory was built on the Bouzaréah site in 1890 during the French occupation (1830 - 1962). Several instruments were soon installed that permitted intense activity at the Algiers-Bouzaréah Observatory during nearly a century. These instruments were a Foucault's telescope, a coudé refractor, a photographic astrograph and a meridian circle (lunette méridienne). The photographic astrograph was the instrument that absorbed a major part of the Observatory's activity. In fact, it was specially built and installed on the site in 1892 to contribute to the French project Carte-du-Ciel (sky map) and to the photographic catalog project. This project involved a network of several astrographs distributed all over the globe (Europe, Australia, Africa and Asia). A precise region of the sky was allotted to each observatory and continuously photographed over several years. Thus, about 5000 plates were recorded by the Algiers photographic astrograph, which operated from 1892 to 1940. Twelve minor planets were also discovered with the instrument during this period. Two of them were named El Djezair (Algeria) and Bouzaréah. The other instruments were used for many programs such as occultation of stars by the Moon, observations of planets and their satellites, of sunspots and of transits of Mercury across the solar disk etc. Another instrument, a Danjon prismatic astrolabe, was later installed on the site (1957) for astrometry programs. It was part of a network program dedicated to study the Earth's rotation and polar movements. The Algiers astrolabe has also participated in the improvement of the fundamental catalog FK4 and the elaboration of the FK5. Planets (Jupiter and its satellites, Saturn, etc.) were also observed with this instrument in order to improve their ephemerides. In the 1960s, all activities at the Algiers Observatory were interrupted by the sudden departure of the French astronomers and the lack of Algerian astronomers at that time.

3. The Present Structure of the Algiers Observatory – CRAAG.

In the early 1980s, several events permitted new astronomy developments at the Algiers Observatory. The earthquake that occurred in 1980 in the west of Algeria (about 200 km from Algiers) made the Algerian government aware of the importance of studying natural phenomena and provided the opportunity to build a new center on the Algiers Observatory site. At this time, new legislation was promulgated for the creation both of centers and of researcher status. This gave total financial autonomy to the centers and defined permanent positions for the researchers. A new center was then created at the Algiers Observatory site which is the present Centre de Recherche en Astronomie Astrophysique et Géophysique (CRAAG). It is composed of three scientific departments. The first is devoted to astronomy and astrophysics researches while the other two are for geophysical studies. Two other departments exist also for scientific support (library, computer center etc.). When CRAAG was created, there were no astronomers. The first Algerian astronomers began to integrate the CRAAG-Algiers Observatory in the middle 1980s. By the late 1980s and early 1990s, there were about ten astronomers in the Center. They were all senior researchers, a large part of them returning from France at the end of their postgraduate studies. After a brief period of research activity in the Observatory, new events in Algeria caused the departure of the half of the total number of the astronomers. It was necessary to replace those who left. The lack of astronomy teaching in the Algerian universities forced the astronomers remaining to develop their own educational program. Thus, many students coming from various physics disciplines continued their postgraduate education by participating in the existing research topics. The group is now composed of 14 astronomers, all having permanent positions. They are part of different research teams working in the two laboratories of the department: (i) the stellar and astrometry laboratory and (ii) the astrophysics laboratory. The projects developed in the laboratories are presented in the following sections.

4. The Stellar and Astrometry Laboratory

Two research teams compose this laboratory. The first one develops studies related to solar astrometry and imaging through atmospheric turbulence while the second one to the stellar instabilities.

4.1. Solar astrometry and atmospheric optics (Dr. A. Irbah)

A small group composed of seven astronomers is involved in these research programs, which are concerned with the variations observed in measurements of the solar diameter and with the effects of atmospheric turbulence on the recorded data. In fact, diameter measurements made with an astrolabe at Calern Observatory (France) during more than two solar cycles show apparent variations. The observed variations are in opposite phase to the solar cycle, defined by the sunspot number. These results have given rise to many scientific works and experimental developments. Thus, several astrolabes in the world were specially adapted for solar observations. The future space mission *Picard* will also be dedicated to this topic. It will consist of the launch in 2002, of a micro-satellite equipped with several solar experiments; one devoted to solar-diameter measurements. This scientific context led to initiation of a solar-radius observation program at Algiers Observatory. An experiment similar to the one at Calern Observatory is now developed for the Algiers astrolabe. This program will be installed in the south of Algeria at the geophysical observatory of Tamanrasset, which is part of CRAAG. Part of this program will be to observe, together with all solar astrolabes around the world, the space mission *Picard* and also to observe in coordination with the new solar observing program at

Calern Observatory (DORAYSOL and PICARDSOL). DORAYSOL (Definition and Observation of the Sun's Radius) is the new version of the solar astrolabe while PICARDSOL is the ground-based counterpart of the space mission. In the framework of this program, the effects of atmospheric turbulence on the measurements are studied. They are developed in order to quantify the errors in diameter measurements with the observing conditions. These are defined by the Fried's parameter, the atmosphere time constant (s), the isoplanatism patch size and the spatial coherence outer scale of the wave-front. Numerical simulations are used to generate random perturbed atmospheric wave-fronts. Solar images as recorded in the program are then simulated in various observing conditions and used to define the behaviour of errors in the diameter measurement. The results deduced from this work led to the development the seeing monitor MISOLFA (French-Algerian Solar Imaging Monitor), which will observe at the same times as the solar observations made at Calern Observatory (DORAYSOL and PICARDSOL) and by the satellite *Picard*. With the Tamanrasset program, observing conditions will be directly deduced from solar-image sequences used for diameter measurement. The third research topic concerns the analysis of visual solar data recorded at Calern Observatory during more than two solar cycles. Special signal-processing methods are needed since diameter measurements are non-uniformly sampled and present some temporal gaps. The methods developed show new periodicities previously undetected, such as the 27-day rotational period, which appears to be close to the Carrington rotation. All these results, which require further study, are currently being developed by the research team. For more details, some references related to all this work are given at the end of the paper (Irbah et al. 1999, Laclare et al. 1996, Lakhal et al. 1999, Moussaoui et al. 2000).

4.2. Stellar instability project (Dr H. Sadsaoud)

The main objective of the stellar group is to develop an automatic photometric observing station in Algeria, and to create an Algerian research team to run and to use it. The observing set-up will enable our researchers to learn the modern techniques of stellar observation and the latest techniques in image and signal processing. The research themes are principally stellar physics and particularly stellar pulsations of hot stars, with the aim of understanding their instability mechanism, functioning and evolution, a subject which the astronomy team of the Bouzaréah Observatory has been working on for several years now. Members of the team have good experience in this observational domain since they often observe abroad using big telescopes equipped with high-performance systems of data acquisition (high-resolution spectrographs). These fields of research need a theoretical and observational study of variable stars and notably the instability they represent. The cause of these stellar instabilities has become more apparent these last two years, thanks to the publication of more precise opacity tables than those previously used. Scientific work reveals that the destabilizing mechanism of these stars is similar to the κ-mechanism operating in the Cepheid instability band. It is, then, with the aim of studying stars in general, and in particular the B-type variable stars, that we are envisaging the installation of the first observing station in Algeria, which would permit us to follow the photometric and spectroscopic evolution of the B-type stars. The station would be installed first in the north, thanks to the help of Algiers Observatory. The final transfer to a much better site would be envisaged after the results of the site survey in Algeria, which is in progress. In recent years, instrumental evolution permits the realization of good observations. Yet, determination of the pulsation parameters by classical methods is not yet on a level of with new data of high quality. By their relative inaccuracy and difficulty of use, these methods are not adapted to interpret all the information held in the high-resolution spectroscopic profiles which are now being obtained. A new method, the moment method, seems very promising because it makes use of the whole observed profile, however, important work is to be done, particularly taking into consideration the Coriolis force or thermal effects. Despite its strength, this new way of analysis rests upon a restricted theoretical basis: the adiabatic linear case, which must be adapted to a high-amplitude case. In fact, the study of the influence of non-linear effects on the pulsations, with the help of an imaging-method development, is thus to be envisaged as an area of an international collaboration.

5. The Astrophysics Laboratory: Solar-Terrestrial Project (Dr T. Abdelatif)

In the Astrophysics Laboratory there is a small group of researchers interested mainly in solar-terrestrial interaction. A few aspects of this very vast domain have been selected for two reasons: (i) the human component is very small and very few astrophysicists have been educated in Algerian universities, (ii) the studies have to be of national interest (e.g. telecommunications). The final goal is to constitute a team of researchers that are able to study most of the solar phenomena that affect the Earth's electromagnetic environment. The first project concerns the study of the propagation properties of the ionosphere. A computer code has been developed in order to simulate the propagation of an electromagnetic pulse through an inhomogeneous plasma. We hope that we will be able to study with this code any observational model of the ionosphere. CRAAG has recently acquired several dual-frequency GPS receivers; we are planning to use the collected ionospheric data to establish a model of the local ionosphere for solar-quiet days. It will then be possible to follow up ionospheric perturbations due to solar events. A theoretical model as well as a computer code will be needed to invert the time-dependent data in order to produce a time-dependent three-dimensional model of the ionosphere. The second study, we are involved in, concerns the effect of sunspots on the total solar irradiance. The total solar irradiance of the Sun is modulated by the solar activity and especially by the sunspots on the surface of the Sun. This irradiance will be computed using the available data of sunspots area versus time and latitude. The computed irradiance will be compared with the measured one using the ACRIM data (also available). An optimization of the fit will be used in order to constrain some of the free parameters like the limb-darkening function, distribution of the blocked energy etc. The third study is related to the study of the solar cycle. The forecasting of solar maximum or minimum is very important for all satellite and telecommunication activities. The study of past solar activity can be a valuable asset for forecasting. The distribution of maxima and minima will be deduced from the raw data using a new technique. Information concerning features of the solar cycle (periodicity, minima and maxima, rate of variation) is extracted using decomposition in terms of Tchebychev polynomials. This new method has the advantage of computing statistical quantities by using a variable window over the data. An ongoing study concerns the propagation of acoustic waves in an atmosphere embedded in a magnetic field, this study has been the subject of many publications in the past. The actual ongoing research concerns the absorption of p-modes by sunspots. The future development of the laboratory concerns the observational aspect; it is intended to acquire a small station for solar observations that will produce daily pictures of the Sun and its activity in different wavelengths. It is clear from the above that our main interest is the Sun, its activity on different time-scales and its effects on the geosphere.

6. International Collaboration

We have much international collaboration with several foreign institutions in various countries (France, Spain and England). These collaborations have been set up in order to help the Department in the development of all the projects. The most important one is, however, the French scientific cooperation that exists up to now. It was initiated in the middle 1980s to reactivate astronomy at Algiers Observatory. Since then, three cooperation programs (CMEP) for astronomy developments at CRAAG-Algiers Observatory have been set up with the French Foreign-Affairs Ministry. There are also annual cooperation programs with the French CNRS. Thus, this collaboration permitted the obtaining of support from French astronomers and also provided useful funds for visiting their laboratories and acquiring some equipment. It also permits young Algerian astronomers to complete their own education. The principal French partners are the Observatoire de la Côte d'Azur and the UMR Astrophysique of the Nice-Sophia-Antipolis University.

7. Difficulties Encountered in the Development of Astronomy in Algeria – Conclusion

Astronomy has existed in Algeria for more than a century. The major activities in this discipline were developed at the Algiers Observatory, which was built in 1890. At this period, several instruments were installed on the site, allowing the Observatory to make considerable contributions to the international astronomical community. However, these intense activities suddenly stopped in the beginning of the 1960s, because of the departure of French astronomers in 1962 and the lack, at that time, of Algerian astronomers to carry on the activities. This situation persisted until the creation of the present CRAAG, twenty years later. It was at this period that the first Algerian astronomers begun to integrate the centre, returning from foreign universities at the end of their postgraduate studies. Several astronomy projects were then initiated and are still developed up to now. However, they encountered many difficulties to develop new astronomy in Algeria. The most important was the very limited funds allotted to the discipline. The consequences for the present astronomers have been difficulties in arranging regular exchanges with foreign laboratories, in acquiring new instrumentation and recent documentation, and in correctly responding to the astronomical needs of Algerian society. Although the astronomer group of the Algiers Observatory is now relatively large, it remains insufficient to allow a full development in all areas of astronomy, taking into account that astronomy is taught neither at school nor in the universities. To avoid these difficulties, the Algerian astronomers set up international collaborations in order to develop research in astronomy. Thus, new efforts will be made in order to pursue the existing projects but also to initiate, in the next few years, new ones in stellar and solar physics principally, in helioseismology and asteroseismology, instrumentation, cosmology etc. This will necessitate however, a noticeable increase in the number of Algerian astronomers and thus the setting up of a better educational program. If the participation of young Algerian astronomers in international schools is very helpful, it still remains insufficient. Introduction of basic programs in schools and the setting up of annual summer universities on specialised topics seem to be the best way for a rapid full development of astronomy in Algeria. International aid for giving shape to this program will be a valauble contribution.

References

Irbah, A., Bouzaria, M., Lakhal, L., Moussaoui, R., Borgnino, J., Laclare, F. and Delmas, C. 1999, *Feature extraction from solar images using wavelet transform: image cleaning for applications to solar astrolabe experiment,* Solar Physics, 185, pp. 255-273.

Laclare, F., Coin, J.P., Delmas, C. and Irbah, A. 1996, *Measurements and long-term changes of the solar diameter,* Solar Physics, 166, pp. 211-229.

Lakhal, L., Irbah, A., Bouzaria, M., Borgnino, J., Laclare, F. andd Delmas, C. 1999, *Error due to turbulence effects on diameter measurements performed with a solar astrolabe,* Astron. Astrophys. Supp. 138, pp. 155-162.

Mpussaoui, R., Irbah, A., Abdelatif, T., Fosat, E., Borgnino, J., Laclare, F., Delmas, C. and Schmider, F.X. 2000, *Analysis of diameter measurements performed at Calern Observatory astrolabe,* Solar Physics, (in press).

Moussaoui, R., Irbah, A., Fossat, E., Borgnino, J., Laclare, F., Delmas, C. and Schmider, F.X. 2000 *Spectral analysis of solar diameter measurements recorded at Calern Observatory astrolabe during two solar cycles,* Submitted to Astron. Astrophys.

Discussion

Wang expressed interest in the studies of atmospheric optics. Have they been published? Irbah replied that some references to the work are given in the paper. The major part of the studies of atmospheric optics are concerned with the effects of atmospheric turbulence on mesasurements of the solar diameter. The studies are developed by numerical simualtions and observational data. Similar work at the Calern Observatory in France will be transferred to the Tamanrasset Observatory in southern Algeria, as part of an international project. We are, however still developing a seeing monitor to rest several sites in Algeria.

Fierro said that México also has an old Carte-du-Ciel instrument. They completed their part of the project and now find the telescope wonderful for educational purposes. Irbah replied that he had not known that México Observatory had been part of the Carte-du-Ciel project, but he agreed about the educational potential of the instrument. It was also still useful for research programmes because of the large stock of plates obtained over a period of about fifty years.

Chamcham remarked that, between Morocco and Algeria there could be uncertainty of up to two days in the visibility of the lunar crescent and astronomers in Morocco hesitated to announce observations of the crescent –are there similar problems in Algeria? Irbah replied that, if the crescent csn be observed in Algeria then all coutries to the west, including Morocco, should also be able to observe it. On the other hand, a crescent seen in Morocco might have been missed in Algeria. The visibility dates between the two countries should not, however, differ by more than a day. In Algeria they make calculations of the visibility and give them to the authorities, who make the necessary decisions. During the last few years, Algerian and Moroccan calculations have been in good agreement, but the dates of celebrations in the two countries have differed.

Astronomy for Developing Countries
IAU Special Session at the 24th General Assembly, 2001
Alan H. Batten, ed.

The Egyptain 1.88-m Telescope

Anas M. I. Osman

National Research Institute of Astronomy and Geophysics, Helwan, Egypt, amiosman46@hotmail.com

Abstract. The Kottamia 1.88-m reflecting telescope in Egypt is the largest in the Middle East and North Africa. An extensive upgrading programe has been undertaken for this telescope to increase its efficiency. A new Zerodur optical system has been delivered by Carl Zeiss, Germany, and a new CCD system including an acquisition Camera Tek 1024 x 1024 pixels, with pixel size 24 x 24 micron and LN cooling, and an offset guiding camera Kodak 1080 x 1024 pixels with pixel size 16 x16 micron and thermoelectric cooling. This CCD system has been attached to the Newtonian focus for direct imaging with scale 22.5 arcsec mm^{-1}. The aluminizing plant has also been refurbished by Balzer, the new pumping system can accommodate mirrors up to 2m. The unit is supplied with a microprocessor, which controls and checks all recoating steps, closes all valves and stops the operation in any emergency. A Cassegrain spectrograph (donation from Okayama Astrophysical Observatory, Japan) will be attached to the telescope, after modification, for use with a CCD camera instead of the image-intensifier used before. This spectrograph will be used to obtain medium-dispersion and low dispersion spectra for faint stars and galaxies down to 15th magnitude.

1. Introduction

Astronomy in Egypt is an old science. The first reasonable-sized telescope used for astronomical observations was the 0.9-m reflector at Helwan Observatory (south of Cairo), which has been used for many decades since 1905 to observe many celestial objects, specially in the declination zone 0° to 30°, and many new objects have been detected. More than 1907 nebulae were observed in that zone and, among them, 505 nebulae were new discoveries. In addition, observations of many comets (e.g. Halley 1910) and planets with their satellites have been collected. In 1962 a new era of astronomy began in Egypt with the construction of the 1.88-m reflecting telescope at Kottamia (70 km due east from Cairo, in the eastern desert of Egypt), see Figure 1. It is constructed on a plateau at about 500 m above see level with geographical coordinates $\phi = 29°\ 55.9'$ N, $\lambda = 31°\ 49.5'$ E. The telescope was manufactured by Grubb Parsons of England and it can be used in the Newtonian ($f/4.9$), Cassegrain ($f/18$) and coudé ($f/28.9$) focii with scales of 22.5, 6 and 3.8 arcsec mm^{-1} respectively. This telescope was provided with: a Newtonian camera ($f/4.9$) for direct imaging with correcting lens (field of view of about one degree); a Cassegrain spectrograph with two cameras of dispersions of 100 A mm^{-1} and 48 A mm^{-1} at 4800 A; and a two-channel photoelectric photometer for stellar photometry with photon counter.

Since the telescope started functioning in 1964, hundreds of observations have been secured for stars, galaxies, star clusters, planets, comets, etc. by Egyptian and foreign astronomers. Based on these observations, a large number of papers have been published and tens of M. Sc. and Ph.D. degrees have been awarded. Due to the successive re-aluminizing of the telescope mirror and the chemical treatment required for the process during about 30 years of work, a serious problem with the surface of the main mirror was detected. The mirror surface would not accept any recoating properly and the resultant coating was dim, poorly reflecting and short-lived. Many attempts were made to clean the surface and to recoat it, without success. Several reports concluded that repeated cleaning of the mirror surface with chemicals had leached it. On the other hand, the detectors provided with the telescope became old and needed to be replaced by new sophisticated ones. After evaluation of the situation, an extensive program for upgrading the telescope was undertaken to provide the telescope with:

- a new Zerodur optical system (primary and Cassegrain mirrors),

- a new large-format CCD camera system with liquid-nitrogen cooling and a CCD camera for offset guiding with thermoelectric cooling,

- a new fast grating spectrograph,

- an updated aluminizing plant for the Kottamia Observatory,

- upgraded the computing facilities.

2.   Upgrading Program for the Kottamia 1.88-m Telescope

2.1.   CCD Camera System

In 1992, the upgrading program was begun by a contract with Astro Cam (Pix Cellent now), U.K., to provide the telescope with a CCD camera system for scientific observations and for guiding purposes, with the following specifications:

A - The Scientific Camera: This is to be used mainly for direct imaging and provided with LN cooling. The main chip is Tektronix with format 1024 x 1024 pixel with pixel size 24x24 microns square, thinned back-illuminated with an anti-reflection coating, to enhance the the chip sensitivity to shorter wavelengths. It is characterized by high dynamic range, excellent charge-transfer rate at low signal level with low dark current and maximum readout flexibility. The quantum efficiency reaches its maximum value (80 %) at room temperatrue between 6000A and 7000A.

B - The Guiding Camera: This camera is to be used for guiding during observations. The main chip is Kodak KAF-1300 with format of 1280x1024 pixels with pixel size 16x16 microns square, cooled thermoelectrically. Its spectral response is from blue to near infrared spectral range with maximum quantum efficiency ($\sim$ 50%) at 8000A. A long pass-band filter RG 610 is installed at the

Figure 1.   The Kottamia 1.88-m Telescope, Egypt

front of the camera to improve the S/N ratio. The two cameras are controlled by the DDE 4202 controller unit and Imager 2 software package. They have been tested and mounted at the Newtonian focus of the telescope (f/4.9, plate scale = 22.53 arcsec mm^{-1}) for direct imaging. This system can be also used at the Cassegrain focus of the telescope (f/18, plate scale = 6 arcsec mm^{-1}) although the long tube of the telescope can affect the optical path to the Cassegrain focus by some disturbance.

2.2.   New Optical system for the Kottamia 1.88-m telescope

In 1994 a contract with Zeiss Jena (Germany) was signed to provide the Kottamia 1.88-m telescope with a new optical system (primary mirror and Cassegrain secondary) made of Zerodur, together with modification of the main-mirror cell and the Cassegrain cell to suit the new optical system. A new design for the supporting system of the main mirror was produced by Zeiss, using the old metal cell of the Kottamia telescope. It consists of 18 axial supports distributed in two rings, the inner ring with six supports and the outer with 12, see Figure 2.

Figure 2.   Distribution of the 18 axial supports in the modified cell

The mirror will be supported by 16 lateral Zerodur pads cemented at the mirror edge, see Figure 3. These pads are fixed on 16 radial counterweight-support units, which are fixed on the inner edge of the mirror cell. The radial position of the mirror in the cell is defined by three units (three fixation points), two of them are hard points (fixed, adjustable stops) and the third point gives a soft pressure to the mirror edge by the use of a spring.

The new primary mirror has the specifcations shown in Table 1.

Table 1.   The New Primary Mirror Specification

Focal length	9138 mm
Weight	1600 kg
Material density	2.53×10^{-3} kg mm^{-3}
Radius of the mirror	965 mm
Radius of inner hole	94 mm
Diameter of encircled 80% of energy	≤ 0.35 arcsec

The new optical system and the modified cell arrived at the site during the summer of 1997. Acceptance tests on site were begun in April 1998, by Zeiss experts, and the results were reasonable but not sufficient. Another test campaign was begun by Zeiss experts on 17 June 1998 and, after few nights, they found that the results were changeable and unsatisfactory. The primary-mirror supporting system was examined and, after the dismantling the cell, Zeiss found that 12 out of the 16 radial glued pads were damaged and had come loose from

Figure 3.   Installation of the new primary mirror in its modified cell

the body of the main mirror. The only explanation offered by Zeiss was that some glued supports fell off because of the material of the glue. After regluing these 12 pads, the Zeiss team began another test during November 1998 and took some frames and analyzed them to evaluate the Zernike Coefficients. The observations were very limited, covering a very short time during two nights (total observing time 6 hours) for *a few stars around the zenith*. Since this test was performed for a very short time and around the zenith, in the absence of any NRIAG consultant, Zeiss was asked to make the telescope available for more observing time to check the accuracy and stability of the whole system before acceptance. During one of the tests by Egyptian astronomers in June 1999, a hard sound of a sliding part below the main mirror was heard, observations were stopped and Zeiss was informed and asked to examine the mirror and the supporting system. After once more dismantling the mirror cell, Zeiss experts found that two lateral pads were loose again and separated from the mirror body (*these two pads were among the twelve reglued pads separated before*). The supporting system and its design became doubtful and Zeiss must find a solution to guarantee the reliability of the supporting system. In the technical proposal of their offer, Zeiss stated that the equilibrium of the mirror would be guaranteed in all positions of the telescope. From that time, the situation became complicated, since Zeiss insists that the main reason for the damage is tilting the telescope tube below horizon by *31* °, although this could not have happened in any way. Accordingly, both sides have gone to arbitration to solve the problem and to finalize the project.

2.3. Updating the aluminizing plant of Kottamia Observatory

The Kottamia aluminizing plant is used for coating the primary and secondary mirror of the telescope, see Figure 4. It can hold mirrors up to two meters in diameter. The unit was delivered by Edwards (England) about 30 years ago, and was used efficiently since that time. Since it was equipped with a semi-automatic pumping system with limited safety facilities, it was important for the safety of the new optical system to upgrade the present unit by replacing the pumping system with a new powerful, automatically controlled unit with accurate measuring facilities. This new pumping system is adapted to the old aluminizing chamber which is still in good condition. At the end of 1997 a contract was signed with Balzers to undertake the upgrading with the aims of:

1 controlling the high-vacuum system by a modern automatic control allowing all operational steps to be programmed,

2 giving the pumping system a high efficiency and equipping it with all necessary safety elements,

3 reducing the manual operations in order to avoid personal errors,

4 preventing back-streaming of diffusion-pump oil and accelerating the evacuation process,

5 providing a very high-speed pumping system for a short processing time,

6 protecting the mirror in case of any failure by safety and control units.

After providing and installing all parts of the new pumping system, the aluminizing plant has been tested for a vacuum of pressure of 10^{-6} mbar inside the chamber and the results were satisfactory.

2.4. New Cassegrain spectrograph

A Cassegrain grating spectrograph has been donated by Okayama Astrophysical Observatory, Japan, to Kottamia Observatory. It will be attached to the telescope after modifications for using a CCD camera instead of the image intensifier used before. This spectrograph covers the wavelength range 4000 A-8000 A, with slit length of 15 mm (1.5 arcmin on the sky). It will be used at middle-dispersion to low-dispersion for faint stars and galaxies down to around 15th magnitude. A new CCD camera head has been ordered from Pix Cellent company to be used with this spectrograph. It is an EEV CCD 42-10 CCD chip with 2048×512 pixels grade 1 with pixel size 13.5 microns square and an active image area of 27.6×6.9 mm^2, with LN cooling. The CCD is back illuminated (thinned) with an extended AR coating. Typical quantum efficiencies are >90% at 500 nm, 50% at 350 nm, > 80% at 4000-6500 A and 30% at 9000 A. The readout noise is three electrons RMS at low pixel rates. The active spectral range of the camera is from 200 to 1060 nm In addition, a CCD camera for tracking has been orderd, it includes a thermoelectric CCD head, of type TE/3 Kodak KAF V0040 grade 1 device with 768×512 pixels, with pixel size 9.0 microns square. The active image area is 6.8×4.5 mm^2. The CCD is front illuminated (thinned) with a typical QE of 35% at 600 nm. Its readout noise is eight electrons RMS at low pixel rates.

Figure 4.   The Kottamia aluminizing plant after updating

2.5.   Upgrading the computing facilities

During the last three years, the Kottamia observatory has been provided with a powerful computing facilities including two Sun Spark workstations and some Pentium PC computers. These systems are used in controlling the CCD systems of the telescope and analyzing the observation by different software packages.

The Kottamia 1.88-m telescope will be back in commission very soon after the problems of the supporting system have been solved. It is a good telescope and can serve all astronomers in the Middle East and North Africa since it is the largest telescope in the region. It needs all the support from all authorities to develop astronomy in many developing countries in the region especially after such upgrading project.

References

El Bassuny, A. 1998, African Skies, No. 2, p. 15.

Hanafy, D. and Heileman, W. 1999, African Skies, No. 4, p. 7.

Hassan, S.M. 1998, African Skies, No. 2, p. 16.

Samaha, A.H. 1964, Helwan Observatory Bulletin, No. 62.

Discussion

In answer to Kochhar's query about the source of funds for upgrading the Kottamia telescope, Osman said that most of the funding came from the budget for the last five-year plan of the National Research Institute of Astronomy and Geophysics, Helwan-Cairo; part also came from the Egyptian Ministry of International Cooperation. In answer to Hingley, Osman said that the mirror supports are not active. Hingley also commented that the Kottamia telescope was one of about half-a-dozen of that size built by Grubbb-Parsons. He was glad to hear of the rebirth of this one and drew attention to a recent article on these telescopes by Richard Jarrell in the *Journal for the History of Astronomy*, (Vol. 30, pp. 359-90, 1999).

Al-Sabti commented that when he went to use the telescope some time ago, the lack of funds was obvious and there was no permanent staff at the Observatory. He was glad to hear that things were improving. Osman said that the financial situation is improving. They had $2.5M from the Egyptian government to upgrade the telescope and to maintain the Observatory. The only permanent staff at the Observatory are the technical workers. Astronomers visit Kottamia only to observe. There are now about 35 astronomers, with Ph.D. degrees, with permanent jobs at the Institute.

Astronomy for Developing Countries
IAU Special Session at the 24th General Assembly, 2001
Alan H. Batten, ed.

Astronomical Research and Education in Tajikistan

Pulat B. Babadzhanov

Institute of Astrophysics, Dushanbe 734042, Tajikistan. e-mail: pulat@astro.tajik.net

Abstract. Astronomical researches in Tajikistan are carried out by the Institute of Astrophysics, Tajik Academy of Sciences. The main scientific fields of investigation are physics and dynamics of asteroids, comets and meteors, variable stars and observations of artificial Earth satellites. The Institute has three observational stations: the Gissar observatory (at an altitude 730 m above sea-level) with photographic fireball and meteor patrols, a 70-cm reflector, a 40-cm Zeiss astrograph, a high-precision astronomical telescope (D=1 m, F=0.75 m), etc.; the Sanglokh observatory (at an altitude 2300 m above sea-level) with a Ritchey-Chretien 1-m telescope and 0.6-m reflector by Carl Zeiss; the Pamir observatory (at an altitude 4350 m above sea-level) with a 70-cm reflector. In 1999 the Department of Astronomy was restored in the Tajik State National University and the first students were admitted to this university for the astronomical profession.

1. Introduction

From the most ancient times astronomy has occupied an important place in the culture of the peoples of Central Asia. *Avesto* contains the earliest items of information on the astronomical knowledge of the Central-Asian peoples. In the ninth to eleventh centuries in Central Asia, alongside mathematics, geography, medicine, philosophy and literature, astronomy flourished greatly. In the works of Aburayhan Beruni (973-1048), the greatest scientist of the tenth to eleventh centuries, detailed evidence of the general astronomical knowledge of the peoples of Central Asia of his period is given. The works of Aburayhan Beruni, Abu Ali ibn Sino (Avicenna, 980-1037), Abumahmud Khujandi (died about 1000), the designer of the sextant (the unique astronomical instrument of that epoch), Omar Khayyam (1048-1131), author of an original solar calendar (and famous poetry), Nasreddin Tusi (1201-1274), Mirzo Ulughbeg (1394-1449) and many talented scientists of the East, were an invaluable contribution in development of world astronomical science. The special development of astronomy in Central Asia was connected with creation of the famous astronomical observatory of Ulughbeg in Samarkand and with the formation of the Samarkand scientific school, where the greatest scientists, such as Qhozizodeh Rumi (1360 - 1437), Jamshed Koshi (died in 1430), Alauddin Ali Qhushchi (died in 1474), and others, worked actively. After Ulughbeg's death astronomical research in Central Asia gradually declined, but the ideas and traditions of the Samarkand scientists were continued by Ali Qhushchi and his pupils in Iran and Turkey. Modern astronomy in Tajikistan began in the 1930s, after the completion of the work of the Tajik-Pamir complex expedition (1928-1932) which recognized the favorable astroclimatic conditions and advantageous geographical location of Tajikistan. In 1932 the Tajikistan government accepted the decision to organize a Tajik astronomical observatory (TAO) and authorized its construction in the outskirts of Dushanbe. The principal directions of scientific research for the Observatory, namely, meteors, comets, and variable stars, were chosen taking into account the geographical location and climatic conditions of Tajikistan. The Observatory was to investigate the astroclimate of other parts of Tajikistan in order to choose the best site for a future high-mountain observatory. These directions, alongside others, remain as the principal ones to this day.

In 1948, TAO was renamed Stalinabad Astronomical Observatory (SAO). In 1958, on the base of the Observatory, the Institute of Astrophysics of the Tajik Academy of Sciences was created. It consisted of three departments:

Department of Meteor Astronomy,

Deparment of Comets,

Department of Variable Stars.

Afterwards the

Department of Theoretical Astrophysics (1962),

Laboratory of Experimental Astrophysics (1972), and

Department of Astrometry (1975)

were created.

Some 30 years after the creation of the Institute, three of its modern observational bases were built:

1. In 1963-1971 the Gissar Astronomical Observatory (GissAO) was built at a distance of 14 km south-west of Dushanbe. Its domes house: a 70-cm reflector supplied with electron-optical, electrophotometric and polarimetric receiving apparatus, intended for observations of variable stars and comets; a 40-cm Zeiss astrograph for observations of asteroids, comets and variable stars; a 20-cm refractor and double astrograph. The observatory is also equipped with a high-precision astronomical camera (D=1m, F=0.75m) for photographic observation of artificial Earth satellites, comets and meteors. One of largest photographic meteor patrols is established in Gissar observatory. It consists of 16 wide-field cameras MK-75 (F=75cm D/F=1:3.5) and 24 cameras MK-25 (F=25cm, D/F=1:2.5). Observations of meteors are carried out also with TV and fireball cameras equipped with "fish-eye" lenses.

2. Sanglokh Observatory, the construction of which was completed in 1980, is located south-west of Dushanbe at a distance of about 90 km. It was built at the top of Sanglokh Mountain, the astroclimatic conditions of which have been widely recognized (Table 1). A Ritchey-Chretien 1-m telescope equipped with a spectrograph, "UAGS", and photometer-polarimeter were set up at this Observatory. A 60-cm Zeiss telescope is being assembled.

3. Pamir High-Mountain Observatory, the so-called "Solar ground-based astronomical complex Pamir" (situated at an altitude of 4350 m above sea level and enjoying 250 clear nights per year). It used to belong to the Main Astronomical Observatory (Pulkovo) of the Russian Academy of Sciences and was nationalized in 1991. It is located in the Murgab district (East Pamir) of the

Figure 1.   Gissar Observatory: The 1-m astronomical camera

Table 1.   Astroclimatic Characteristics of the Astronomical Observatories of Tajikistan and some other Countries

Observatory	Longitude	Latitude	Height[1]	N[2]	A''[3]	Op[4]
Sanglokh	69.0	38.2	2300	1700	0.54	0.78
Pamir	74..	38..	4350	1820	0.54	0.86
GissAO	68.6	38.5	730	1620	2.10	0.72
La Silla	-70.7	-29.3	2347	2100	0.60	0.78
Crimea	34.0	44.6	600	1219	2.12	0.73
Kitt Peak	-111.6	32.0	2120	2100	1.20	0.78
Palomar	-116.9	33.4	1706	1706	2.00	0.80

[1] in metres above sea level; [2] number of astronomicla hours per year; [3] astroclimatic index (image quality in arcsec); [4] optical transmission factor of the atmosphere.

Figure 2.   Sanglokh Observatory: Dome of the 1-m telescope

Gorno-Badakhshan Autonomous Region of Tajikistan. There a 70-cm telescope RT 700 (with Cassegrain optical system) and a solar telescope are installed. Pamir observatory, with its unique astroclimate, is an excellent long-term site for astronomical submillimeter, IR and optical observations. There the average annual precipitation is less than 100 mm per year; monthly average wind-speed is about 6 m sec^{-1}; the brightness of the night sky in the V-band is less than 22 mg sec^{-2} (Kanaev et al., 1983). The basic astroclimatic characteristics of the astronomical observatories of Tajikistan are given in Table 1 in comparison to the data for some other observatories (Shevchenko, 1972).

Below we list some important scientific results obtained in the Institute of Astrophysics, Tajik Academy of Sciences, lately.

2. Investigation of Meteors and the Earth's High Atmosphere

The nature of meteoroids and the phenomena accompanying their flight in the Earth's atmosphere, the atmospheric trajectories, meteor radiation and ionization, the heliocentric orbits of meteoroids, distribution of meteor matter in the near-Earth space and, finally, the origin and evolution of meteoroid streams and meteor showers, are problems in the area of interest to scientists of the Institute of Astrophysics of the Tajik Academy of sciences.

Long-term complex (photographic, spectrographic, radar and TV) observations of meteors give extensive information about radiants and velocities, masses and densities, and orbits of many thousands of meteors. On the base of these observations the physical parameters and wind mode in the meteor zone of the Earth's atmosphere (heights 60km-120km) have been determined; the contribution of meteors to the ionization of the high atmosphere and the influx of meteoric matter on to the Earth, Venus and Mars are studied; new meteor associations (i.e. groups of meteors with close orbital elements and radiants) have been revealed (Babadzhanov et al. 1990). From the results of photographic observations of meteors made from Dushanbe, catalogues of their orbits were compiled, which have been included in the data base of the IAU Meteor Data Center in Lund (Sweden).

In 1964-1965, in Dushanbe, a method of "instantaneous exposure" was devised. This method allows us, for the first time in the world, to obtain meteor photographs with exposures of 0.00056s. and, so, to establish the fact of fragmentation of meteoroids in the atmosphere. Later this method was used to obtain "instantaneous" meteor spectra as well (Babadzhanov, 1994).

Simultaneous photographic and radar observations of meteors, which have been carried out at the Gissar observatory, allow the specification of the correlation between intensity of meteor radiation and electron line-density of its train. According to radar observations of hundreds of thousands of meteors the spatial distribution of air flow in the meteor zone of the atmosphere and the processes of deionization of meteor trains were investigated.

During two years, (1968-1970), the Soviet Equatorial Meteor Expedition organized by the Institute of Astrophysics of the Tajik Academy of Sciences and the Kharkov Politechnic Institute worked in Somalia (East Africa) and, as results, the radiants and orbits of meteors of both the northern and southern hemispheres and the wind modes of the high atmosphere of the Earth's equatorial area were determined. Processing of the observations obtained has provided fundamental astronomical and geophysical results, namely, a value for the influx of meteoric matter onto the Earth, the regularity of global circulation of the Earth's atmosphere at the heights of 80-120 km, and a catalogue of orbits of several thousand faint meteors (Babadzhanov et al. 1974).

In the activity of scientists of the Institute of Astrophysics during the last 12 years, an important place was taken up by the large-scale investigation of the evolution of meteoroid streams and short-periodic near-Earth objects (comets and asteroids). As a result, a completely new conception of the evolution of both the shape of the meteoroid streams and the dynamics of meteor showers has been obtained. It was shown that the number of meteor showers produced by a comet depends on the Earth-crossing class of the cometary orbit, and that the majority of short-periodic comets are quadruple crossers of the Earth's orbit and, hence, each of them might produce four meteor showers. So, it was shown, that according to "meteor signs" a number of near-Earth asteroids, representing a danger to the Earth, are extinct cometary nuclei (Babadzhanov, 1998). Basic principles of meteoroid-stream formation and evolution developed in Dushanbe have served as the basis for a large series of scientific works carried out in other countries in recent years.

3.   Cometary Investigations

The physics of comets is another important direction of scientific research at the Institute of Astrophysics, Tajik Academy of Sciences. These researches cover all sections of cometary physics and extensive observational and experimental material on comets has been obtained there. In 1955, A.M. Bakharev discovered a comet which became known as the Bakharev-McFarlane-Crinke comet. The expansion of the observational base and the use of telescopes installed in Gissar and Sanglokh observatories allowed spectroscopic, photoelectric, and polarimetric observations of comets to be carried out. In the 1950s a mechanical theory of cometary forms was developed, and problems of the interaction of corpuscular solar streams with cometary atmospheres have been investigated. A theory of plasma cometary tails was developed and non-stationary processes in comets have been also studied. It was shown that many such processes, for example cometary outbursts, are associated with solar activity (Dobrovolsky, 1966).

Our scientists have also developed a theory of the disintegration of cometary nuclei. The disintegration of cometary nuclei modeled with H_2O, CO, NH and CH ices, and the snow model of a cometary nucleus, have been studied at the institute. The heat regimes of such nuclei were estimated. Theoretical data were tested under laboratory conditions which simulated the cosmic ones. The process of disintegration of dust matrices, the velocities of ejection of matrix fragments, and so on were studied in the laboratory. It was established that the nuclei of some short-periodic comets develop a refractory crust and they might evolve into asteroid-like bodies (Ibadinov, 1993).

Polarimetric and colorimetric investigations of comets have revealed negative polarization of the continuum radiation of comets at small phase angles (angle Sun-comet-Earth). Non-linear growth of cometary brightness in the region of small phase-angles has been found ("oppositional effect"). The bases of the theory of shock waves arising from the interaction of magnetized corpuscular solar wind with cometary plasma was worked out. The interaction of interplanetary dust particles with dust particles of the cometary atmosphere are being investigated and, a mechanism of X-ray radiation of comets and the occurrence of multicharged ions is proposed (Ibadov, 1996).

The Institute of Astrophysics of the Tajik Academy of Sciences actively collaborated in working up and in the realization of the International Halley Watch programs. Astrophysical and positional observations of a number of asteroids have been carried out and the orbital evolution of some Near-Earth asteroids associated with meteoroid streams, including the Taurid Complex asteroids, are being investigated.

4. Astrometric Tasks

In connection with the precise determination of the coordinates of the celestial objects and phenomena (meteors and comets, stars and asteroids, etc) astrometric methods were widely used and developed in the Institute. Furthere development of astrometic work in Dushanbe was connected with the launching of the first artificial Earth satellite (AES) (1957) and observation of such objects. A high-precision astronomical camera of the Gissar Observatory allows the making of photographic observations over a wide range of visible speeds.

5. Investigation of Non-Stationary Stars and Galaxies

Observations of variable stars in Tajikistan were made from the first days of the astronomical observatory in Dushanbe. Over the past years a "Sky Survey" has been created, unique photo-archives consisting of almost 70,000 sky negatives. From these some novae and more than 100 variable stars have been discovered in T-associations. In addition, some features of light-curves, the variability of periods of variables of different types, and oscillations in brightness of novae were studied. With the coming into operation of the Sanglokh 1-m telescope, photometric, polarimetric, and spectrographic observations of faint objects, extremely young stars, such as T Tauri and Ae/Be Herbig types, began to be made. Carrying out simultaneous polarimetric and photometric observations in different colors on some stars of the T-Tauri type, scientists of the Institute study physical conditions in the gas-dust shells of these stars (Borisov & Minikulov 1997, Borisov & Redkina, 1997; Redkina & Zausaeva, 1997; Kiselev et al., 1995).

Theoretical investigations of the dynamical phenomena of the collective gravitational interactions of stars in galaxies have been widely developed at the Institute. Investigation of star formation in galaxies has shown that spiral waves of density in galaxies were active in producing not only the features of motion of stars and interstellar gas, but also the peculiarities of late star-formation processes.

6.   Publications

Since 1952, the Institute has issued the *Bulletins of the Institute of Astrophysics* of the Tajik Academy of Sciences. Up to the present time, 83 numbers have been issued. From 1958 till 1990 the Institute also issued the Soviet journal *Comets and Meteors*, up to its 40th number. Besides 12 monographs and thousands of scientific papers by the Institute scientists have been published in various places.

7.   Personnel and Training of Specialists

The disintegration of the U.S.S.R., the social-political instability during 1991-1997 and the economic crisis in Tajikistan negatively affected the functioning of the Institute of Astrophysics: some buildings and structures of its Observatory were damaged seriously; in particular, the structures of Sanglokh Observatory. Now, none of the telescopes are supplied either with new radiation detectors (e.g., CCDs), or with old detectors (photographic plates and films). The scientific and technical personnel of the Institute have also suffered serious changes. Many highly qualified research workers have left Tajikistan, or transferred to more well-paid positions. Now, out the Institute's staff of 70, 20 are scientists, including four with the Dr.Sci degree and seven with the Ph.D. degree. Therefore, the Institute undertakes measures to restore its personnel.

Until recently there were two types of curricula for training astronomers: (1) in 1972-1988, an "astrophysics" specialization within the "Physics" curriculum in the Tajik State University; (2) beginning 1986, a "Physics and Astronomy" curriculum for schoolteachers in the Dushanbe Teachers' Training University.

On the initiative of the Institute of Astrophysics in 1999, a specialty of "Astronomy" and a Department of Astronomy were organized in the Tajik State National University. Twenty first-year students have finished the 1999-2000 academic year. The Institute of Astrophysics (and its observatories) has become the base research establishment for this specialization and expects an influx of capable youth into astronomy.

It is proposed to use the multistep system (Bachelor, Master, postgraduate studies) for training astronomers in Tajikistan. The structure of the astronomy bachelor curriculum is shown in Table 2.

Table 2.   Astronomy Bachelor Curriculum

Series of Subjects	Planned hours
General humanitarian, economic and social disciplines	984
Mathematical and natural sciences disciplines	1416
General vocational and special disciplines	2797
Total	5197

General vocational and special disciplines consist of the following courses:
General astronomy
General astrophysics
Practical astrophysics

Theoretical astrophysics
General astrometry
Celestial mechanics
Physics of the Sun and planets
Small bodies of the Solar system
Physics of stars
Physics of the atmosphere
History of astronomy
Galactic and extragalctic astronomy
Cosmology
Radioastronomy
Special laboratories and Practical astronomy

Now, for the training of personnel we need astronomical educational materials, such as sky atlases and maps, astronomical softwares, computers, etc.

8. Prospects

The prospect for astronomical research in Tajikistan depends mainly on the existing scientific and technical base, on training highly skilled scientific and engineering staff, on qualitative improvement of the scientific and technical base and on normal financing of the Institute.

In the next five years in the Institute and its observatories complex research on physics and dynamics of the Near-Earth Object (asteroids, comets, meteoroid streams and meteor showers), investigation of non-stationary stars etc. is planned for the Institute. The program of scientific research at Sanglokh Observatory, after its repair, envisages the following tasks:

- detection and investigation of extremely faint variables, members of T-associations in the Galaxy and Cepheid variables in M31;

- complex photometric, polarimetric and spectroscopic investigation of the gas-dust shell of non-stationary stars of the T-Tauri type;

- observations of comets, near-Earth asteroids, meteor showers and fireballs.

The Institute has experience of work on international scientific programs and projects. In particular, important results were obtained there for the programs of the International Geophysical Year (IGY), the International Year of Quiet Sun, the IHW, etc. Together with Russian and Slovak astronomers a catalogue of orbital evolution of short-period comets was created. The Institute of Astrophysics of the Tajik Academy of Sciences and its observatories are open for scientific collaboration with scientists of other countries.

References

Babadzhanov, P.B. 1994, *Asteroids, Comets, Meteors* 1993, 45-54.
Babadzhanov, P.B. 1998, Cel.Mech. & Dynam.Astronomy, 69, p. 221.

Babadzhanov, P.B., Bibarsov, R.Sh., & Kolmakov, V.M. 1990, *Mass Distribution and the Flux of Sporadic Meteoroids*, Moscow, 56 pp.

Babadzhanov, P.B., Kashcheyev, B.L., Nechitaylenko, V.A., & Fedynskiy, V.V. 1974, *Radar-meteor investigation of the high-atmosphere circulation*, Publishing house "Donish", Dushanbe, 172 pp.

Borisov, Yu.V. & Minikulov, N.K. 1997, Bull. Inst. Astrophys., Tajik Acad.Sci., No 83, p.19.

Borisov, Yu.B. & Redkina N.P. 1997, Bull. Inst. Astrophys., Tajik Academy of Sci., No 83, p. 3.

Dobrovolsky, O.V. 1966, *Comets*, Moscow, Nauka, 288 pp.

Ibadinov, Kh.I. 1993, Doklady Academii nauk Resp. Tajikistan, XXXVI, No 3, p.182.

Ibadov, S. 1996, *Physical Processes in Comets and Related Objects*, Moscow, Kosmoinform Publ. Comp., 181 pp.

Kanaev, I.I., Sholomitsky, G.B., Maslov, I.A., & Grozdilov, V.M. 1983, Review of Science and Technics (Series of "Astronomy"), Vol. 22, p. 286.

Kiselev, N.N., Minikulov, N.K., & Chernova, G.P. 1995, Bull. Inst. Astrophys., Tajik Acad. Sci., No 82, p. 42.

Redkina, N.P. & Zausaeva O.G. 1997, Bull. Inst. Astrophys, Tajik Acad. Sci., No 83, p. 7.

Shevchenko, V.S. 1972, *Young Star Complexes*. Astroclimat, Publ. house "Fan", Uzbekistan, p. 84.

Discussion

In answer to a question, Babadzhanov stated that salaries of astronomers in Tajikistan are not in arrears.

Work at Bosscha Observatory

Moedji Raharto

Bosscha Observatiry, Lembang 40391, Java, Indonesia. e-mail: moedji@sirius.as.itb.ac.id

Abstract. An overview is presented of work at Bosscha Observatory during the last ten years. Future developments are also discussed.

1. Introduction

Bosscha Observatory is one of the oldest observatories in Southeast Asia. It belongs to the Republic of Indonesia and operates under the Bandung Institute of Technology (ITB), Department of Education and Culture. The Observatory is located at a hill in Lembang with altitude 1310 m above sea level, geographical latitude -6° 49' and longitude of 107° 30' East (or $7^h 10^m$ East); it is north of Bandung. The Observatory was founded in 1923, so it is 22 years older than the Republic of Indonesia. In the period 1923-1958 the Dutch astronomers contributed work at the Observatory and Indonesian astronomers have continued to develop the work since 1958. The population of the country is more than 200 millions and the area is about 2 million square degrees, spread over geographical longitude of 95° E to 140° E and latitude of 5° N to 10° S; one third of the area is land and the rest is sea. For more detail about astronomy in Indonesia see Hidayat (1996) and van Albada-van Dien (1995).

ITB has three missions in higher education: to develop scientific and technical knowledge, to transmit such knowledge and to integrate it in society. Within the mission of ITB, Bosscha Observatory concentrates on contributing to the development of astronomy, but the Observatory also provides facilities for astronomical education, as well as facilities for education in science and technology.

An additional job for staff members at the Bosscha Observatory is to give popular lectures for visitors. "Astronomy for non-astronomers" may be an appropriate term. How we explain astronomy in popular ways depends on the educational level and background experiences of the visitors.

2. Contribution to Astronomy

The main instrumentation consists of 60-cm Zeiss double-refractors (focal lengths 10.78 m (in visual wavelengths) and 10.72 m (in photographic), a Schmidt telescope (51/71/127 cm), a GOTO Cassegrain (45 cm aperture and 54 m focal length), the 37-cm Bamberg refractor with a focal length of 7 m and a Unitron 10.2-cm refractor with a focal length of 1.5 m.

The Zeiss double-refractors (visual and photographic) have been extensively used for observation of visual double stars. The visual telescope is dedicated to observing visual double stars with photographic plates. About 9000 plates of astrometric images of visual double stars with angular separations in the range of 2″-20″ have been obtained through the visual telescope. The recent published data for visual double stars can be found in Jasinta et al. (1994, 1999), Jasinta and Hidayat (1999).

Observations of Mars have also been made with the visual telescope equipped with a planetary camera during the Mars opposition period in 1995 (Iwasaki et al., 1995).

In addition, the 37-cm Bamberg refractor has been used for photometric observations of variable stars (Malasan, 1996) and recently has been used for public education by showing lunar craters, planets and stars. After a new GOTO 45-cm Cassegrain telescope was installed in 1989, it was used for practically all the photoelectric photometry, see, for example, Okazaki et al., (1995). The 45-cm Cassegrain telescope is a computer-controlled telescope donated by the Japanese government donated under its cultural agreement. The telescope is equipped with photometer (Johnson UBV) and spectrograph.

The smaller Unitron refractor is suitable for lunar and solar observations. Imaging of eclipses of the Sun and Moon is also among its targets. The telescope is used for studying the proper motion of sunspots by some students (Soegiyono 2000, Herdiwijaya et al. 1997).

In 1958, UNESCO donated a Schmidt telescope with a size of 51/71/127 cm and field of view of $5° \times 5°$. The telescope is equipped with 6°-prism with a dispersion of 187 nm mm^{-1} at the A band (λ 759.4 nm). The survey telescope (plate scale 162″ mm^{-1} and f-ratio of about 2.5) is used for the M-star survey (Ichikawa et al., 1982). Observations of emission-line stars, of bright novae such as V382 Velorum, and astrometry of comets were also made with photographic plates (Hidayat et al., 1999 and Ikbal et al., 1995)

Several astronomical meetings have been held in Indonesia, for example IAU Colloquium No. 80 in 1983 on the occasion of the sixtieth anniversary of Bosscha Observatory (Hidayat et al. 1984), an IAU Symposium on Stellar Photometry and Spectral Classification in 1963 and IAU Symposium 143 on Wolf-Rayet Stars in 1991, the second IAU Regional Asia-Pacific Meeting on Astronomy in 1981 and IAU Schools for Young Astronomers in 1963 and 1983. Most recently, IAU Colloquium No. 148 on The Future Utilization of Schmidt Telescopes was held in Bandung in 1994 (Chapman et al. 1995).

3.   Some Notes on Maintenance and Development

Staff from the department of astronomy in ITB contribute to the administrative work of the Observatory as well as doing some astronomical research with the Observatory's facilities. Economical and political crises began in Indonesia in 1997. The direct psychological impact of the crises made the younger generation think about survival and forget the development of science. It is not easy to find bright members of the younger generation willing to become astronomers.

The uncontrolled prices and undervalued currency have made life harder. The currency is stabilizing now at about one-third of its previous value, so many things are about three times more expensive than they were before the crisis. Nobody thought about the serious impact of the economical and political crises on astronomy. Direct impacts will be on maintenance costs; the government cut our budget because most of the budget is now devoted to a great effort to help students, with scholarships, to continue their studies. The second direct impact comes from adjustment of the currency, which implies higher prices and higher operating costs in terms of the domestic currency.

It was good fortune for Bosscha Observatory that some supporting units, a PC computer network for Internet and a ST6B CCD camera were donated by a well-known company, Schlumberger, in 1999. The CCD camera is used to observe visual double stars with the Zeiss double refractor, as well as for stellar photometry. The continuation of support from LKBF for astronomical journals keeps the library up to date and the Internet connection runs well. We face a general problem that there is no budget for research and development; I hope that in the future a rich man or company will come to help.

Our experiences show that it is necessary to have a dedicated optical-electronic engineer for astronomy as well as some budgetary provision for maintenance costs. At present, such human resources are less available, although there may be such people in other Departments (Physics, Engineering Physics, Electrical Engineering and Mechanical Engineering) in the Bandung Institute of Technology. Friendly relations with these Departments enable us to get help to identify or to solve instrumentation problems, but the time that the expert has avaialable may not be enough and weak commitment will be the result. Research and development, the backbone of the institution may stall or may not be well planned because of the lack of engineer and of an environment in which astronomers and engineers work together. We are not well prepared to maintain modern facilities like the GOTO telescope. After five years we face many problems with that telescope; some can be identified and solved by replacement with new material, but now the situation has become serious because the telescope-control system does not work. Photometric observations of dwarf novae with the GOTO telescope equipped with a CCD ST6B photometer, reported by Kunjaya (2000), and spectroscopy were in progress.

Work on astrometry, photometry, imaging and spectroscopy will be transferred from photographic plates to electronic detectors. It is difficult to set up a competitive program with a small telescope. Recently Le Poole (2000), a guest visitor from Leiden Observatory, proposed simultaneous spectroscopic work using the Bosscha Schmidt telescope with fiber optics.

Access to a new telescope is always an issue for Indonesian astronomers, as for other astronomers in Southeast Asia. Swarup (1981) and Swarup, Hidayat and Sukumar (1984) proposed collaboration on GERT (Giant Equatorial Radio Telescope). Politically, the proposal had been endorsed and supported by the Indonesian government, but now this project remains a plan; we have learned that without international collaboration, the Indonesian government will not be able to provide the whole budget needed.

Van der Hucht (1984) explored the possibilities of a plan for a new generation 2-m class telescope in Indonesia under the umbrella of INA (Indonesia-

Netherlands Astrophysics). The light-pollution level at Bosscha Observatory is about 4 mag brighter than the dark sky (Malasan et al., 2000). A new site in Indonesia will be sought for the new telescope of 2-3 m size and joint efforts among ASEAN countries should be important for funding the project.

International cooperation and help for the young staff who continue their studies in astronomy and astrophysics are also important for the future of astronomy in Indonesia. We are optimistic that we shall survive for the next century, with different features.

4. Exploring Public Awareness for an Active Role in the Development of Astronomy and Space Science

The library is also one of the sources for research and education in astronomy in Indonesia. This library consists of the main astronomical journals and can be accessed by students in the Department of Astronomy. In addition, computer facilities, PCs with Internet connections, are used also for education. Basic observational science may be introduced through the telescopes. Students from various backgrounds, like architecture, cinema and art, often consult us about their thesis work concenred with space-science museums, observatories etc. In July 2000 we organized a month of training for 20 people from Indonesia, Brunei and Malaysia, to enhance astronomical knowledge of the Islamic calendar system. The Department of Religious Affairs in Indonesia sponsored that activity.

The informal Colloquium held at the Bosscha Observatory is a kind of melting-pot among scientists with different fields of interest. These activities are also an interactive process among scientists, as well as a process to educate new young scientists.

The potential synergy between space and ground-based astronomy will push frontiers of knowledge in astronomy and space science further. For example Raharto (1996) using optical data for M-type stars, obtained from a near-IR survey with the (relatively small) Bosscha Schmidt Telescope and from IRAS (InfraRed Astronomical Satellite) point sources, discovered optically luminous IR (InfraRed) sources which can be identified as M-type supergiants with large IR excesses, or AGB stars. Homogenous IR photometry of IR point sources with a high degree of completeness in almost all directions of the sky were obtained by IRAS in 1984. The data open the opportunity to study the overall structure of the Galaxy. The experience of the successful IRAS mission thus led to progress in space science and IR astronomy.

Many things remain to be explored from space and space exploration (for example, planetary or cometary exploration) needs more long-term investments and persistence of activity in research and education through several generations. It is necessary to enhance the impact of new knowledge and progress on the wider society. It is impportant to have local people always available to lecture in the local language and this warrants our continuing research activity. Human resources can be cultivated through higher education in astronomy, space science and related subjects. The total number of students in the Department of Astronomy is often directed by the availability of jobs in astronomy or space science, which is usually very limited. In Indonesia, graduates work in the satellite division or the computer-analysts division of the Aircraft Company

IPTN, in Planetarium Jakarta, Aeronautic and Space Science (LAPAN), the Department of Astronomy in Indonesia or abroad etc. Some graduates work as bankers or journalists or in other areas. In Indonesia, the intake of students for Department of Astronomy is very small (for example, the intake of students for the undergraduate course in the Department of Astronomy in Bandung Institute of Technology is about 15 among the total intake at ITB of 1250 (compared to a total Indonesian population of 215 millions, spread over 2 million km^2 in the equatorial region). The small number of students makes it difficult for the Department to be self-sustained; some additional funds are necessary to pay for the courses offered. The economical crisis, that began in 1997, made the situation worse. It became literally a matter of survival to fight critics of the budget for continuing astronomical research and education, confronted as we were by the basic needs of life, to increase of live salaries and to support education in other fields. A new concept of astronomical education is needed, that includes basic education in space science and recognizes that graduates can also fill the need for trained people with expertise in related areas of high technology. If that is accepted, then the solution to the problem of preparing the next generation of astronomers and space scientists could be found, through preparation of human resources for modern society. From the economic point of view only, it is hard to develop astronomy and space science in an underdeveloped country. The development of astronomy and space science in Indonesia will take longer due to these additional problems arising from the economical and political crises.

Another way to develop astronomy and space science may be through promoting the need for, and benefits, of space science for the future. It is frustrating to have to advertise astronomy and space science frequently but, in the long run, such a campaign may influence the opinion of politicians and decision-makers so that they make better provision for the development of space science. Unfortunately, the budget for high technology in Indonesia was recently cut, the reason given being that high technology spends too much money and the public does not get enough direct benefit from these activities. The politicians and decision-makers do not have enough time to discuss and to see even a little into the future. It is not easy to erase a wrong image of "luxurious" space science with little benefit to society. This situation is a kind of bottleneck in attempts to make a synergy between society, government, industry and education or research in space science.

Astronomy is closely related to space science; all astronomical objects are in outer space and observation of astronomical objects from space has opened a new horizon of astronomical knowledge. Nowadays ground-based and space-based astronomy and astrophysics are very well developed in modern countries while, on the contrary, in underdeveloped countries they languish. This situation makes a big gap, in the activities of space exploration and dissemination of space science, between developing and developed countries. This gap gives the impression that astronomy and activities of space exploration are luxurious activities, and can only be a dream for people in underdeveloped countries. I suggest that we need to emphasize a policy that the worldwide progress of space science should also support existing bodies that are still struggling to survive and to develop in their difficult circumstances. Collaboration between countries in educational activities, exchanging knowledge and data, would be worthwhile and would overcome some of the difficulties of underdeveloped countries.

The gap in knowledge between society and expert space-scientists becomes wider, for example, in the field of space-technology as more sophisticated and advanced technology, harder to understand, is used. The results of HST (Hubble Space Telescope), the Voyager and Apollo missions and other space missions are wonderful. How do we keep the public interest in them? How do we deepen the public interest so that people even begin to understand astrophysics? How do we give the public a sense that astronomy belongs to them? The public needs more information as the rate of progress of knowledge in astronomy and space-science accelerates. About 10,000 visitors come to Bosscha Observatory in Lembang every year and about 200,000 visit the Planetarium Jakarta. The visitors interact with astronomers through popular lectures on astronomy. These potential resources should be used to move the role of public from passive to active participants in the mainstream of development, or at least of transmission of new knowledge in astronomy and space-science to society.

That more than 10,000 people visit Bosscha Observatory's program of "Astronomy for Non-astronomers" every year provides another chance that our society will contribute directly to the Bosscha Observatory.

Acknowledgments.   I would like to thank the IAU for providing a travel grant to attend the IAU GA 24 in Manchester. I appreciate very much the provision of hospitality during my stay in Manchester by the LOC of IAU GA 24. Finally I would like to thank Dr Alan Batten for valuable comments, suggestions and revision during editing the manuscript.

References

Chapman, J., Cannon, R., Harrizon, S. and Hidayat, B. (editors), 1995, *The Future Utilization of Schmidt Telescopes*, IAU Colloquium 148, ASP Conference Series No. 84.

Herdiwijaya, D., Makita, M., Anwar, B., 1997, PASJ vol. 49, 235-248

Hidayat, B., Kopal, Z. and Rahe, J. (editors), 1984, *Double Stars, Physical Properties and Generic Relations* Proceedings of IAU Colloquium No. 80 held at Lembang, Java.

Hidayat, B., 1995, in Chapman et al. (eds.) pp.

Hidayat, B., 1996, Journal of the Korean Astronomical Society 29, S455-S457

Hidayat, B., Ikbal, M., Wiramihardja, S.D., Raharto, M., 1999, *Astronomissche Gesellschaft Tagung*, Germany

Ichikawa, T., Hamajima, K., Ishida, K., Hidayat, B., and Raharto, M., 1982, PASJ 34, 231

Ikbal, M.A., Irfan, M., Raharto, M., Hidayat, B., 1995, Astronomische Nachrichten 317, 1581

Iwasaki, K., Sagar, R., Ghosh, K.K., Raharto, M. and Dirghantara, F., 1995, *Lunar and Planetary Symposium* 1-4

Jasinta, D.M.D., Soegiartini, E., 1994, Astron. Astrophy. Suppl. Ser. 107, 234-241

Jasinta, D.M.D., 1999, *Proceedings of Fourth EAMA, Yunnan Obs.*, pp. 235-238

Jasinta, D.M.D., Raharto M., Soegiartini, E., 1999, Astron. & Astrophy. Suppl. Ser. 134, 87

Jasinta, D.M.D., Hidayat, B., 1999, Astron. & Astrophy. Ser. 136, 293-295

Kogure, T. and Hidayat, B. (editors), 1985, *Galactic Structure and Variable Stars*, Proc. Of the Three Year Cooperation in Astronomy between Indonesia and Japan 1979-1984

Kunjaya, C., 2000, *Observation of Dwarf Nova* (in preparation)

Le Poole, R., 2000, private communication

Malasan, H.L., 1996, *Instrumentation and Research Programmes for Small Telescopes*, IAU Symposium vol. 118, J.B. Hearnshaw and P. L. Cottrell eds. pp. 303-304, Reidel, Dordrecht.

Malasan, H.L., Senja, M.A., Hidayat, B. and Raharto, M., 2000, *Preserving the Astronomical Sky*, IAU Symposium 196, R. J. Cohen and W. T. Sullivan eds., pp. 1-4, ASP Publications.

Okazaki, A., Mahasenaputra and Hidayat, B., 1995, Southern Stars, vol. 36,65

Raharto, M., 1996, *Study of Galactic Structure Based on M-Type Stars* (RON-PAKU - Thesis)

Soegiyono, 2000, (a thesis for Sarjana Science degree in Indonesian)

Swarup, G., 1984, *Proceedings of Second Asian-Pacific Regional Meeting on Astronomy in 1981*, 22-31

Swarup, G., Hidayat, B. and Sukumar, S., 1984, in Proceedings of IAU Colloquium no 80 Double Stars, *Physical Properties and Generic Relations* held at Lembang Java edited by B. Hidayat, Z. Kopal, and J. Rahe, 403-407

van Albada-van Dien, E., 1995, in *The Future Utilization of Schmidt Telescopes*, IAU Colloquium 148, 15-18.

van der Hucht, K.A., 1984, in Proceedings of IAU Colloquium no 80 *Double Stars, Physical Properties and Generic Relations* held at Lembang Java edited by B. Hidayat, Z. Kopal, and J. Rahe, 409-410

Discussion

Orchiston asked if Indonesian astronomers had considered trying to establish collaborations with astronomers in other countries who could provide access to telescopes or data. Alternatively, had Indonesian astronomers thought of applying for service observing (which would not require their presence at the observatory) on, for example, the 3.0-m Anglo-Australian Telescope?

Rijsdijk suggested that, rather than training local people to use the Indonesian telescopes, those people sholud be sent abroad for training so that they could bring expertise back to Indonesia. With that added expertise, they could think of looking for new sites and telescopes. They could not expect foreign aid for sustaining their astronomy program to continue indefinitely.

Hearnshaw asked if there were any good dark sites in Indonesia, for telescopes in the 2m to 3m range. Raharto repeated that the sky brightness at Bosscha Observatory is about 4 magnitudes brighter than the darkest skies but he thought that there were several dark sites in Indonesia, for example in northern Sumatra, close to Toba Lake, or in the eastern part of the country, close to Timor Island. He thought that a telescope in the 2m to 3m range was very necessary for the future of astronomy in Southeast Asia.

Kitamura asked if there was sufficient support by technicians from other Departments. Raharto replied that there was no fixed arrangement for the Observatory to use technicians from other Departments and those who came were more committed to the institutions in which they worked. Observatory projects tended, therfore, to have low priority.

Astronomy in Venezuela

Patricia Rosenzweig

Universidad de Los Andes, Facultad de Ciencias, Departamento de Física, Grupo de Astrofísica Teórica (GAT), Mérida, Venezuela.
e-mail: patricia@ciens.ula.ve

Abstract. Since the installation of the Observatorio Cagigal in Caracas, astronomy in Venezuela has developed steadily, and, in the last few decades, has been strong. Both theoretical and observational astronomy now flourish in Venezuela. A research group, Grupo de Astrofísica (GA) at the Universidad de Los Andes (ULA) in Mérida, started with few members but now has increased its numbers and undergone many transformations, promoting the creation of the Grupo de Astrofísica Teórica (GAT), the Grupo de Astronomía, the Centro de Astrofísica Teórica (CAT), and with other collaborators initiated the creation of a graduate study program (that offers master's and doctor's degrees) in the Postgrado de Física Fundamental of ULA. With the financial support of domestic Science Foundations such as CONICIT, CDCHT, Fundacite, and individual and collective grants, many research projects have been started and many others are planned. Venezuelan astronomy has benefitted from the interest of researchers in other countries, who have helped to improve our scientific output and instrumentation. With the important collaboration of national and foreign institutions, astronomy is becoming one of the strongest disciplines of the next decade in Venezuela.

1. Introduction

The history and development of the astronomy in Venezuela has been reviewed by some authors in Spanish (Stock, 1981; Olivares, 1986; Hubschmann, 1988; Quintero, 1989; Chalbaud, 1990). However the reader is referred to Inglis (1961) and Alvarez and MacConnell (1978) who give partial accounts in English. On the basis of these works, a brief account will be given in this short communication.

It can be said that astronomy began in Venezuela when the first observatory of the country was installed in Caracas in 1888 by Juan M. Cagigal (Hubschmann, 1988), today known as Cagigal Observatory after its founder. At that time, Caracas was a suitable site for an observatory, not only because the minimal amount of pollution and light contamination provided appropriate atmospheric conditions, but also because the city is the capital. Since then, this observatory has been in operation, serving mainly for astronomical observations, legal time, and, in addition, for meteorological measurements. From 1888 to 1958 the observatory was under the administration of the Ministry of Education; since then and up to the present, it has been under the administration of the Venezuelan Navy.

To improve the Observatory's equipment, in 1953 it was proposed to buy sophisticated telescopes from the Askania and Zeiss companies. At that time, the Director of the Observatory was Eduardo Röhl who was the principal supporter of the Observatory's development. The major instruments were the 1.5-m Schmidt camera, the 1-m coudé-Cassegrain reflector, a refractor of 65-cm aperture (10.5 m in length), and the double astrograph (each of 50-cm aperture). Unfortunately, these instruments could not be installed in Cagigal Observatory because light contamination and atmospheric pollution in Caracas began to be a major problem from the 1960s onwards. Instead, it was recommended to locate these instruments in another observatory to be built in a suitable place with better conditions. Two major factors contributed to modify this initial project, namely, the death of Röhl in 1959 and the political situation of the nation, which led to a change in the government system in 1958.

The construction of a second and bigger observatory was necessary to install the main instruments mentioned above. In 1962 a commission of AURA (Edmondson, 2000) recommended the installation of this new observatory in the Venezuelan Andes, specifically in Mérida State, taking into account two major factors: (i) the Mérida sky presents good seeing compared to other places in the country, and (ii) the logistical support of a large university (the Universidad de Los Andes) that could assist in the creation of astronomy and astrophysics studies in cooperation with the observatory.

It was not until 1970 that the construction of this second observatory in Venezuela began, operated this time independently from the Cagigal Observatory, and under the supervision and support of the National Council for Scientific and Technological Research (CONICIT). Based on this introductory historical review, the purpose of this paper, as mentioned before, is to give a report of the present status of Venezuelan astronomy and future national and international projects.

2. Present Status of Venezuelan Astronomical Research and Teaching

This second, modern observatory was built next to a small village in Mérida state called Llano del Hato, from which the name "Observatorio Nacional de Llano del Hato" was derived. In order to make a final decision, the Universidad de Los Andes (ULA) provided the premises for the storage of the instruments and construction of four telescope buildings. Also, ULA provided qualified personnel to examine whether or not the ground could hold the weight of the instruments mentioned (Stock, 1981). At the same time of the creation of the Observatorio Nacional de Llano del Hato, the Fundación Centro de Investigaciones en Astronomía "Francisco J. Duarte" (CIDA) was also created to manage the Observatory. On the other hand, in the Universidad de Los Andes, the birth of the Astrophysics Research Group within the Physics Department of the Faculty of Science took place.

This Group, consisting of students, instructors, and professors, worked closely with Dr. J. Stock, the Director-in-Chief of the Observatory project to complete the remaining work. With the purpose of starting a teaching program in astronomy, courses in different subjects of this discipline were given by foreign and national astronomers. Also, some theses were offered and consequently, in 1975, students began to obtain their undergraduate degrees in physics with a major in astronomy. Many of these students had to travel to other countries in order to obtain a third-level degree. When they returned, interests of several sorts emerged. Thus the study of theoretical astrophysics was considered to be of high interest. In this way, the original Astrophysics Group gave rise to another group that focused on this discipline (the Group in Theoretical Astrophysics, GAT). The interest grew more and more and a "critical mass" of well-trained specialists in astronomy stimulated the creation of graduate studies in astronomy. In 1991 the Graduate Program called "Postgrado en Astronomía y Astrofísica (PAAS)" was created and offered the master's degree. Then, in 1995 the doctoral degree was offered for the first time. Because this program included many specific areas, some of which were only marginally related to astronomy, and because it started to involve collaboration from other educational institutes of the same level, this graduate program changed its name in 1997 to Postgrado en Física Fundamental (PFF). This program comprises 21 professors from ULA and from other institutions. Since the creation of the PAAS (now PFF), 13 master's and 1 doctoral degrees have been awarded. Now the program has an enrolment of approximately 19 master's and 10 doctoral students.

Parallel to the creation of the Graduate Study Program, the creation of a Center of Theoretical Astrophysics (CAT) was established in the Faculty of Science of the Universidad de Los Andes (ULA). The Group of Theoretical Astrophysics (GAT), together with the CAT, involve the collaboration of approximately 22 members, not only from ULA but also from other universities and research institutes, such as Instituto Venezolano de Investigaciones Científicas (IVIC), Universidad Simon Bolívar (USB), and Universidad Central de Venezuela (UCV).

The lines of research are very diversified. Some of them, included either in the PFF, in the GAT, or in the CAT are, astrophysical plasmas, the interstellar medium, solar system, observational astronomy (spectroscopy, photometry, astrometry, etc.), relativistic astrophysics, quantum theory of fields, chaos and non-linear dynamics, chemistry and physics.

3. Ongoing Scientific Projects

Venezuelan astronomers have opened different collaborations with institutions abroad. These collaborations have also improved the instruments in the Observatory. For instance, the QUEST project is a collaborative effort of the Physics and Astronomy Departments at Yale University, Indiana University, CIDA and the University of Los Andes, to study gravitational lensing of quasars. At this time, QUEST has designed and built a large-area CCD mosaic camera with 16 CCDs, This camera has been installed on the Schmidt telescope of the Observatory and is now routinely taking high-quality data. Another important collaboration has been established with the University of Canterbury in New

Zealand. Specifically, the spectroscopic study of novae and supergiant stars has been based on spectra taken in the Observatory and the Mt. John University Observatory in New Zealand. Additionally, collaboration has been established with UNAM in México in which studies of Cepheid stars are based on photometric data from the Observatory and the San Pedro Mártir Observatory in México.

Many other collaborations with European observatories and research centers have also been established. For example, theoretical work in physics is being undertaken with the Abdus Salam International Center for Theoretical Physics. There are many more projects and all of them promising high-quality researche to be published in several scientific periodicals.

4. Future Perspectives

In spite of the economic crisis, which Venezuela has been facing for a long time, astronomers, as well as other researchers, can obtain financial support from several organizations. The most important national agency that supports science is CONICIT along with its local regional offices called Fundacite. Also, in each university there are internal agencies such as, the CDCHTs, the University Scientific Exchange Program, and the International Inter-Institutional Relations Office that provides a substantial financial assistance as well. Besides, each qualified researcher can count on grants provided by their respective institution.

5. Final Comments

From my point of view, even though Venezuela is a developing country and thereby subjected to many crises (especially of the financial kind), the organizations involved in the progress of science have given, and are still giving, substantial priority to the consolidation of the astronomical heritage. Above all, taking into account that even though any scientific enterprise is undertaken with great difficulty, the scientific output is of a high standard as is demonstrated by the different scientific articles and publications found in recognized periodicals. In part, this high standard of production results from the efforts by several faculty members who had the opportunity to obtain their doctorates abroad and returned to the country some time ago; also from many projects that are active at the present time with other colleagues overseas, and from the work of colleagues who have obtained their degrees from our graduate programs as well.

Acknowledgments. Special thanks to the IAU for their support and kind invitation to the Special Session. I am also grateful to Dr Alan Batten and Prof. John Hearnshaw for their guidance throughout this work and the Session. I would like also to thank Dr. Marcos Peñaloza for assistance in preparing the manuscript and useful discussions. Finally, to the CDCHT-ULA and CONICIT, both in Venezuela, for their financial support.

References

Alvarez, H., and MacConnell, D.J. 1978. *Astronomy in Venezuela Today*, Sky and Telescope, 55, 103.

Chalbaud, P. 1990, *Indicios de la Astronomía Moderna en Venezuela: El Proyecto de Eduardo Röhl Aportes para su Estudio*, Thesis, Bachelor Degree, Universidad de Los Andes.

Edmondson, F.K. 2000, private communication.

Hubschmann, K., 1988, *Observatorio Cagigal: Cien Años de Historia y Ciencia*, Colección: Cuadernos Lagoven, Caracas, Editora Venegráfica.

Inglis, S.J. (1961). *Astronomy in Venezuela*, Sky and Telescope, 22, 251.

Olivares, A. E. 1986, *Dr. Luis Ugueto, Ingeniero, Astrónomo y Profesor*, Caracas, Academia de Ciencias Físicas, Matemáticas y Naturales.

Quintero, I. 1989, *Del Observatorio Cagigal al CIDA: Dos Instituciones Científicas en la Historia de la Ciencia en Venezuela*, Caracas, Fondo Editorial Acta Científica Venezolana

Stock, J. 1981, *Astronomía en Venezuela*, Revista Mexicana de Astronomía y Astrofísica, 6, 13.

Discussion

Mahoney asked about the policy of the new Venezuelan government for astronomy and Rosenzweig replied that it was to continue with the budgetary plans. Tancredi asked if the QUEST instrument on the Schmidt telescope was available for programs other than the microlensing search. Rosenzweig replied that it was also used, among other programs, to search for minor planets, supernovae, and to study variable stars. Fierro asked if Venezuelan degree courses were open to students from other countries (especially in Latin America) and Rosenzweig confirmed that they are.

Astronomy for Developing Countries
IAU Special Session at the 24th General Assembly, 2001
Alan H. Batten, ed.

Astronomy Research in China

Jingxiu Wang

National Astronomical Observatories, Chinese Academy of Sciences, Beijing 100012, China. e-mail: wjx@ourstar.bao.ac.cn

Abstract. Decades of efforts made by Chinese astronomers have established some basic facilities for astronomy observations, such as the 2.16-m optical telescope, the solar magnetic-field telescope, the 13.7-m millimeter-wave radio telescope etc. One mega-science project, the Large Sky Area Multi-Object Fiber Spectroscopic Telescope (LAMOST), intended for astronomical and astrophysical studies requiring wide fields and large samples, has been initiated and funded.

To concentrate the efforts on mega-science projects, to operate and open the national astronomical facilities in a more effective way, and to foster the best astronomers and research groups, the National Astronomical Observatories (NAOs) has been coordinated and organizated. Four research centers, jointly sponsored by observatories of the Chinese Academy of Sciences and universities, have been established. Nine principal research fields have received enhanced support at NAOs. They are: large-scale structure of universe, formation and evolution of galaxies, high-energy and cataclysmic processes in astrophysics, star formation and evolution, solar magnetic activity and heliogeospace environment, astrogeodynamics, dynamics of celestial bodies in the solar system and artificial bodies, space-astronomy technology, and new astronomical techniques and methods.

1. Introduction

At this time when we are seeing the dawn of the new century, Chinese astronomers are facing great opportunities and challenges, as well as great difficulties.

Generally speaking, astronomy is an observational science. Its glamour and influence lie in new observations and observational discoveries. Progress in astronomy research heavily depends on the advanced instrumentation and technology, which are expensive and require continuous innovation. It is not appropriate to talk about astronomy research in China without mentioning its weaknesses. The ground-based instruments, developed by Chinese astronomers, are of small and, at most, medium sizes. The biggest optical telescope in China, 2.16 m, was put into operation in early 1990s forty years after the 5.7-m Hale Telescope on Palomar Mountain started to work in the U.S. So far, no Chinese satellite has yet been launched for astronomical observations. The funding level for Chinese astronomy research is less than one tenth of that in developed countries. There are few world-renowned Chinese astronomers. Some of our brilliant young graduate students and qualified astronomers continue to drain into developed countries. A detailed analysis of Chinese astronomy can be found in a report by Su et al. (1997).

With such a difficult situation, our long-term goal in astronomy research raises several questions: what can Chinese astronomers do in the foreseeable future? what are their strategic considerations? what would they like to share with, and contribute to the international astronomy family? With regard to these questions, two events are indicative. First a mega-science project, Large Sky Area Multi-Object Fiber Spectroscopic Telescope (LAMOST), intended for astronomical and astrophysical studies requiring wide fields and large samples, has been initiated and funded. Secondly, the National Astronomical Observatories (NAOs) of China have been coordinated and established by the Chinese Academy of Sciences. Chinese astronomers have set up their goals, and are poised to make their contribution to a more comprehensive understanding of our universe and all of its components.

2. Basic Observational Facilities

Chinese astronomers have dreamed and worked generation by generation to lay their own foundation of astronomical observations. They clearly understand that, although the facilities they established might be very limited, they are most important not only for frontier studies of carefully-selected topics, but also for training astronomers who will be ready to work later with more advanced facilities. Indeed, developing one's own instrumentation creates a group of astronomers who gain knowledge of both the science behind the observations and the techniques used for the observations, is created. Decades of efforts made by Chinese astronomers have established some basic facilities, mostly for common purposes (see Table 1).

Table 1.   Key Commonly-Used Astronomical Instruments in China

Wavelength	Diameter	Location	Main science
Optical	2.16	*Xinglong (NAOs)	AGN, stars, galaxies
Optical	1.56	Sheshan (ShAO)	astrometry, Galaxy
Optical	1.00	Kunming (NAOs)	AGN, stars
Infrared	1.20	Xinglong (NAOs)	infrared photometry
Millimeter	13.70	*Delinha (NAOs & PMO)	star formation
Meter	28×9.00	Miyun (NAOs)	meter source survey
Radio	25.00	*Sheshan (NAOs & ShAO)	VLBI observations
Radio	25.00	*Urumqi (NAOs)	VLBI observations
solar	.35	*Huairou (NAOs)	vector magnetograph

* Observing base of NAOs, Southern Base will be at Lijiang in Yunnan
PMO – Purple Mountain Astronomical Observatory
ShAO – Shanghai Astronomical Observatory

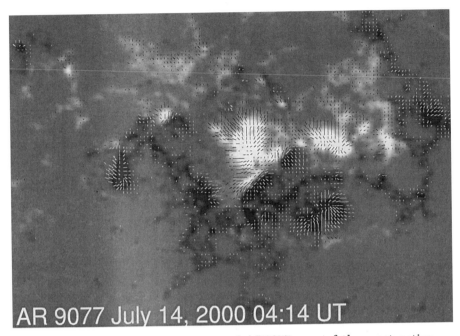

AR 9077 July 14, 2000 04:14 UT

Figure 1.   Vector magnetogram of AR9077, one of the most active regions in this solar cycle. The line-of-sight component is represented by brightness, with darker (lighter) color for negative (positive) polarity; the transverse components are presented by short line segments with their length proportional to field strength and alignment parallel to the field direction. Courtesy J. Zhang and Y. Deng

The majority of these instruments are operated by the National Observatories in the Chinese Academy of Sciences (CAS). They are accessible to researchers from both CAS and universities. The scientific output of the facilities is considered satisfactory. For example, 30 papers were published in SCI source journals based on the observations with the 2.16-m telescope in 1999. The 2.16-m telescope is actively involved in international campaigns on AGN and pulsating stars. The majority of its observing proposals focus on AGN, starburst galaxies, normal galaxies, supernova, metal-poor stars, pre-main sequence stars, pulsating stars, and coronally active stars. Based on observations from the 1-m telescope at Kunming, great progress has been made on the observational studies of BL Lac objects. The Solar Magnetic Field Telescope at Huairou has produced some of the best vector magnetograms for solar activity studies. One vector magnetogram is shown in Figure 1 for a superactive region, AR9077, observed in July, 2000.

Except for the key instruments of common purpose, some specific instruments are operated by research groups. To list a few, an old 60/90 Schmidt telescope equipped with 2K×2K CCD is effectively used for multiple-color surveys of large-red-shift galaxies by a leading group from NAOs; a 60-cm reflector, adapted as a specialized telescope for supernova (SN) monitoring at Xinglong

Figure 2. A image of the well studied galactic open cluster M67. The image is made by composition of three images at 8490A (red), 6075A (green) and 3890A (blue)Å. Courtesy L. Deng and the survey team

Station of NAOs – one of the best systems for SN survey – has captured 35 bright supernovae from early bursts in the last three years; A 50-cm solar-spectrum telescope is working on Stokes polarimetry in another research group of NAOs at Kunming. Those telescopes are really small, but specialized for a particular scientific topic and combined with intelligence from leading scientists. Therefore the scientific returns from these small telescopes are impressive. Figure 2 is an example of the deep-sky survey images of the galactic open cluster M67, the deepest view so far for this open cluster.

So far, in Chinese universities there has been one advanced telescope, the 60-cm solar-tower telescope equipped with an imaging spectrograph of multiple wavelengths. A leading group from the Astronomical Department of Nanjing University has made important progress in the studies of semi-empirical models of solar flare and sunspots based on the observations from this telescope.

3. Major Initiatives

As a developing country, China certainly lacks the capability to invest heavily in astronomy. The Chinese Academy of Sciences has worked out a way for major initiatives. It has been suggested to the country to sum up the available funds, and in each 5 years to open an opportunity for a mega-science project. Chinese astronomers seized a chance in early part of 1990s, and had a national mega-science project, Large Sky Area Multi-Object Fiber Spectroscopic Telescope

(LAMOST), approved by the State Planning Committee of China in 1997. The concept of LAMOST is described by Wang et al. (1996).

LAMOST is a meridian reflecting Schmidt telescope laid on the ground with its optical axis fixed in the meridian plane. Since it is aimed at the scientific goal of "wide field-of-view and large samples", it adopts a simple meridional structure, so that the telescope, being stationary on the ground, can be very long and thus, since it is a Schmidt-type telescope, have a broad field-of-view. The other two essential factors for the telescope, correction of spherical aberration and tracking corrections, are effected by the ingenious introduction of an active reflecting corrector. The aperture of LAMOST is 4 m, enabling it to obtain the spectra of objects as faint as 20.5 magnitude with an exposure of 1.5 hours. Its focal plane is 1.75 m in diameter, corresponding to a 5° of field-of-view, and may accommodate as many as 4000 optical fibers. So the light from 4000 celestial objects will be led into a number of spectrographs simultaneously. Thus, the telescope, when complete, will have the highest spectrum-acquiring rate in the world. In technique and technology, the active reflecting corrector not only plays a key role in the design of LAMOST, but also presents itself as a significant innovation for the further development of modern astronomical instruments.

LAMOST will be located at Xinglong Station of NAOs. It will be open to the world-wide astronomy community. The budget for this project is RMB 235 M Yuan (about U.S. $28 million. It will come into operation at the end of 2004. LAMOST will bring Chinese astronomy to a frontier position in large-scale observations of optical spectra, and in wide-field astronomy.

Several other major projects are under active study and preparation for future selection as national mega-science projects. Among them, an astronomy satellite, e.g., Space Solar Telescope (SST) and a ground-based Five-hundred meter Aperture Spherical Telescope (FAST) are most attractive and promising. SST is aimed at observations of solar magnetic activity at the fundamental spatial scale, i.e., 75 km on the solar surface. FAST will be the largest single-disc radio telescope. The ultimate goal of the major initiatives is greatly to strengthen the observating facilities for Chinese astronomy.

Coordinated with the LAMOST project, the Ministry of Science and Technology of China has established a National Key Basic Research Development Program – Formation and Evolution of Galaxies. The funding level of such a national key program amounts to 20 M Chinese Yuan. The other two National key programs, Modulated Hard X-ray Telescope and Solar Activity and Space Weather, are approved and under coordination; they are considered as scientific supporting programs for mega-science projects.

4.   The National Astronomical Observatories of China

Several key factors made the Chinese astronomy community feel a great need to have their National Astronomical Observatories. As a developing country, China has neither sufficient investment in astronomy, nor enough leading scientists. To realize a national mega-science project there must be a national organization which can concentrate and coordinate the best efforts from Chinese astronomers and engineers. It can only be done if NAOs maintain and fully open the key national facilities to all scientists, not only in CAS, but also in the universities

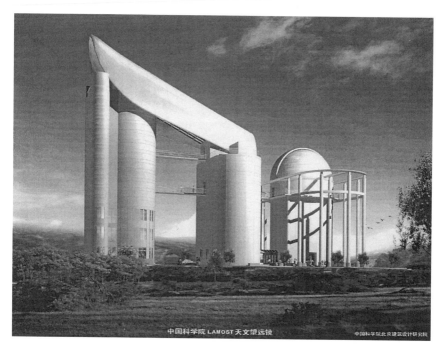

Figure 3.   An artist's-eye view of LAMOST. The reflecting Schmidt plate is located in the lower dome on the right-hand side of the plot (south). The spherical primary mirror is on top of the higher tower, which is about 44 m from the ground. The central block is for the focal surface and the spectrograph room. Courtesy LAMOST team

and other institutions, to optimize scientific output. To reduce the low-level repetition in research and technical development, the NAOs can also play an important role. To have the best astronomers and engineers, and to foster leading research groups in the world, some long-term support, and an easy and comfortable working environment need to be created in NAOs.

The National Astronomical Observatories of the Chinese Academy of Sciences was established in April of 1999 on the basis of the existing astronomical observatories and stations. It consists of five observing bases (see Table 1), six common laboratories (see Section 5), 30 research groups composed into a research network, and four combined research centers, (Section 5) with some leading universities in China.

The top priority of the observing bases is to operate effectively and open fully the key national facilities (see Table 1) to domestic and overseas astronomers. At each base there are a few resident astronomers. They carry out the scientific work, in addition to supervising graduate students and astronomers who come to the observing base. Laboratories of NAOs work on keeping the existing facilities productive, at the same time, developing the new techniques necessary for future initiatives.

Current astronomy research in China covers a diversity of areas, such as the deep survey of large-red-shift galaxies, supernovae and supernovae remnants, dark matter and gravitational lensing, galaxy interaction and star-burst galaxies, active galactic nuclei, γ-ray bursts, molecular clouds and star formation, stellar convection and pulsation, star evolution, solar magnetism and magnetic activity, astro-geodynamics, dynamics of celestial bodies and artificial bodies, astronomical optics, modern studies of ancient Chinese astronomical literature, and so on. In each working area there are some Chinese astronomers whose work has been highly regarded by international colleagues. Based on an analysis of the general trend of astronomy research in the coming new century and the status quo of astronomy research in China, nine research fields have been set up as priority for support from the NAOs. In the frontier research area, the large-scale structure of universe, formation and evolution of galaxies, high-energy and cataclysmic processes in astrophysics, formation and evolution of stars, solar magnetic activity and heliogeospace environment receive enhanced support; in applied astronomy studies (Section 5), astrogeodynamics, dynamics of celestial bodies in the solar system and artificial bodies are selected as principal fields; in technology studies, space observation and exploration, and new astronomical techniques and methods are taken as top priority.

The principal professors together with their research groups in NAOs were selected in the priority areas, by peer review and open defense. Under the direction of the principal professors, the research groups were organized. There are altogether 30 groups which were established and have received strong support from NAOs of CAS. They have become active centers of astronomy research in China, and are very attractive to young graduate students.

The NAOs look forward to the future of Chinese astronomy. To further strengthen the observation facilities, The Southern Base of NAOs is under construction in Lijiang, Yunnan Province. Currently, a 2.3-m optical telescope is under consideration, 3-m to 4-m telescopes have been identified as appropriate for this site in the future. The Chinese astronomy community continues to search for good sites for next-generation observatories. The first site survey in Tibet was made in the late 1980s. A more recent site survey in Tibet was made last June (2000) for both solar and night-time astronomy. The site-survey observations are under analysis. The observations appear encouraging. Further survey observations will be made. Figure 4 is a star-trail picture taken in Tibet at an altitude over 5,000 m.

5.   Key Relations

For a developing country, it becomes most important to optimize the supporting system, and to maximize the benefits of astronomy studies to the country and the society. It is recognized that the following six relations should be properly dealt with.

1. The National Astronomical Observatories and Other Observatories.

   So far, all the Chinese observatories are national. They are institutes of CAS. The key function of NAOs is developing national facilities and having them efficiently operated and open to all active domestic astronomers

Figure 4.   Site survey in Tibet in June 2000, courtesy K.X. Chang

and international colleagues; while the other observatories carry out advanced research based on the performance of the national facilities as well as their home instrumentation which is specialized for particular scientific topics. The other observatories seem more to concentrate on their own distinctive and strong points in research. All the observatories work in a complementary way to achieve the highest output of the astronomical observations. Purple Mountain Astronomical Observatory focuses on theoretical astrophysics, millimeter and sub-millimeter astronomy, and celestial dynamics; Shanghai Astronomical Observatory concentrates on the Galaxy and extra-galactic astrophysics and astrogeodynamics.

2. Observatories and Universities.

Astronomy researchers in Chinese universities have established themselves by advanced theoretical studies and comprehensive data analyses. Most of their work is in the frontier areas of astrophysics. A few current working directions are listed below for four top universities in astronomy.

- Nanjing University (NU) – γ-ray bursts and high-energy astrophysics; supernova remnants, galaxy dynamics, solar active-region physics, nonlinear celestial dynamics;

- Beijing University (BU) – pulsars, star formation;

- University of Science and Technology in China (USTC) – early universe and cosmology, large-scale structure in universe, AGN, accretion theory, solar magneto-hydrodynamics;

- Beijing Normal University (BNU) – quasars, stellar physics.

More universities are producing high-quality research in astronomy. It has been urged that NAOs and other CAS observatories work in concert with universities to promote strongly astronomy research in China, and to improve the education of graduate students, the new generation of astronomers (Fang and Tang, 2001). To realize this, four research centers have been established in NAOs, supported both by NAOs and universities. They are Beijing Astrophysics Center in BU, East China Center of Astronomy and Astrophysics with NU, Astrophysics Center of USTC , and Astrogeodynamics Center in Shanghai with several universities and institutes. Among them, Beijing Astrophysics Center (BAC) jointly sponsored by NAOs of CAS and Beijing University was established first. It has enjoyed great success (Annual Report of BAC, 1998, 1999), and will grow up as a new astronomy department.

3. Frontier Studies and Applied Astronomy

By "applied astronomy" is meant developing and applying astronomical knowledge and methods to other disciplines in the natural sciences and to the national economy and security enterprises. For a developing country, frontier studies are fascinating and we like to see the country catch up in natural science; it helps public appreciation of science; on the other hand, applied astronomy is also important to benefit the country and people in a more direct way. The NAOs suggest a ratio of 7:3 in supporting frontier studies and applied astronomy. Currently, the applied studies are concentrated on astrogeodynamics, detection of near-Earth asteroids and comets, dynamics of the artificial bodies, solar and space-environment prediction, astronomical factors in global changes and natural disasters, modern studies of historic astronomy materials and so on. Great progress has been made in astrogeodynamics; for example, the new nutation model of a non-rigid Earth and an initial tectonic block-motion model of China have been established.

4. Scientific and Technical Work

Astronomical research depends heavily on innovations in technology and instrumentation. Therefore technical work should be considered an ingredient of astronomical investigations. The ratio of astronomers and experts in technical work is maintained at 1:1 in NAOs. Six laboratories were built in NAOs for common usage. They focus on techniques which are closely related to the key existing facilities and major projects under development, as well as on what is potentially important for future studies. Six laboratories are devoted to astronomical optics, millimeter and sub-millimeter techniques, optical and infrared detectors, VLBI technique, space-astronomy technology, and techniques for large radio-telescopes, respectively. They are distributed in Nanjing Astronomical Instrument Research Center and a few observatories.

5. Observational and Theoretical Research

Astronomy never stops at the stage of new discoveries. Generally speaking, astronomy research includes, at least, five steps: observations, discoveries, physical understanding, mathematical descriptions, and scientific predictions. Theoretical studies are tied not only to observations and data interpretations, but often have an independent function in predicting and guiding new observations and creating new knowledge in the natural sciences more generally. To maintain a relatively independent position for theoretical astronomy is essential for astronomy research.

Chinese astronomers have made advanced theoretical studies on stellar convection and evolution; but, generally speaking, theoretical research is a weak link in Chinese astronomy. To promote theoretical research in China more fully is of fundamental importance for Chinese astronomy in the coming new century. It has been proposed to establish a Theoretical Astrophysics Center in NAOs. Inviting scientists from theoretical physics, high-energy physics, mathematics, plasma physics and other branches of physics to work with astronomers in this center is encouraging. It is not appropriate for original theoretical research only to count papers. A small but very excellent theoretical group of astronomers is urgently needed for Chinese astronomy.

6. Competition and collaboration

Competition always goes with collaboration in astronomy research. Astronomy is a science that demands extensive international collaboration. Chinese national observatories and universities are now open to our international colleagues. The Chinese Academy of Sciences has opened a Guest Senior Investigator Program. Under this program NAOs and all the observatories are able to provide fine working conditions and environments for their guests. Chinese astronomers are encouraged to join actively in international collaboration, observing with world-class instruments, joining in international campaigns, making collaborative data analysis and theoretical studies. This will certainly benefit Chinese astronomers by bringing them into frontier research areas. On the other hand, in such collaborations Chinese astronomers share their experiences and intelligence with the international astronomy community. The real challenge is to work on big science with only small instruments. This will raise a generation of capable astronomers. In this sense Chinese astronomers are creative. To establish a brotherly and sisterly relationship with leading astronomy groups, to join some selected international projects are essential for deeper and closer collaborations. The Asia-Pacific Space Geodynamics (APSG) Program was initiated in 1996 to coordinate research projects in plate-tectonic, crustal motion and deformation, and to study the sea level in the Asia-Pacific area, by space techniques. A Max-Planck and CAS partner-research group in astronomy was established in Shanghai Astronomical Observatory last June. Many more close collaborations are projected.

Because of the author's limited knowledge, astronomy research in Hong Kong, Aomen, and Taiwan has not been mentioned in this paper.

Acknowledgments. The author is grateful to many colleagues for their kind offers of the materials and ideas which have been presented in this article. They are professors S.Wang, J.Chen, S.Ye, F.Cheng, G.Ai, T.Lu, Z.Wang, H.Su, G.Lo, Y.Chu, F.Yang, C.Huang, L.Deng, J.Wei, Y.Zhao, and J.Zhang, The author is also indebted to Drs. Alan Batten and W.-P. Chen for their suggestions to improve the paper.

References

Annual Report 1998, 1999, Beijing Astrophysics Center.

Fang, C. and Tang, Y. 2001, this volume, pp. 80-86.

Su H.J. et al. 1997, Report on Development of National Sciences: Astronomy, Science Press, Beijing.

Wang, S.G., Su, D.Q., Chu, Y.Q., Cui, X.Q., and Wang, Y.N., 1996, Applied Optics, 35(25), 5155.

Discussion

Anandaram and Kozai both asked if Chinese observatories are open to foreign astronomers. Wang replied that they are; in particular, both the Huairou and Xinglong stations of Beijing AO have been open to foreign astronomers for several years. Applications from foreign colleagues are much encouraged because of the contacts and opportunities for interaction that they give. Applications should be sent to the Principal Astronomer of the observing base concerned. Announcements of visits will be made on the home page of the National Astronomical Observatories of the Chinese Academy of Sciences: (http://www.bao.ac.cn)

Astronomy for Developing Countries
IAU Special Session at the 24th General Assembly, 2001
Alan H. Batten, ed.

SALT as an African Facility

Peter Martinez

South African Astronomical Observatory, P.O. Box 9, Observatory 7935
South Africa. e-mail: peter@saao.ac.za

Abstract. Over the next five years, South Africa and its international partners will construct the Southern African Large Telescope (SALT), a 10-m optical/near-infrared spectroscopic survey telescope. This paper traces the arguments used to justify the construction of such a facility in Africa. The collateral benefits expected from this project are discussed, along with suggestions of how to realize these benefits. SALT has the potential to promote the development of astronomy on the continent. The creation of an African Network for Astronomy Education and Research is proposed as a way to stimulate such development.

1. A Large African Telescope

Since the early 1970s the major facility for optical/infrared astronomy in sub-Saharan Africa has been the South African Astronomical Observatory (SAAO) in Sutherland. This facility has made many major contributions to southern hemisphere astronomy. The largest of the SAAO telescopes is the 1.9-m Radcliffe reflector. By exploiting advances in detector quantum-efficiency and in computing power, this telescope has remained a significant facility for modern southern hemisphere astronomy. However, by the 1980s it became clear that South African astronomers needed a successor to the aging 1.9-m telescope at Sutherland.

In the early 1990s a case was made to construct a 3.5-m telescope similar to ESO's New Technology Telescope. This was a modest aspiration, coming at a time when other countries were planning to build 8-m to 10-m class telescopes, but it seemed to be the best that South Africa could afford. This case for a large national telescope was being made in parallel with the great political changes taking place in South Africa at the time. The right political context for a large national telescope project did not exist at the time, and it remained a notional project.

Then, in 1994, a remarkable thing happened; South Africa experienced a peaceful transition to a democratically elected, majority government. One of the processes that the new government set in train was an audit of South Africa's science and technology capability. The resulting policy document concluded that *"scientific endeavour is not purely utilitarian in its objectives and has important associated cultural and social values ... Not to offer 'flagship' sciences (such as physics and astronomy) would be to take a negative view of our future — the view that we are a second-class nation, chained forever to the treadmill of*

221

Figure 1.   Artist's conception of SALT as it will appear among the existing telescopes in Sutherland.

feeding and clothing ourselves." In other words, this newly elected government had recognized the value of science and technology as an agent for national transformation.

Concurrently with these political developments, but half a world away, a consortium of American and German astronomers was constructing the Hobby-Eberly Telescope (HET), in Texas. This telescope represents a radical departure from conventional telescope design, and offers much of the functionality of a conventional telescope at a fraction of the cost. By the mid-1990s, the builders of the HET were thinking of a southern-hemisphere twin of their telescope, and they approached their South-African colleagues, with whom they already had strong ties. The idea was enthusiastically adopted by South-African astronomers as this offered the possibility of building a 10-m class telescope for a price similar to that of the more modest 3.5-m telescope project, which seemed to be going nowhere. Moreover, with SAAO's well-developed infrastructure in Sutherland, it seemed likely that South Africa would succeed in attracting international partners to join a 10-m telescope project in the southern hemisphere. The project soon attracted the interest and support of the country's science administration, and lobbying for the Southern African Large Telescope (SALT) at top government level began in earnest in 1997.

On 1 June 1998 the South-African Minister of Arts, Culture, Science and Technology announced in Parliament that the South African Cabinet had approved the construction of SALT on condition that the South African Government would fund 50% of the cost and the balance would have to be funded by international partners. There was unanimous support from all political parties for this announcement. By November 1999, the Minister was satisfied that suffi-

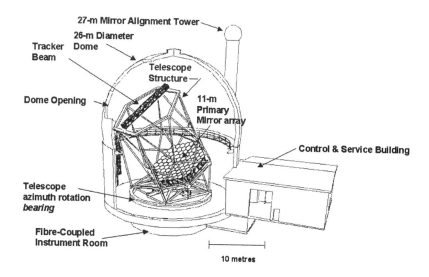

Figure 2.   Schematic view of the SALT facility. (Image courtesy of the HET Board.)

cient international funding had been committed to authorize the commencement of expenditure of the South African funds allocated for this project.

As of August 2000, the SALT partnership included South Africa, the HET Board, Poland, Rutgers University, Göttingen University, the University of Wisconsin-Madison, Carnegie Mellon University, Canterbury University (New Zealand) and a consortium of U.K. universities comprising the University of Central Lancashire, Keele University, Nottingham University and Southampton University, and Armagh Observatory in Northern Ireland. As of this writing, SALT is about 90% funded for construction and the first 10 years of operations. Prospective additional partners are in the process of raising funds to join SALT.

How does South Africa justify building one of the world's largest telescopes when it has so many other pressing needs? This paper outlines the ideas that have emerged (and continue to emerge) in response to this question. Section 2 contains a brief description of the design of SALT. Section 3 discusses current plans for a programme of collateral benefits to be implemented during the construction and operation phases of SALT. Section 4 discusses how to promote the use of SALT as an African facility.

2.   The Design of SALT

SALT is based on the design of the Hobby-Eberly Telescope (HET), which was built at McDonald Observatory, in Texas. For a description of the HET project, refer to the paper by Ramsey et al. (1998) and references therein. Briefly, the HET design concept is as follows. The telescope has a segmented spherical primary mirror array 11 metres across. The mirror array comprises 91 hexagonal mirrors, all 1 metre across. This mirror array is supported in an open framework structure inclined at a fixed tilt (35 degrees, in the case of SALT). The telescope structure moves only in azimuth (not elevation), and only when slewing. Rotation in azimuth is accomplished by means of floating the telescope structure on an air cushion. During an observation, the telescope structure remains stationary, and tracking motions are accomplished by a tracker beam that moves across the top hexagon of the telescope structure.

In one full rotation, the telescope sweeps out a 12°-wide annulus on the sky, centred on the telescope's latitude. Astronomical objects are accessible to this telescope only when passing through this 'annulus of visibility'. The amount of time it takes an object to cross the annulus of visibility is declination-dependent. At the worst extreme, near the celestial equator, objects can be tracked for about 48 minutes, while at the southern extreme of the annulus of visibility, objects can be tracked for up to 2.4 hours. This feature of the telescope's operation dictates that observations are done in a queue-scheduled mode. That is, unlike a conventional telescope, which is dedicated to one programme for a whole night, SALT will typically observe targets from many different programmes during the same night.

The fixed tilt of the telescope simplifies the structure considerably since the gravity load on all structural members and the mirror array is constant. Likewise, the use of a spherical primary means that all 91 mirror segments are identically figured. Together, these cost-saving features allow access to 70% of the sky accessible to a fully steerable conventional telescope at 20% of the cost.

SALT will operate in the wavelength region of 0.3 to 2.5 microns. The thermal background increases rapidly redward of 2.5 microns, and the telescope is not efficient at those longer wavelengths. Moreover, the altitude of Sutherland (1757 m) makes it uncompetitive in the thermal infrared, compared to other high-altitude sites.

One of SALT's most complex subsystems is the tracker beam, which moves across the top hexagon of the telescope structure to accomplish tracking. This tracker beam weighs 4.5 tons and must be capable of 9 degrees of freedom to move accurately on a spherical focal surface with a linear precision of 6 microns. The tracker beam will support the Prime Focus Instrument Platform (PFIP), a package housing the spherical-aberration corrector, atmospheric-dispersion corrector, exit-pupil baffle system, acquisition/guide CCD camera, a Prime Focus Imaging Spectrograph (PFIS), and a fibre feed to medium-resolution and high-resolution spectrographs in an environmentally controlled room under the telescope. For a more detailed discussion of SALT instrumentation plans, refer to the paper by Buckley et al. (2000).

SALT is a copy of the HET, but not an identical copy. The HET Board is a partner in the SALT project, and the SALT project team is benefiting from the hindsight of the team that built the HET. The tilt of SALT has been changed from HET's 35° tilt to 37° to allow SALT to access most of the Small Magellanic Cloud. SALT will also incorporate edge-sensors to facilitate alignment of the primary mirror array, a modification soon to be applied to the HET as well. The most significant departure from the HET design is in the manner in which spherical aberration is corrected in SALT. A new design by O'Donoghue (2000) promises to deliver substantially better-quality images, an 8-arcmin field (double that of the HET), somewhat reduced vignetting, and an increased pupil size (pending implications of redesigning the top hexagon). The differences between SALT and HET are discussed in more detail by Stobie et al. (2000).

3. The SALT Collateral Benefits Plan

The decision to build SALT was taken at the highest levels in South Africa. By funding SALT, the South-African government is making the statement that it sees South Africa as a significant role player in the international scientific community. SALT will enable South-African scientists to remain internationally competitive in astronomy well into the 21st century. However, SALT has an importance to South Africa (and Africa) far beyond its astronomical research mission. SALT will provide distinct and substantial benefits in the development of people, technology, and the economy. The SALT Collateral Benefits Plan is designed to maximize the benefits from the investment of public funds in the construction and operation of SALT. There are three main thrusts to this plan: industrial empowerment, educational empowerment, and public outreach. The Collateral Benefits Plan is being drafted in consultation with stakeholders from government, industry and education. The discussion in the remainder of this section draws heavily on the discussion document 'SALT Collateral Benefits Plan' (1999).

3.1. SALT industrial empowerment

The SALT project is being planned in such a manner as to maximize the industrial pay-offs from the development of components, subsystems, equipment, and labour for the construction and provision of SALT. Specific elements include a procurement policy with emphasis on empowerment and capacity building of people, firms and companies from previously disadvantaged communities. Foreign companies contracted to work on SALT could host South-African technical personnel at the contractor's facility to collaborate in the design and manufacturing efforts. Careful selection of personnel will maximise this benefit.

The result of these activities will be industrial collaborations joining South African and foreign companies in new ways to increase technical capabilities. This will result in technology transfer in the areas of optics, electronics, micropositioning, dynamic control systems, precision sensing, and others. These companies will be able to compete for additional business within and outside of South Africa.

3.2. SALT educational empowerment

One of the biggest educational challenges confronting South Africa is to broaden the science, education and technology training base, particularly in the black community. SALT can play a significant role in meeting this challenge by providing educational and training opportunities for astronomers, physicists, computer scientists, and engineers during the construction (5 yr) and operational (25+ yr) phases of the project. This will specifically involve student participation in the SALT project team, student placement with SALT contractors, student participation in collaborations with SALT partners, and student involvement in instrumentation development. Since these students all have to be registered at their own academic institutions, this will promote closer ties between SAAO and those institutions, and empower them in new technologies.

The involvement of students in all phases of the SALT project will equip them with the experience of participation in a large, complex, high-technology project, which experience will be carried into later careers. Knowledge of the technology, interactions and techniques for development of such projects is vital, as South Africa must address ever more ambitious technical undertakings. Experience with SALT will provide confidence in using SALT's constituent technologies in future projects.

Another component of this educational thrust is to promote the development of astronomy at the historically black universities in South Africa, where the subject has not featured in the past. This will have the advantage of increasing the numbers of scientists and engineers participating in the construction and use of SALT, and will also make SALT more accessible to the wider community in South Africa. The achievement of research success in astronomy at these institutions will have long-term spin-offs in terms of encouraging success in other fields at those universities by stimulating both students and faculty. The SAAO will play a key role in implementing these policies.

The astronomical activities generated by the SALT project will provide excellent training opportunities in astronomy. However, the nature of the training is such that students will be well equipped to pursue challenging careers in other disciplines should they wish to leave astronomy. There is a growing trend internationally for students trained in astronomy and astrophysics (particularly at post-graduate level) to be sought after for employment in business and industry for their analytic and problem-solving skills.

3.3. Public outreach and direct education benefits

The South-African government recognizes that scientific literacy is the key to ensuring the technological competitiveness of the nation. Until recently, black people in South Africa had very limited access to science, engineering and technology education, training and careers, with the result that the majority of the labour pool is unskilled in this vital area. South Africa requires a sustained campaign to promote awareness and understanding of science, engineering and technology at all levels in society. SALT will contribute to this process by stimulating excitement about science and technology in general.

A careful programme of educational and public outreach benefits is being compiled in collaboration with other stakeholders to capitalize on the interest generated by SALT. This programme includes activities such as training of educators in astronomy, a 'Starbus' science roadshow for outreach to rural communities, development of educational resources, etc. To accommodate the large numbers of visitors expected to visit SALT, a Visitor Centre will be constructed in Sutherland. The Visitor Centre will contain interactive hands-on displays on SALT, astronomy, and other scientific aspects related to the Sutherland/Karoo region. For operational reasons the Visitor Centre will be in a separate building from SALT, but the SALT telescope dome will incorporate a special viewing gallery where visitors will be able to witness the full scale of the telescope at close quarters. This facility is discussed in detail by Rijsdijk (2001) elsewhere in this volume.

4.   SALT as an African facility

Capability in the space sciences will be a vital factor in the development of Africa in the 21st century. The applied space sciences are underpinned by the basic space sciences, such as astronomy and astrophysics. One of the indicators of activity in basic space science is membership of the International Astronomical Union (IAU). This indicator reveals that Africa is very poorly represented in the space sciences, compared to other regions in the world. Currently, only four out of 53 African states are members of the IAU. As of 1999, the total IAU membership among African countries amounted to 95 persons, or 1.1% of the total IAU membership. The African membership of the IAU is dominated by South Africa (48%) and Egypt (41%), followed by Nigeria, Algeria, Morocco and Mauritius, each with 4% or fewer members. Not surprisingly, these countries are among the more prosperous nations on the continent.

One should bear in mind that the above statistics reflect the presence of nationally organized astronomical communities. There are many individual scientists distributed throughout Africa who are involved in astronomy education, and in some cases research also, who are not members of the IAU. Isolation, and lack of access to world-class facilities, are the two factors that retard the development of basic space science in Africa.

At present there is a dearth of large-scale astronomical facilities in Africa, and African astronomers are forced to develop their careers elsewhere. The creation of new large-scale facilities like SALT will contribute to encouraging highly trained professionals to remain in their own cultural and political environment. As the development agenda of the region would likely preclude the construction of another facility comparable to SALT in the foreseeable future, this project was conceptualised from the start as one that would serve the needs of the region. SALT has been intentionally named the Southern African Large Telescope, as opposed to the South-African Large Telescope, to underscore the fact that this is seen as a regional project. However, if there is to be an African user community for SALT by the time that the telescope is commissioned in 2005, then the first steps have to be taken now to develop that community.

In order to promote the development of astronomy in Africa, one has to provide adequate opportunities for African astronomers to practice their science in Africa, rather than pursuing careers elsewhere. Networking is the key to establishing a critical mass of African astronomers working in Africa. Networking breaks the isolation of scientists by involving them in a distributed community of scientists working towards a common goal in the same field, thereby giving them access to the collective expertise and experience of that whole group.

The advent of the Internet is revolutionizing science by giving scientists everywhere unprecedented access to information and data. Although African scientists may not enjoy the same standards of connectivity as their European or North-American colleagues, many now have access to electronic mail, and limited access to the worldwide web. This access provides the foundation upon which to build an *African Network for Education and Research in Astronomy*, whose hub could be the optical and radio observatories in South Africa. Such a network would allow astronomers from a variety of African countries to access instruments at one the best astronomical sites on the continent, without crippling investments in their own countries.

The Working Group on Space Sciences in Africa (Martinez 2000) and the South-African Astronomical Observatory are currently planning a pilot project which could form the basis of an African Network for Education and Research in Astronomy. This is to be implemented through a Fellowship programme, hosted by SAAO. Fellows from different African countries would work together at SAAO for a period of several months, during which time they would develop research skills in astronomy, as well as the personal acquaintances so important in scientific collaboration. At the same time they would collaborate on the development of educational resources which could be used upon their return to their home institutions.

The programme for the Fellowship should be devised in such a way as to ensure the long-term goal of establishing a sustainable network after Fellows return to their home institutions. The scientific programme must be developed in such a manner that it is topical, yet accessible to physicists who may have little or no background in astrophysical techniques. The networked approach demands that the participants should work on a common project, or 'key theme'. The key theme should be chosen with regard to scientific relevance, modest computing hardware and software requirements, and modest data storage and data transfer requirements. In time, the network will reach a critical mass, after which it may be capable of addressing other key themes.

It is important to structure the network in such a way that scientists may fruitfully collaborate with each other from their home institutions. This will be accomplished through providing Fellows with a common set of software tools and teaching resources. Some measure of remote access (via email) to a robotic telescope at SAAO could also be included. In this way the participating scientists could continue to obtain new data of excellent quality for their own research projects.

If this pilot project is successful, a new group of Fellows will be selected each year, and the network will grow with time. In this way, there could be a number of potential users of SALT by the time the telescope is commissioned in 2005.

Acknowledgments. The author gratefully acknowledges financial support from the IAU and the South African Astronomical Observatory for participation in this Special Session.

References

Buckley, D.A.H., O'Donoghue, D.O., Sessions, N.J., Nordsieck, K.H. 2000, *Instrumentation options for the Southern African Large Telescope*, Proc. SPIE, in press

Martinez, P. 2000, *Bulletin on Teaching of Astronomy in the Asia-Pacific Region*, (See also http://www.saao.ac.za/~wgssa)

O'Donoghue, D.O. 2000, *The correction of spherical aberration in the Southern African Large Telescope (SALT)*, Proc. SPIE, in press

Ramsey, L.W., et al. 1998, *The early performance and status of the Hobby-Eberly telescope*, in *Advanced Technology Optical/IR Telescopes VI*, Proc. SPIE, 3352, 34-42

Rijsdijk, C. 2001, this volume, pp. 117-130

SALT Collateral Benefits Plan 1999, South African Astronomical Observatory/National Research Foundation

Stobie, R.S., Meiring, K., Buckley, D.A.H. 2000, *Design of the Southern African Large Telescope*, Proc. SPIE, in press

Discussion Batten noted that Martinez had mentioned five African countries are members of the IAU. At this Session we had representatives from seven African countries – probably a record for the IAU and a tribute to the energy that Martinez had put into the UN-sponsored Working Group on Space Sciences in Africa. Hemenway asked if the pilot fellowship scheme described by Martinez was confined to university level, or whether it was also open to primary-school and secondary-school teachers. Martinez replied that it does not at present include school teachers but thanked her for the suggestion.

Abstracts of Poster Papers

Abstracts of poster papers relevant to this section of the Special Session are presented below.

The Error in Solar-Diameter Measurements Induced by Atmospheric Turbulence

Lyes Lakhal et al., Observatoire d'Alger-CRAAG, BP 63 Bouzaréah 16340 Alger, Algeria

Diameter measurements performed with the solar astrolabe are affected by atmospheric effects. In order to deduce significant astrophysical results, it is necessary to know how these effects contribute to the error of diameter measurement. For this purpose, a numerical simulation was developed leading to a law of error behaviour with observing conditions. In order to validate these results, recorded data obtained at the Calern observatory astrolabe (France) during the period 1996 - 1997 are used. A good agreement is observed with the simulation, which permit salso to give access to daily atmospheric time constants. (Co-author is A. Irbah, Algeria.)

Astronomy Research in Bolivia

Dmitry D. Polojentsev et al., Pulkovo Astronomical Observatory, St Petersburg, Russia

1. An astronomical expedition from Pulkovo observatory in Bolivia, near Tarija was organized in 1982. The first telscope was an astrograph (D=23 cm, F=230 cm, field = 5×5 degrees). Sucsessful observations on this instrument are still being made. In all 7 astronomical instuments were installed. Now they are the National Bolivian Observatory.
2. The main results of astrophysical investigations were devoted to 4-color photometry of supernova 1987A and the creation of a Spectrophotometric Catalogue of 60 Selected Southern Stars.
3. The main results of astrometrical investigations were made on two catalogue problems: Photographical Catalogue for Southern Star (FOCAT-S) and Equatorial Catalogue (ECAT). The first was the foundation for southern part of PPM Catalogue.
4. A time service was organized in 1988 in Tarija at the National Astronomical Observatory "Santa Ana". In 1997 Pulkovo observatory assisted to reconstract it.
5. The only Planetarium in Bolivia "Max Schreider"in La Paz was founded in 1976.
6. The Associacion Boliviana de Astronomia (ABA) was organized in 1969 in accordance with a Goverment Resolution. It has branches in Potosy, Santa Cruz, Sucre, Tarija etc.
7. The development of the astronomy in Bolivia depends directly on cooperation with the astronomically developed countries.
(Co-author is R. Zalles, Brazil.)

Sofia Sky-Archive Data Center: Photographic Plate Collections for Developing Countries

Milcho. K. Tsvetkov et al., Institute of Astronomy, 72 Tsarigradsko Blvd, BG 1784 Sofia, Bulgaria.

The Sofia Sky Archive Data Center (SSADC) was developed on the base of the Wide-Field Plate Database (WFPDB - http://www.skyarchive.org) as a project of the Working Group on Sky Surveys, Commission 9 of the IAU, and is dedicated to saving plate collections. The center manages 12 PCs connected in a local computer network and a PDS 1010 microdensitometer donated by the European Southern Observatory. The main field of operation is the WFPDB development, plate digitization and image processing for different astronomical tasks in South- and East Europe (Russia, Ukraine, Armenia, etc.), and as a regional coordinator especially for the neighbour countries - Romania, Yugoslavia, Macedonia, Greece and Turkey. The main problem in the way of the WFPDB development is the creation of the computer-readable plate catalogues of the original logbooks because the speed of converting the logbooks in a computer-readable form is very low. The message is: we have to find the way to accelerate this important part of the project where the role of the developing countries in this direction should be very important. (Co-authors are: K. Tsvetkova, K. Stavrev V. Popov, H. Lukarski, A. Borisova, M-E. S. Michailov and G. Borisov of Sofia, Bulgaria, and S. Christov, Bulgarian South-West University).

CCD Observations with Small Telescopes of Moving Bodies

Oleg P.Bykov, Pulkovo Astronomical Observatory, St Petersburg, Russia.

Astronomy for developing countries must be simple, cheap and attractive. Advanced amateurs with small astronomical CCD instruments could be its base in these regions. Together with the astronomical community and professional astronomers of other countries, amateurs can solve a lot of practical tasks connected with the CCD observations of moving celestial bodies.

The author has analyzed the CCD observations of numbered and unnumbered Minor Planets made in 1998-1999 by amateur astronomers around the world and published in the MPC. The accuracy of their observations is sufficiently high and their contribution to the MPC database is considerable. Amateurs are discovering unknown celestial objects and have a right to name the discovered minor planets.

Pulkovo observatory could take part in Astronomical education and Software creation for the professional astronomers and amateurs from Developing Countries.

Astronomy in Uzbekistan

Sabit P. Ilyasov et al., Ulugh Beg Astronomical Institute, Uzbek Academy of Sciences, Astronomicheskaya Ul 33, Tashkent 700052, Uzbekistan

Ulugh Beg Astronomical Institute (UBAI) of the Uzbek Academy of Sciences is one of the oldest scientific institutions not only in Uzbekistan, but in the whole of Central Asia as well. There are five departments in the institute. The main directions of research are solar physics, young non-stationary and close binary stars in star formation regions, satellite geodynamics, non-linear and non-stationary evolution of galaxies. Helioseismology studies carried out in the frame of the IRIS (International Research on the Interior of the Sun) and TON (Taiwan Oscillation Network) projects. Astrophysical programmes such as a search for periodicity in star-formation regions, study of close binary stars in the same regions, as well as in open clusters, CCD photometry of extra-galactic objects as gravitation lenses have been made at the Maidanak Observatory, which is located in the south-east of Uzbekistan. Monitoring of the seeing at Mt. Maidanak from 1996 to 1999, using ESO Differential Image Motion Monitor, showed that its atmospheric conditions are comparable with the best international observatories. The present status of the main fields of research and prospects are discussed. (Co-author is Shurat A. Ehgamberdiev, Uzbekistan.)

Astronomy in the Republic of Macedonia
Mijat Mijatovic, Institue of Physics, Skopje, Republic of Macedonia

Astronomy in the territory of today's Republic of Macedonia has a century-long history. This history is presented in the essay, begining with M. Trpković's suggestions to reform the orthodox calendar in 1900s, through the foundation of the first faculty of Macedonian language in Skopje in 1946, until today's situation in astronomy.

In second half of the twentieth century, the development of astronomy in the Republic of Macedonia is divided in two different periods: before and after the big earthquake in Skopje in 1963. The first period is characterized by hope and enthusiasm, and a little observatory was started to be built, but it was destroyed in the earthquake. The last ten years a new upsurge is seen in Astronomy in the Republic of Macedonia, which is founded on Balkan and international collaboration.

Astronomy without Astronomers?
Magdalena Stavinschi, Astronomical Institute, Romanian Academy of Sciences, Cutitul de Argint 5, RO 75212 Bucharest, Romania

Astronomy in Romania has an old tradition. After half a century of privations and isolation from the rest of the world, we believed that the changes undergone by our country in 1989 (and by the neighbour countries, as well) will be benefit for the Romanian astronomy, too. Indeed, it was, but for a very short period. The young people left the country, one by one, and others cannot accept the low salary offered by a research institute. The economy doesn't allow us to enrich the astronomical endowment. Of course, we cannot close the observatories. We have to find other ways to save the astronomy in this part of Europe, especially in the epoch of the space astronomy.

A 100-Year Astronomical-Data Bank: Collaboration Possibilities and Some Problems

Salakhutdin N. Nuritdinov, Ulugh-Beg Astronomical Institute of Uzbek Academy of Sciences, Astronomicheskaya Ul 33, 700052 Tashkent, Uzbekistan

Our Department of Galactic astronomy and Cosmogony of the Astronomical Institude of Uzbek Academy of Sciences has rich photo meterials obtained over more than 100 years by two telescopes (Normal Astrograph with F= 3500 mm, D= 330 mm and Zeiss Double Astrograph with F= 3000 mm, D= 400 mm). The main objects studied are:

1 Open and Globular star clusters,

2 Regions of the Milky Way and the Pulkovo Observatory program.

3 Comets and Asteroids

4 Planets and their Satellities.

There are also astronomical data received in the framework of a number of International programmes. Now we are working out some research-complex programmes of these objects. We are ready to collaborate on these programmes.

Galactic Cluster Studies and Emission-Line Star Surveys with the Schmidt Telescope at Bosscha Observatory

Suhardja D. Wiramihardja, Bosscha Observatory and Institut Teknologi Bandung, Lembang 40391, Indonesia

Galactic cluster studies and emission-line star surveys have conducted at the Bosscha Observatory since the establishment of Schmidt telescope in 1959. Though the mirror diameter is moderate, its coverage of 25 square degree sky area makes this telescope capable for survey works. Some results and plans will be given in the poster

Development of Astronomy in Ecuador

Ericson Lopez, Observatorio Astronomico de Quito, Ecuador

Section 5: Small Telescopes or Internet Access?

Astronomy for Developing Countries
IAU Special Session at the 24th General Assembly, 2001
Alan H. Batten, ed.

The Choice of Small Telescopes

David L. Crawford

GNAT, Inc. Tucson, Arizona, 85716, U.S.A. www.gnat.org . e-mail: crawford@darksky.org

Abstract. Small telescopes can be powerful tools for astronomical research. Many are being used by professional and amateur astronomers to do important, even frontier, research. They are also extremely valuable tools for education. This paper discusses the characteristics of such telescopes, and it makes recommendations about items such as field of view, focal length, and so on. It also discusses a few small telescopes representative of what is currently being done. Astronomy needs such facilities as much as it needs the giant telescopes. They complement each other very well.

1. Introduction

New-generation small telescopes, ones using modern CCD imaging detectors and taking full advantage of computer control and of communication networking, are truly frontier instruments for astronomical research and for education. These new-generation aspects apply just as much to small telescopes as they do to the much larger (and much more expensive) ones. It is a fact that the automatic and networked use of such small telescopes could provide many more quality observing hours to the astronomical community worldwide. A non-profit organization, GNAT, Inc., has been created with the goals of developing and operating such a global network of astronomical telescopes and in being a catalyst for all those interested in the effective use of small telescopes for research and for education. GNAT will be a relatively low-cost operation, with low overhead and a small staff, but with many members, allies, and partners. We believe that the program has very little risk, but with nearly unlimited upside potential. This paper can only summarize some of the problems and potentials.

2. Special Problems for Developing Countries

1 *Hardware and other facilities.* These often do not exist in developing countries and, if they do, may be outmoded or poorly maintained. It is difficult to build, to operate and to maintain facilities with low funding and few trained staff. In general, of course, the staff are more interested in do-

ing research, or education, than in developing and maintaining a viable facility.

2 *Bureaucracy and other issues.* Unfortunately many developing countries suffer from these problems as much as, or even more than, do the more technically advanced countries.

3 *Poor communications.* These are a big problem in almost all developing countries. Astronomers in such countries need a great deal of help with the Internet and staying in touch.

4 *Exchange of people.* This is needed for interaction and cultivating allies, but is often a problem for developing countries.

3.   Research and Education Drivers

In all countries, there are good people and there are good ideas. We have heard here at this meeting of many potentials for research and education, and of many places where realization of these potentials is possible, and much needed. The people in these places want to do good things, interesting and useful research, and to use astronomy as an educational tool for students. There are many other examples that we have not heard about here. Valuable techniques with small telescopes include: imaging, photometry, spectroscopy, and others. There is a growing array of good instrumentation at lower costs than in previous years. Small telescopes complement very well the research being done with large telescopes and with space facilities. Just like the operator of a fleet of trucks, we need resources of all sizes. The small ones cost much less than the large ones; both are much needed tools. There is, of course, other needed infra-structure, such as libraries, technical facilities and help, and allies and mentors. Finally, we note that it is essential to fit the local situations and potentials. All countries are different. Viable solutions must take that fact into account.

4.   Telescopes and Instrumentation Issues: Types, Specifications and Sources

There are three ways to obtain access to these tools. Astronomers can develop and operate an observatory within their own country (perhaps of particular interest for those countries with good observing conditions), or they can become involved in the operation of a remote observatory, located elsewhere in excellent observing conditions, or they can "observe" by accessing existing data bases. We will discuss these options below. But wherever telescopes are located, some items are in common. First, what type of telescope? What mounting, size and instrumentation? For a number of reasons, we will consider here only "small telescopes." By these, we mean ones with an aperture of 0.5 m to 1.0 m. Instruments in this size range are of relatively low cost, easier to construct, to ship, to install, to operate, to maintain. These modest-aperture telescopes, if quality ones, can be used for very viable and exciting research, at the frontier of many scientific problems. Any type of mounting will be acceptable, and many of the problems of design are much simpler than for larger telescopes. However, the telescopes should be good ones, of high quality. What about design specifications? GNAT has published the following specifications, after much thought about needs and current and past designs, and after a much discussion with observers and with telescope makers:

1 Imaging and photometry will be the main roles for a small telescope, although spectroscopy and other techniques could be used.

2 An imaging CCD photometer is the main instrument, but others could be used.

3 *Value per cost* is a key element.

4 The use of a common design for many telescopes results in the sharing of design and fabrication costs by many users, thus maximizing *value per cost*.

5 Users working together can help to improve quality and to lower costs.

6 Reliability of the instrument is critical: low operating costs are as important as low capital costs.

7 The primary focal ratio should be in the range 1.5 to 2.0; the secondary focal ratio in the range 6 to 9.

8 The field of view should be designed to handle a 2000 square CCD chip, although telescopes could operate with smaller chips. The are potential uses for telescopes with larger fields of view but these would cost considerably more.

9 Pixel size, seeing and field of view should all be matched but some compromise will probably be necessary. Note: in this area small telescopes are better than larger ones.

10 Image quality should be between 0.6 and 0.8 arc-sec at full-width half-maximum (FWHM).

11 Pointing accuracy should be approximately 10 arcsec (open loop) or 1 arcsec (closed loop)

12 Tracking should be accurate to within about 0.1 arcsec over several minutes.

13 The telescope control system should offer the possibility of fully automatic and remote operation.

14 Communication to the telescope to be via ATIS (Automatic Telescope Instruction Set).

15 Telescope scheduling must take account of the needs of many users of many telescopes at many sites.

16 Good documentation is critical.

17 The housing and site of the telescope are dictated by local considerations and are not in the standardized design.

18 Sites should be of high quality with good seeing and many clear hours, but it is not essential to have the site of highest quality if developing it would be too costly.

4.1. On-site facilities in one's own country

How does one go about buiding an observatory? Among the issues to be considered are funding, observing conditions in the country concerned, technical issues (avaiability of electric power etc.), capital costs, operating costs, maintenance issues and upgrading: all are important and all need very careful consideration. It is often easier to get the capital costs covered than to obtain the operating costs. Many examples exist of real problems due to lack of operating funds, even in developed countries.

What is "On site"? Pride of ownership is not enough. Remember that even a site within the country is remote for most users.

4.2. Remote facilities

Do astronomers in a coountry set up a remote observatory as a single group or join in a consortium? Several possibilities exist; the choice will depend on circumstances. Similar technical considerations to those already mentioned must be taken into account. There are again many possibilities for doing the research. Scientists can work as individuals, as members of a team or on observatory projects. There are also many ways to be involved in the technical development of hardware and software. Access to the telescope can be by Internet or even by mail. GNAT for example is not a real-time operation, but nearly 100 percent queue scheduling. Pluses and minuses need to be thoroughly understood and discussed, in each individual case. Most remote operations can and will be very adaptable to individual needs and constraints. Overall operational costs can and should be relatively low.

4.3. Observing data bases

Problems and potentials of access are the same as in remote observing. Solving the access problem, whether by Internet access or otherwise, will be one of the first things that must be successfully addressed by any country. Other papers at this meeting have discussed this issue. The potential is great for anyone and any country. But note that the data in the database must be of high quality, and anyone (observatory or individual) inputting data must ensure that they are only high-quality data. One must understand what makes for high-quality data. Unfortunately, too many examples exist of low-quality data.

5. Can a Remote-Observatory Concept Work?

Of course. The needs and the potential are there, and the technology is ripe. The implementation of a successful project will need help from many individuals, from many countries. It can happen. It will. The only questions are "When?" and "How?" and "How Cost Effective?" The key elements and needs are:

i Funding. Note that not only can the observatory be distributed globally but so can the funding sources.

ii People, in the countries concerned and partners.

iii The facilities themselves, both in the home country and the remote site.

iv Education, in the home country, of those involved in research and technology, and of teachers and students. Some edication can be abroad, through partners, and "meetings."

v Travel funds.

One example of the many with potentials is GNAT. GNAT's Goal is to develop and operate a global network of small telescopes. At least two 0.8-m aperture telescopes at a minimum of each of six worldwide quality sites, three in the northern hemisphere and three in the south. In addition, there would be a home base, to be the catalyst of the operation and the communications center. Besides the 0.8-meter (and larger apertures ones in the future), at the same sites, GNAT will develop and implement "three-shooter telescopes." These are small-aperture instruments, on the same mounting, directed at a fixed altitude and azimuth for months at a time. They operate in a drift-scan mode and scan the sky in a strip of 48 arcminutes wide at a particular declination. Comparison of the images from night to night will permit the recognition of slow moving objects and of variable brightness objects.

Access to GNAT should be possible by almost anyone, anywhere. Usage of the facility can be via observatory programs, consortia programs, or individual research projects. Many types of programs are possible, ranging from variable stars through the search for extra-solar planetary occultations. The data will be accessible to all via on-line data bases. GNAT is really an electronic, distributed, world-wide observatory, in facilities and in staff. To be successful, it needs funding, of course, and the involvement of many individuals, as unpaid, part-time staff members and as observers.

GNAT is now operating a prototype 0.5-m telescope, in a fully automatic and efficient mode, in the Tucson, Arizona, area, every clear night. We are using a 1000 square pixel CCD imaging photometer and ATIS communication software. The telescope control software, by Don Epand, is working very well. The telescope has been obtained from SciTech, a California company. There are other suppliers, but we have been dealing only with SciTech. Some of the suppliers come and go, and some seem to deal mostly with "virtual telescopes," not really yet having quality operating telescopes. We will be negotiating a lease/option arrangement for the first 0.8-m telescope later this year. It will be located in the San Diego area, and allow linked operation between the two telescopes, both operated remotely and automatically.

GNAT will operate many of its programs through Working Groups, whose job will be to develop and implement the observing programs. The first such, for Photometric Systems and Standards, is now being formed. Both BVRI and Strömgren standard systems will be in operation on the 0.5-m telescope later this year, with a program of observing standard stars on a regular basis. A second working group, on Open Cluster Photometry, will be operating shortly. Others will follow next calendar year.

A prototype of the "three-shooter" is now being operated to verify system operation and performance. Work has begun on the first full-sized system, and software development is well underway. Costs of such a "three-shooter" facility will be remarkably low, and the data output rate remarkably high.

6. Conclusions

There is a real need for a networked remote observatory of small telescopes. It is a very viable, cost effective concept, one that can supply a lot of good-quality photometric data to the astronomical community, world-wide. For example, with GNAT, astronomers do research and GNAT takes care of the rest: development, implementation, operation, liabilities, and all the non-research aspects of owning and operating a major observatory. All "members" can and would be active partners in the efforts. It has a very high *value per cost* ratio. There are other similar proposed networked facilities. Such facilities will be of great benefit to astronomical research and education worldwide in the coming years. Small telescopes will remain a needed and valuable resource to the worldwide astronomy indefinitely.

Discussion
Hearnshaw thought GNAT a great concept but pointed out that Crawford had been advocating it for many years. Why was it taking so long to get started? Crawford replied that the start was bound to be slow in the absence of funding and permanent staff, but progress is being made and the protoype telescope is performing very well. Rijsdijk asked if existing small telescopes at "mothballed" observatories could be incorporated into GNAT. Crawford thought not, in general, since GNAT is based on having "new-generation" telescopes. Other organizations have tried taking over the telescopes to which Rijsdijk referred, with mixed success. Hingley drew attention to the New Zealand conference in 1985 in which Warner analyzed "numbers of research papers per megabuck" which favoured small apertures. Warner also spoke of "aperturism" in astronomy – which kills telescopes. Crawford said that many astronomers catch the disease of "Aperture fever". Metaxa appealed for all astronomers to fight light pollution and similar problems in other parts of the electromagnetic spectrum, before we lose the night sky.

Astronomy for Developing Countries
IAU Special Session at the 24th General Assembly, 2001
Alan H. Batten, ed.

What Can be Done with Small Telescopes?

Boonraksar Soonthornthum

Sirindhorn Observatory, Department of Physics, Faculty of Science, Chiang Mai University, Chiang Mai 50200, Thailand. e-mail: boonraks@cmu.chiangmai.ac.th

Abstract. Modern astronomy has developed rapidly in the last few decades. Large telescopes, several metres in diameter, have been constructed and installed at many sites around the world, in order to carry out high-quality research work at the frontiers of astronomy. In many parts of the world, only small telescopes are available, because of budgetary limitations and the stage of development of science in the country. However, much effort has been put into using these small telescopes to the best of their capabilities. A small telescope with modern detectors can do good astronomical research. Sirindhorn Observatory, Chiang Mai University, is the only observatory in northern Thailand which plays an active role in astronomical research. The major instrument is the 40-cm Cassegrain reflecting telescope with standard wide-band and intermediate band photoelectric photometers, CCD photometers and a CCD spectrograph. Because of Thailand's hot and humid climate, the telescopes and detectors at Sirindhorn Observatory need regular maintenance. Major research at the Observatory emphasizes the study of physical propoerties and evolution of close binary systems, especially near-contact and contact binaries, by photometric techniques. Networks in this field of research have been established through national and international collaboration with some astronomical institutes in the region. A larger telescope is being developed for more efficiency in operation; it is expected to be able to serve the future development of astronomical research in Thailand.

1. The Situation in Thailand

Thailand is one of the countries with very little science tradition in the past. Astronomy and other branches of science, in their original local form, were on modest levels. Astronomy was not practised as a pure science based on observation or theoretical background, but rather to discover items of immediate usefulness. Knowledge was mainly obtained from outside the country and research was hardly ever undertaken. For this reason, the development of astronomy in Thailand was slow.

In the last twenty years, astronomy became more popularized and interesting to the public. Both government and private sectors promoted many astronomical activities. However, higher education and research activities in astronomy are still growing rather slowly in Thailand. Some universities initiated astronomy as a field of studies in the school of science, both at undergraduate and graduate levels. A few astronomical observatories were founded to serve educational, research and social needs.

However, due to the limited budget for the development of astronomy in Thailand, only a modest-sized telescope was provided for each observatory. Many local astronomers have put their efforts into developing astronomical facilities. Various kinds of detectors e.g. photoelectric photometer, CCD photometer, CCD spectrograph etc. were provided as equipment for existing telescopes.

2.   What can be done with Small Telescopes?

Small telescopes 40 cm to 60 cm in diameter, with modern detectors, can do good astronomical research. If all the correct procedures of data collection are followed, data of good quality can be obtained from these telescopes.

Chiang Mai University, the main educational institute in northern Thailand, possesses an observatory named "Sirindhorn Observatory", founded in January 1977. The site of this observatory is in Doi Suthep-Pui National Park in Chiang Mai province at an altitude of 784 metres above sea level, at latitude $18°$ $47'$ $19''.5$ N and longitude $98°$ $55'$ $29''.9$ E.

From the beginning, the major instrument has been a 40-cm Cassegrain reflecting telescope with an SSP-3A solid-state photometer with standard BVRI filter system. The main purpose is to use this telescope to support astronomical research work, teaching and social public service at the Chiang Mai University.

The main concern after completion of the Observatory was how to design good projects for this small telescope and solid-state photometer. We first realized that if we worked on sufficiently bright stars, a small telescope is capable of the same photometric accuracy as a large telescope.

The next problem was: what types of celestial objects should be selected for observation with the telescope in order to gain the maximum outcome? We finally came to the conclusion that there are many areas, particularly in binary-star research, where powerful contributions can be made, even with only a small telescope.

Several astronomical projects on binary stars were then planned for the 40-cm Cassegrain reflecting telescope and solid-state photometer. Finally, many close binary systems e.g. RZ Cas, IU Aur etc. were observed and their light-curves were obtained. Times of minima and contact times were determined for U Cep, during the hours of totality, with this small telescope, as shown in Figure 1. Absolute photometry was also undertaken; extinction coefficients were obtained at Sirindhorn Observatory.

However, observations on faint celestial objects, normally stars fainter than eighth magnitude, showed that, for this small telescope and solid-state photometer, the fluctuations in light-curves were too large, under the sky conditions of Chiang Mai, for good-quality data to be obtained.

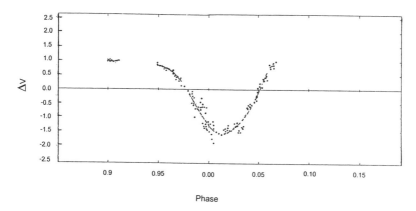

Figure 1.   Observations of U Cephei at primary minimum, using the 40-cm telescope and SSP-3A photometer.

3.   Improvements in using a Small Telescope at Sirindhorn Observatory

Since 1992, astronomical collaboration between Thailand and the Peoples' Republic of China has been initiated through government support from both countries. Two major research works were proposed. The first proposal emphasized photometric and spectroscopic studies of starspots on RS CVn binary systems, in collaboration with Beijing Astronomical Observatory. The second proposal emphasized photometric studies of some near-contact and contact binary systems, in collaboration with the Yunnan Observatory.

By these collaborations we received opportunities to use 60-cm, 1.00-m and 2.14-m telescopes in China for photometric and spectroscopic observations. More experience was gained in observational techniques with CCD photometers and spectrographs. We also gained more experience in using some software for data reduction and analysis, e.g. IRAF image-reduction program, Wilson-Devinney program etc.

The astronomical facilities at Sirindhorn Observatory were also improved during these collaborations. A Star I CCD camera system and CompuScope CCD 800 integrating camera were provided for use with the 40-cm Cassegrain reflecting telescope as CCD photometers. Observations on faint celestial objects are now easily made with the small telescope and CCD photometer. The IRAF image-reduction program has been installed under the LINUX operating system at Chiang Mai University.

A CCD spectrograph with a dispersion of 120 A mm^{-1} has been provided to upgrade astronomical research with the 40-cm Cassegrain reflecting telescope at Sirindhorn Observatory as shown in Figure 2. Several spectroscopic projects can be undertaken with this combination. Typical projects fall into two main categories: measurement of spectral properties to determine chemical and physical processes in celestial objects and measurement of radial velocities. A sample CCD spectrogram is shown in Figure 3.

Figure 2.   The 40-cm Cassegrain reflecting telescope and CCD spectrograph.

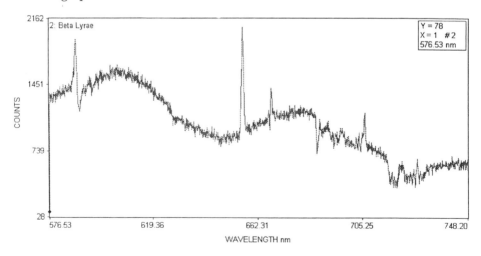

Figure 3.   Spectrogram of β Lyrae obtained with the small telescope and CCD spectrograph.

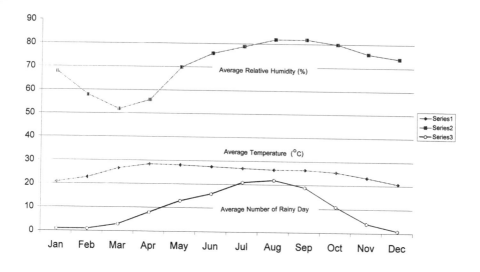

Figure 4.   Climatic data for Chiang Mai averaged during 1988-1997.

4.   Working in a Hot and Humid Climate

Chiang Mai is the major city in northern Thailand. The city lies in a fertile valley region between terraced mountain ranges. The climate of Chiang Mai is slightly different from those of other parts of Thailand, mainly due to the highland structure in this region. The altitude of Chiang Mai is about 300 metres above sea level. Climatic data for Chiang Mai, averaged for the period 1988-1997, are shown in Figure 4. From these data, the average climate in Chiang Mai is seen to be rather hot and humid, with average temperature and relative humidity of 25°.4C and 71%, respectively.

The cool season lasts from late October to the end of February. The average daytime temperature is 21°C; nights are much cooler, the temperature can go as low as 0°C in many areas. The coolest months are December and January. The hot season is from early March until the end of May, when the average daytime temperature is 30°C. The hottest month is April. The rainy season usually begins in early June and lasts until the end of October. The average temperature then is 25°C. The wettest month is September.

Heat, humidity and rain all jeopardize telescopes and other astronomical equipment. Therefore, regular maintenance is needed at Sirindhorn Observatory. In the hot climate of Chiang Mai, where the relative humidity remains above 60%, lenses, mirrors and other equipment have to be regularly checked and kept in a dry place. Crystals of silica gel are also used to protect lenses and mirrors from moisture. As the wet season approaches, the wet and humid weather presents an obstacle to any observation. The Observatory is always closed all through the wet season and maintenance of the telescope is done annually at the beginning of the cool season.

The sky in Chiang Mai is always clear in the cool season. There are about 130-160 nights in a year available for observations at Sirindhorn Observatory. Some spectroscopic observations can be made during the hot season.

5. Future Development

The construction of a new 50-cm Ritchey-Chretien reflecting telescope in a new housing at Sirindhorn Observatory is now in progress. It is expected that the new telescope will be installed at the Observatory by the end of the year 2000. It is planned to use this telescope for CCD photometric observations of celestial objects, especially binary stars, in multi-wavelength bands.

Computing facilites with inter-networking are now developed and widely used in Thailand; astronomers in Thailand also have opportunities to access astronomical data archives, information and services, which are widely distributed through the internet. Software, catalogs, data, journals and updated astronomical information can be accessed and freely downloaded from many WWW and ftp sites. This provides opportunities for astronomers in Thailand to obtain high-quality data and standard software packages which are very important in supporting future research and international collaboration for the staff at Sirindhorn Observatory, even though only small telescopes are available to them.

Acknowledgments. The author would like to express his thanks to Dr Alan Batten for giving him the opportunity to participate and to give an oral presentation in the Special Session of the XXIV General Assembly of the IAU in Manchester, U.K. Financial support from both the International Astronomical Union and Chiang Mai University are gratefully appreciated.

References

The following documents, although not specifically cited in the text, give further background information:

Model 10c Cassegrain Spectrograph, Optomechanics Research, Inc. 1997.

Model SSP-3A Solid State Stellar Photometer: Technical Manual for Theory of Operation and Operating Procedures, Optec., Inc, 1988.

Star I CCD Camera System: Hardware Reference Manual, Photometrics LTD., 1989.

The Compuscope CCD 800/1600 Integrating Camera: User's Hardware Guide, Compuscope Precision Instrument and Software, 1993.

Kannapan, S. and Fabricant, D., 2000, Sky and Telescope 100, No. 1, pp.125-132.

Percy, J.R., 1986, in *The Study of Variable Stars Using Small Telescopes*, Cambridge University Press, Cambridge.

Discussion

Dworetsky pointed out that the Starlink software distribution is available free of charge on CD-ROM, from the Starlink Project at the Rutherford-Appleton Laboratory in the U.K. It includes many types of extraction and reduction programs and can be obtained in LINUX, Solaris and Dec Alpha programs.

Astronomy for Developing Countries
IAU Special Session at the 24th General Assembly, 2001
Alan H. Batten, ed.

Simple Science, Quality Science

John R. Percy

Department of Astronomy, Erindale Campus, University of Toronto,
Mississauga ON, Canada L5L 1C6. e-mail: jpercy@erin.utoronto.ca

Abstract. This paper examines ways in which those in astronomically-developing countries can begin to take part in meaningful astronomical research. It is based on a discussion paper, prepared by the author for a special session of IAU Commission 46 on 12 August 2000, and on the many comments contributed by the audience after the paper.

How can those in the astronomically-developing countries (ADCs) begin to take part in meaningful astronomical research? In this context, research is defined as "producing useful astronomical data or results, leading to presentations at conferences, and publications in journals by the observers, or by others". The answer to the question will differ from one ADC to another. It will depend on the local economic, structural, and human capacity. Possible answers should be the result of careful, ongoing discussions between the local astronomical community, and those from "outside" who wish to help. These discussions, and the follow-up assistance and advice, may have to be continued over 3-5 years, to overcome the initial "learning curve", and to build up the necessary momentum to reach the goal. Good research cannot begin instantly; there may have to be a transition period between "no research" and "front-line research". The IAU Commission 46 Program Groups on Advance Development, and on Teaching for Astronomical Development, are appropriate bodies to engage in these discussions.

It is assumed that the local researchers will begin with small astronomical telescopes — as small as 20-cm aperture. They may then progress to telescopes in the 40-cm to 60-cm range. The capabilities of small telescopes have recently (Irion 2000) been discussed in *Science* — one of the two most prestigious general science journals in the world. The government of Japan has very generously donated telescopes, of approximately 0.5-m aperture, to several ADC's, and has provided training in Japan for some of the local astronomers. In addition to the research (and education) capabilities of such telescopes, they can serve as "attractors" of students, and they can give credibility to the local astronomy activities in the eyes of other academics, administrators, and government.

The issue of "quality science" is important. The quality can usually be assessed by the local observers in some way, but it is also useful to have an external mentor or collaborator who can assess the quality of the data being produced, and offer advice and instruction on how to improve. This is one advantage of participating in an international research collaboration. It takes a number of weeks or months to be sure that the equipment and instruments are working satisfactorily, and producing good data. Snowden (2001) has questioned whether astronomy research can be done in an ADC without extensive direction from "outside". It is true that *some* of the local astronomers must have a sense of how science works, and what research is all about. It is especially important for research to be part of the local university teaching curriculum, so that students will learn that knowledge is developed through research, not through textbooks. They must learn — preferably from experience — that research is exciting, and fun! Education also serves to heighten the profile of astronomy in the ADC, both within the universities, and among politicians and the public. There are many practical considerations to beginning astronomical research, and Kochhar (1996) has mentioned four:

- When a new astronomical facility is set up, it should be at a level consistent with the workshop facilities and infrastructural support available. The equipment should not overwhelm the user.

- For the initial training of manpower, co-operation should preferably be sought from countries which are culturally akin to the host country.

- Attempts should be made to integrate astronomical facilities with the teaching program.

- For best results, observational programs should be chosen so as to form a part of an international campaign.

Because of the small number of potential researchers in the ADCs, it would be most efficient and effective if the research was carried out collaboratively by several members of the local "astronomical community" — professional astronomers (if any), scientists and technologists from related disciplines (physics, engineering, math, statistics, geology), advanced amateur astronomers, undergraduate and graduate students, science teachers etc. Each of these groups brings special skills and enthusiasm to the research enterprise. On the other hand, there may be problems of hierarchy — such as lack of respect for the absolutely essential role that technicians play in making the equipment work effectively, or for the capabilities of students as researchers.

There is also the practical matter that astronomical research is usually done at night. That may raise issues of security, of accommodating mixed observing groups, or simply the tradition of "normal working hours".

Small telescopes can also be used for teaching at the high-school and university level, for public education, for obtaining images which can be used by the local media etc. These applications will not be discussed here, but they should certainly be integrated into the program of the telescope as a way to build up public support for astronomy.

The most obvious field of research (in part because it is my own) is the measurement of variable stars. This can be carried out at various levels of expense, technical difficulty, and sophistication: visual, photographic, photoelectric (PEP) and charge-coupled device (CCD) photometry. These require increasing levels of instrumentation for data acquisition, reduction, and analysis. All are useful for research. There are several organizations which support and co-ordinate such work — notably the American Association of Variable Star Observers (AAVSO). The demand for AAVSO visual observations, from the professional astronomical community, for instance, has increased by a factor of 20 since the 1970s. The AAVSO also has an education project *Hands-On Astrophysics* which provides a basic introduction to variable star observing and analysis, suitable for senior high school students, undergraduates, and amateur astronomers.

At a more advanced level, using a telescope of 20-cm aperture or larger, researchers can do PEP or CCD photometry. The cost of a PEP photometer is about U.S. $1500; a computer is useful but not absolutely essential. The cost of a CCD photometer is probably U.S. $5000, and a computer (486 or better) is essential. The AAVSO co-ordinates both PEP and CCD photometry, and there is also the International Amateur-Professional Photoelectric Photometry (IAPPP) organization which supports such work internationally. At a slightly higher level, there is the international Center for Backyard Astrophysics (CBA), co-ordinated by Professor Joseph Patterson of Columbia University. The CBA uses an international network of CCD-equipped small telescopes to measure the rapid variations in cataclysmic variables. This has resulted in a steady stream of important papers. The typical cost of a CBA "station" is U.S. $12,000 for a telescope, CCD camera, and computer. (This, of course, is a large amount of money for an ADC, but it is the kind of sum which might be provided by an external grant or donation. The goal is to make the best possible use of such funds.) There is a wide variety of other photometric projects which could be done through existing international collaborations: such as monitoring variable Be stars, RS CVn binaries, and other types of stars (Percy 1998).

Another interesting possibility in variable-star research is the analysis and interpretation of on-line archival data — from the HIPPARCOS epoch photometry database of 120,000 stars, for instance. HIPPARCOS maintains an education web page, and there is software available on the web (from the AAVSO, for instance) which can be used for the analysis.

There are other fields of astronomy which lend themselves to "simple, quality science". They are traditionally associated with advanced amateur astronomers, because the instrumental and technical requirements are modest (Percy & Wilson 2000). [For those who are unconvinced of the value of amateur science, I recommend the eloquent article in *Science* by Mims (1999)] The fields include:

- Monitoring sunspots and sudden ionospheric disturbances, both caused by solar activity; both are co-ordinated by the AAVSO. The results are used by US government agencies as one measure of solar activity.

- Timing of occultations of stars by the Moon or other solar system objects; this is co-ordinated by the International Occultation Timing Association. Since the events are geographically specific, astronomers in the ADCs can provide useful results by timing events in their vicinities. This work provides information on the precise positions and motions of stars, and on their duplicity, and (in the case of grazing occultations) on the profile of the lunar limb.

- Meteor observing; this is co-ordinated by the American Meteor Organization, and other groups. Recent observations of the Leonid meteor shower by astronomers in Jordan and Morocco have illustrated the potential of such work. It can provide important information about the structure of the cometary debris which causes meteor showers.

There are additional, more challenging fields in which small telescopes with CCD or video-cameras are being used. These fields are opening up because of the availability of good small telescopes, CCD cameras, and powerful computers, at "affordable" (for amateur astronomers in the industrialized countries) prices — typically at least U.S. $10,000 in total. Paul Boltwood, an amateur astronomer in climatically-underprivileged Ottawa, Canada, recently reached visual magnitude 23 with a 40-cm telescope (the same level achieved by the Palomar 5-m telescope a generation ago). Ron Dantowitz, in Boston USA, has obtained ultra-high resolution images of the planets with a video-camera on a small telescope. These technologies have opened up fields of research such as (i) searching for optical after-glows from gamma-ray bursters; (ii) discovering and measuring faint asteroids and Kuiper Belt Objects, both photometrically and astrometrically; (iii) discovering and measuring distant supernovae; (iv) monitoring the photometric variations of Active Galactic Nuclei, etc.

Spectroscopy is also a suitable field for small telescopes. Simple, inexpensive spectrographs are available, using either photographic or CCD cameras, and they can be used for long-term monitoring of variable emission-line stars, for instance. Very few large, professional observatories carry out these kinds of observations. For further ideas, see Hearnshaw & Cottrell (1986), Percy & Wilson (2000), and many articles in *Sky & Telescope*.

Infrastructures are being set up to facilitate and co-ordinate partnerships between amateur and professional astronomers — the American Astronomical Society's (AAS) Working Group on Professional-Amateur Co-operation, for instance, and a series of conferences and scientific sessions (see *Sky & Telescope*, June 2000).

All of these projects, of course, require training — whether for astronomers from developed countries, or ADCs. This may be accomplished by having the trainees work with an experienced astronomer at an established observatory for as much time as is necessary, or by having the experienced astronomer spend some time with the astronomers in the ADCs. It may also be accomplished through hands-on workshops — as long as these are true workshops, and not just series of lectures. Hands-on workshops may be one of the projects to be organized by the AAS Working Group mentioned above. The UN Office of Basic Space Science, or the IAU, or organizations such as the AAVSO, might be willing and able to organize such workshops.

Again, there are practical problems. It may be difficult to find an experienced astronomer who is willing to spend several weeks in an ADC. The cost of bringing an astronomer from an ADC to an established observatory, or to a lengthy workshop, is not trivial. IAU Commission 46's Program Group on the

Exchange of Astronomers (formerly Commission 38) must address these problems. How can workshops be funded?

There are other possibilities for research, which are not as directly linked with amateur astronomers. It is possible that an astronomer in an ADC has developed a line of research already, or has a collaboration with a professional astronomer elsewhere, which could be continued. The possibilities are endless; data could be provided from an archive, or by a collaborator; major observatories might be willing to provide opportunities for an astronomer in an ADC to work at the telescope. The research might involve a solar telescope, or a radio telescope, or it may involve computer simulation or theory. In general, young astronomers need opportunities to travel, and be immersed in a research atmosphere. They need opportunities to return to positions in the ADC's, and to continue their research. International travel cannot always be limited to senior astronomers from ADC's, who may no longer be active in research because of the pressure of administrative and education duties. Young astronomers are the hope of the future. The international development of astronomical research needs the help of all IAU members who might be willing to include the ADC's in their research programs. It needs the help of organizations such as the AAVSO and its counterparts in other countries and in other fields of astronomy. It needs the help of astronomical observatories and institutes, both large and small. I hope that these proceedings will open the eyes of individuals and organizations to the research needs and opportunities of the astronomically-developing countries.

References

Hearnshaw, J.B. & Cottrell, P.L. 1986, *Instrumentation and Research Programmes for Small Telescopes*, D. Reidel, Dordrecht, The Netherlands.

Irion, R. 2000, it Science, **289**, 7 July, 32.

Kochhar, R. 1996, paper presented at the Sixth UN/ESA Workshop on Basic Space Science, Bonn, Germany.

Mims, F.M. 1999, *Science*, **284**, 2 April, 55.

Percy, J.R. 1998, *Astrophys. Space Science*, **258**, 357.

Percy, J.R. & Wilson, J.B. (editors) 2000, *Amateur-Professional Partnerships in Astronomical Rsearch and Education*, ASP Conf. Series, in press.

Snowden, M.S. 2001, this volume, pp. 266-275.

Astronomy for Developing Countries
IAU Special Session at the 24th General Assembly, 2001
Alan H. Batten, ed.

Simple Instruments in Radio Astronomy

Nguyen Quang Rieu

Observatoire de Paris, Departement DEMIRM, 61 Avenue de l'Observatoire, 75014 Paris, France. e-mail: Nguyen-Quang.Rieu@obspm.fr

Abstract. Radio astronomy has a major role in the study of the universe. The spiral structure of our Galaxy and the cosmic background radiation were first detected, and the dense component of interstellar gas is studied, at radio wavelengths. COBE revealed very weak temperature fluctuations in the microwave background, considered to be the seeds of galaxies and clusters of galaxies. Most electromagnetic radiation from outer space is absorbed or reflected by the Earth's atmosphere, except in two narrow spectral windows: the visible-near-infrared and the radio, which are nearly transparent. Centimetre and longer radio waves propagate almost freely in space; observations of them are practically independent of weather. Turbulence in our atmosphere does not distort the wavefront, which simplifies the building of radio telescopes, because no devices are needed to correct for it. Observations at these wavelengths can be made in high atmospheric humidity, or where the sky is not clear enough for optical telescopes.

Simple instruments operating at radio wavelengths can be built at low cost in tropical countries, to teach students and to familiarize them with radio astronomy. We describe a two-antennae radio interferometer and a single-dish radio telescope operating at centimetre wavelengths. The Sun and strong synchrotron radio-sources, like Cassiopeia A and Cygnus A, are potential targets.

1. Introduction

Thousands of bright stars in the Milky Way can be seen with the naked eye. Light from remote galaxies is detected with large optical telescopes. Astronomical objects emit not only in the visible but also in the whole electromagnetic spectrum. The investigation of the physical conditions of astronomical objects requires observations in as many wavebands as possible, from gamma and X-rays through the ultraviolet and the visible to infrared and radio waves. The frequency and the spectral extent of the cosmic radiation depend on the radiation mechanisms and on the physical conditions, in particular the temperature, the density and the magnetic field in the object. Radiation from outer space is, however, absorbed or reflected by the Earth's atmosphere in a large part of the electromagnetic spectrum. There exist only two narrow spectral windows which are almost transparent to cosmic radiation, namely the visible-near infrared and the radio windows. Outside these windows, the observations must be made by telescopes on board balloons or from satellites launched above the terrestrial atmosphere.

2.   Some Salient Results in Radio Astronomy

Radio-astronomical observations play a major role in the investigation of the universe. The frequency band observable with ground-based radio telescopes ranges from a few Megahertz to $\sim$800 Gigahertz. The cosmic background radiation was discovered at radio wavelengths (Penzias and Wilson, 1965). Its spectrum can be fit accurately by a 2.735 K black body curve peaking in the millimetre spectral region. The observations with the satellite COBE (Smoot et al. 1991) and balloon-borne experiments (de Bernardis et al. 2000; Hanany et al. 2000) have revealed that the background radiation exhibits tiny temperature fluctuations which are the imprint of the large scale structures in the primordial universe. These temperature anisotropies contain basic information allowing astronomers to discriminate between cosmological models.

The spiral structure of the Milky Way was first discovered by the extensive mapping using the 21-cm line of atomic hydrogen, which is also used to map the grand-design spiral pattern of other galaxies (e.g. van de Hulst et al., 1954). Figure 1 shows the image of the spiral galaxy NGC 6946 observed in the 21-cm line.

Furthermore, observations of the 21-cm line have shown that the rotation curves of several galaxies are much flatter beyond the optical disks than predicted by Kepler's law, suggesting that some dark matter may exist in the halos of galaxies (e.g. Sancisi and van Albada, 1987). All these radio data should provide valuable constraints on a number of cosmological parameters.

Gas in dense dark clouds is essentially in molecular form, thereby precluding observations in the visible, namely the Hα line, and in the 21-cm line of atomic hydrogen. The investigation of this dense gas component has been made possible thanks to the radio observations of interstellar molecules. The envelopes of evolved stars and the remnants of supernovae are rich in molecules, which are reprocessed within the stars and injected into the interstellar medium. Molecular-line observations are, therefore, useful for studying the chemical evolution of our Galaxy. Furthermore, spectacular maser action was discovered in molecular clouds capable of amplifying the background radiation by several orders of magnitude.

Observations of highly redshifted radiation from remote galaxies and quasars are becoming a favourite field of research for instruments operating at long radio wavelengths. Since long radio waves propagate almost freely in space, radio observations do not depend strongly on the weather conditions. The presence of turbulence in the Earth's atmosphere does not distort the wavefront of radio signals from outer space. This fact simplifies the building of radio telescopes which do not require sophisticated devices to correct the turbulence effect.

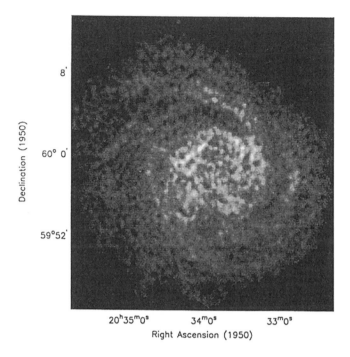

Figure 1.   The spiral structure of NGC 6946 observed in the hydrogen line at λ=21 cm with the Westerbork Supersynthesis Radio Telescope in the Netherlands (Boulanger and Viallefond, 1992).

3.   Instruments in Radio Astronomy

An astronomical instrument is characterized by its sensitivity and its resolving power. It should have a large collecting area to achieve this double performance. The number of photons collected by the instrument increases with its size. The angular resolution is $\alpha \sim \lambda/D$, λ is the observing wavelength and D the diameter of the instrument. In order to obtain angular resolutions similar to those achieved in optical astronomy, radio astronomers have to build antennae whose sizes are larger than those of optical telescopes by a factor of $\lambda_{radio}/\lambda_{optical}$. The resolution of a 10-m optical telescope built with advanced technology working at $\lambda_{optical} \sim 0.6$ μm is theoretically ~ 0.015 arcsec. The diameter of a radio telescope working at $\lambda_{radio} \sim 10$ cm with the same resolution would be ~ 1670 km. Building such a huge single antenna is beyond the range of present technology. But radio interferometers consisting of a network of small antennae separated from each other by thousands of kilometers are currently used in radio astronomy to achieve high resolution.

Astronomical observations at centimetre wavelengths or at longer wavelengths are quite appropriate for countries where the humidity of the atmosphere is significant and the sky is not sufficiently clear to use optical telescopes. This is the choice of Indian scientists who built their Giant Metrewave Radio Telescope

(GMRT) near Pune. This instrument consists of 30 parabolic antennae, each of 45-m diameter, functioning as an interferometer in the form of the letter 'Y'. Each arm extends to ~15 km from the centre of the system. With a collecting area of ~30,000 square metres, the GMRT is the largest interferometric array in the world. The GMRT which operates at low radio frequencies, from 150 MHz ($\lambda = 2$ m) to 1420 MHz ($\lambda = 21$ cm), is devoted especially to the detection of embryonic galaxies present in the early universe, through the 21-cm hydrogen line expected to be redshifted to metre wavelengths because of the expansion of the universe. The angular resolution of the GMRT is as high as ~2 arcsec at $\lambda = 21$ cm.

Great progress has also been made in the development of detectors, by the use of superconductor material. All the receiving system is cooled to a very low temperature, ~4 K by liquid helium in order to minimize the receiver noise.

In countries where the research in astronomy is not yet well-developed, equipments at low cost are appropriate. A simple instrument can be used by teachers to train students in physics and to familiarize them with radio astronomy by considering the universe as a laboratory. The radio emission from the closest star, the Sun, can be easily detected. Strong synchrotron radio sources, like Cassiopeia A and Cygnus A can also be possible targets. A small instrument equipped with an uncooled receiver is most suitable for this purpose.

4. The Basic Concept of Interferometry

Let us consider the one-dimensional case with two antennae A_1 and A_2 separated by a distance L (Figure 2). The radio signal of frequency ν coming from a point source is received at A_1. The output voltage is $U_1 = Esin(2\pi\nu t)$. The signal reaches antenna A_2 with a time delay $\tau = (Lsin\theta)/c$, resulting in a phase shift $\phi = 2\pi\nu\tau$; c is the velocity of light and θ denotes the direction of the incoming signal, that is the angle between the direction of the radio source with the perpendicular to the baseline.

The output voltage at antenna 2 is then, $U_2 = Esin(2\pi\nu t + \phi)$. If we now add the two signals by connecting the two antennae to the same receiver, we get the interferometric signal S:

$$S = U_1 + U_2 = Esin(2\pi\nu t) + Esin(2\pi\nu t + \phi)$$

This expression is equivalent to:

$$S = 2Ecos(\phi/2)sin(2\pi\nu t + \phi/2)$$

After detection, we get an intensity I proportional to the square of the "interferometric term" $cos(\phi/2)$, i.e.:

$$I = S^2 = 4E^2cos^2(\phi/2) = 2E^2(1 + cos\phi) \tag{1}$$

with $\phi = (2\pi Lsin\theta)/\lambda$

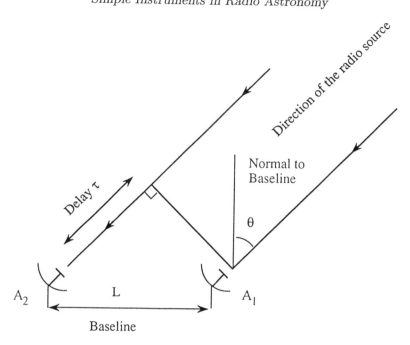

Figure 2.   A two-antennae interferometer

As the radio source, S, is rising towards the zenith, θ varies, as well as ϕ, as a function of time. For a baseline oriented east-west, it can be shown that (by using formulae of spherical trigonometry in the spherical triangle PSE):

$$sin\theta = cos\delta sinh \qquad (2)$$

where δ is the declination of the radio source and h is the hour angle.

Eqs.1 and 2 show that the intensity I of the signal of the source varies periodically with maxima occurring at hour angles h such that:

$$sinh = n\lambda/(Lcos\delta) \qquad (3)$$

where n is an integer, varying from 0 to n.

From Eq.3 we derive the time interval between the two nearest maxima:

$$\Delta h = \lambda/(Lcos\delta cosh) \qquad (4)$$

Figure 4 shows the fringe pattern calculated around $h = 0$, with $\lambda = 0.5$ m, $L = 25$ m and $\delta = 15°$ (Thai-Q-Tung, 1997).

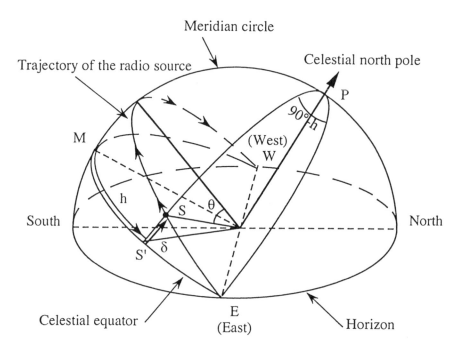

Figure 3.　　　Observation of a radiosource

5.　Extended Sources

Equation (1) can be generalized to calculate the response of an extended (one dimensional) source whose size is D (Thai-Q-Tung, 1997). The intensity of the signal is:

$$I = 2 \int_{-D/2}^{D/2} E_x{}^2 [(1 + cos(2\pi L sin(\theta + x)/\lambda)] dx$$

For the sake of simplicity, we assume that x is small and the intensity distribution is uniform and symmetrical with respect to the x axis. We then get:

$$I = 2E_0^2 D + 2E_0^2 cos(2\pi L sin\theta/\lambda) sin(\pi L D cos\theta/\lambda)/(\pi L cos\theta/\lambda) \qquad (5)$$

Figure 4 shows the fringe pattern (dotted lines). The fringe amplitude V(u) now changes with θ:

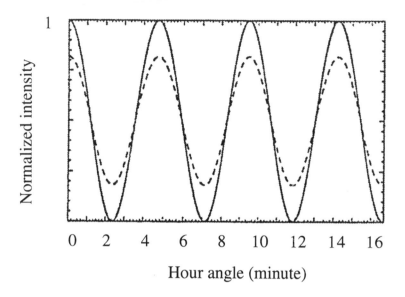

Figure 4.   Fringe pattern: point source (full line); uniform extended source (dotted line).

$$V(u) = 2E_0^2 sin(\pi Du)/\pi u$$

with the spatial frequency u $= Lcos\theta/\lambda$

The brightness distribution is restored by Fourier transform:

$$B(x) = \int_{-\infty}^{\infty} V(u)exp(-i2\pi ux)du \qquad (6)$$

The restoration of the brightness using Eq.5 is illustrated in Figure 5.

6.   A Two-Element Radio Interferometer

In the following we describe a simple radio interferometer. The two antennae are ordinary TV dipole Yagi aerials fixed to two flagpoles. The radio signals coming from the antennae are added and sent to a video-recorder which is used as a receiver. After detection, the radio signal is sent to an analog-digital converter and then to a computer which records the signal and displays the fringe pattern (Biraud, 1985; Biraud and Darchy, 1990).

High-gain antennae (gain $\geq$ 10 dB) are required to obtain a large equivalent collecting area. A gain of 16 dB corresponds to a peak directivity of the antenna main beam G = 40. The collecting area is $A = \lambda^2 G/(4\pi)$ and the main beam is $\alpha = 1.2\lambda/D$, with an antenna size $D = 2\sqrt{A/\pi}$. At λ =0.5 m, one finds

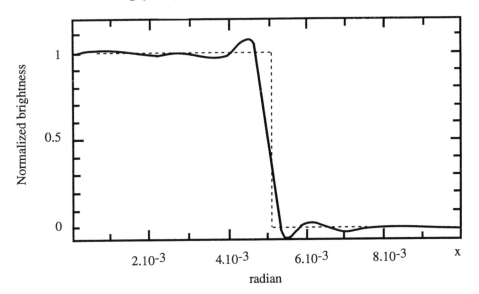

Figure 5.   Signal restoration : model (dotted line), restored bright-ness (full line).

A= 0.8 m^2 and α = 34°. The beam is large enough to maintain the fringe am-plitude of the radio source practically constant during several hours, facilitating the pointing procedure. The pointing of the antennae can be made manually. A baseline $L \sim$ 25 m gives a fringe interval (interval between two nearest maxima) of $\sim$ 1.5°. This value, much larger than the solar diameter ($\sim$ 32 arcmin), is suitable to detect fringes from the Sun. Such an instrument with an east-west baseline $L = 25$ m was used to observe the Sun during the solar eclipse in Phan Thiet (Vietnam) on October 24, 1995 (Figure 6).

Figure 7 shows the fringe pattern in different phases of the eclipse. The fringe amplitude decreases notably when the Sun was totally hidden, between 11 h 14 m and 11 h 16 m (local time). The Sun which had a declination $\delta = -11.5°$, culminated at 11 h 32 m.

Observation started at 09 h 10m (hour angle h = -2 h 22 m) and stopped at 12 h 00 m (h = +0 h 28 m). The fringe interval is 5 m 45 s at the beginning and 4 m 43 s at the end of the observation (Eq.4).

The observations suffered from the interference due to the presence of many eclipse watchers riding motorcycles and driving cars. Furthermore, a station for radio telephones was installed near the observation site. The record represented in Figure 7 results from a cleaning procedure to eliminate the interference peaks.

The solar flux F is on average $\sim$ 2.10^{-21} Wm^{-2}Hz^{-1}, corresponding to a peak antenna temperature $T_a \sim$ 45 K, at λ =0.5 m. The antenna is electrically equivalent to a resistor. The antenna temperature is the temperature of the antenna radiation resistance, which depends on the temperature of the emitting region in the sky. Figure 7 shows that the signal-to-noise ratio is excellent. The fringe amplitude decreases when the baseline increases. The fringes eventually disappear when the baseline becomes too large. The baseline of the interferom-

Figure 6.   The two antennae of the interferometer (visible in the fore-
ground and background) used to observe the solar eclipse in Phan Thiet
(Vietnam).

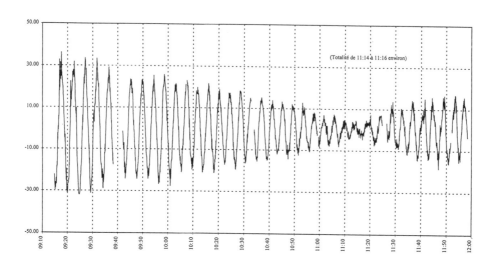

Figure 7.   Fringe pattern during the different phases of the solar
eclipse in Phan Thiet, Vietnam. B. Darchy, Observatoire de Paris

eter can be varied and oriented in different directions to measure the diameter or to map the emission of the radio source (Nguyen-Q-Rieu, 1993).It is possible to observe a few synchrotron radio sources such as the supernova remnant Cassiopeia A. This source is $\sim$ 45 times weaker than the Sun at $\lambda = 0.5$ m. The antenna temperature of Cassiopeia A would be $\sim$ 1 K. It is possible to detect this source and strong radio galaxies like Cygnus A, if we allow for a long integration time, consisting of adding several tens of fringes together.

Solar astronomers observe sunspots at $\lambda \sim 10$ cm to estimate the activity of the Sun. A radio interferometer with an east-west baseline of 350 m consisting of two 5-m dishes, gives an angular resolution $\sim$ 1 arcminute, which is adequate for the purpose. With such an instrument which has a primary beam (field of view) $\sim 1°$, we can detect sunspots whose fluxes are 10^{-22} - 10^{-21} Wm^{-2}Hz^{-1}.

7. A Single-Dish Radio Telescope

A simple single dish radio telescope can be built with an 0.7-m parabolic antenna designed to receive television signals from satellites (Darchy and Flouret, 1999). This type of antenna operates around frequencies of 11 GHz ($\lambda = 0.027m$). The antenna beam, $\alpha_b \sim 2.6°$, is narrow enough to require a pointing accuracy which should be better than $1°$. For the Sun, its optical image formed at the antenna focus can be used to check the pointing. The radio emission profile appears approximately as a gaussian curve, while the Sun drifts along the beam as the result of the diurnal and solar proper motions. At $\lambda = 0.027$ m, the observed antenna temperature T_a of the Sun is $\sim$ 360 K. The solar brightness temperature, $T_b \approx T_a(\alpha_b/\alpha_{Sun})^2$, is 8500 K, if the angular diameter of the Sun α_{Sun} is $32°$.

The flux F of the synchrotron emission from radio sources decreases at short wavelengths, according to the power law $F \propto \lambda^\beta$. For Cas A, the spectral index β is 0.77. The flux of Cas A at $\lambda = 0.027$ m is $\sim$ 4.6 10^{-24} Wm^{-2}Hz^{-1}. The collecting area of a 0.7-m parabola is $A = 0.38$ m^2. The antenna temperature is given by $T_a = F \times A/(2k)$, where k is the Boltzmann constant. At $\lambda = 0.027$ m, the antenna temperature of Cas A is only $T_a \sim 0.06$ K, which is $\sim$ 6000 times weaker than that of the Sun. Furthermore, at $\lambda = 0.027$ m, the observation is rather sensitive to the weather conditions. The sky must be as clear as possible. In order to detect Cas A, we should equip the antenna with a tracking system to observe the source during hours. We should also use a larger antenna together with a high sensitivity receiver to detect other radio sources.

8. Conclusion

Radio telescopes operating at decimetre and centimetre wavelengths may be built in developing countries where the weather conditions are not favourable for optical observations. The antennae and most of the electronic equipments can be bought at relatively low cost in the market, especially to build solar instruments. A solar radio interferometer was built to observe the 1995 solar eclipse in Vietnam. It was offered to the National University of Vietnam in Hanoi to train students in physics. This kind of instrument can be used to observe strong radio sources. Simple interferometers consisting of a pair of 5-m antennae can be built to observe sunspots, but the building technique does not require any special expertise.

References

de Bernardis, P., et al., 2000, *Nature* **404**, 955.

Biraud, F., 1985, *l'Astronomie*, November Issue, p. 529.

Biraud, F., and Darchy, B., 1990, March, Report of Station de Radio Astronomie de Nancay.

Boulanger, F, Viallefond, F., 1992, *Astron.Astrophys.* **266**, 37.

Darchy, B., and Flouret, B., 1999, January, Report of Station de Radio Astronomie de Nançay.

Hanany, S., et al., 2000, *Astrophys. J.* submitted.

Hulst, van de, H.C., Muller, C.A., and Oort, J.H., 1954. *Bull. Astr. Netherlands*, **12**, 117.

Nguyen-Q-Rieu, 1993, in *Microwave Engineering Handbook*, vol.3, Ed. B. Smith and M-H. Carpentier (Chapman and Hall), p. 511.

Penzias, A.A., and Wilson, R.W., 1965, *Astrophys. J.* **142**, 419.

Sancisi, R., and van Albada, T.S., 1987, in *Dark Matter in the Universe* (IAU Symp. Nr. 117), Eds. J Kormendy and G.R. Knapp, p. 67, D. Reidel, Dordrecht, The Netherlands.

Smoot, G.F. et al., 1991, *Astrophys. J.* **371**, L1.

Thai-Q-Tung, 1997, Report, Observatoire de Paris.

Discussion

In answer to Fierro, who asked how many antennae had been built and are functioning, Rieu said that a few had been built at the Radio Astronomy Centre of the Paris Observatory and are functioning. As stated in the paper, one was used to observe the 1995 solar eclipse in Vietnam and it was offered to the National University of Hanoi, to train physics students. Wentzel observed that a small radio telescope had been built near Calcutta, as a student project in an engineering course supervised by Ramesh Sinha. Rieu remarked that, apart from its value for training physics students, such an instrument could be built by amateur astronomers. Schreuder stated that although radio astronomy is very important in The Netherlands, Dutch radio astronomers tended to think it too complicated for amateurs. Nevertheless, a group of Dutch amateurs, with professional guidance, had built and successfully operated an instrument very similar to the one described by Rieu. Rieu said that the group in question lived about 30 km from Westerbork and one of the examples he had shown came from Westerbork. Kochhar observed that local people living near the GMRT in Pune, India, felt their job opportunities restricted because of the opportunities created for radio astronomers. Rieu replied that this was a social problem for which Indian radio astronomers are not to blame. The GMRT is an excellent low-frequency interferometer.

Astronomy for Developing Countries
IAU Special Session at the 24th General Assembly, 2001
Alan H. Batten, ed.

Is Astronomical Research Appropriate for Developing Countries?

Michael. S. Snowden

P.O. Box 44, MacPherson Rd, Singapore 913402. e-mail: mss@sri.lanka.net

Abstract. An unproductive 45-cm astronomical telescope, given by JICA (Japan) to Sri Lanka, raises general questions as to the reasons for unproductive pure science in developing countries. Before installation, site, maintenance, and scientific objectives were discussed. The facility was launched with a conference organised by the UN Office for Outer Space Affairs. Unfortunately, no research or significant education has resulted after four years. The annual operating cost is U.S. $5000 per year, including salary for a trainee, maintenance, and a modest promotional programme. Comparison with a similar installation in Auckland suggests lack of funding or technical competence do not explain the failure in Sri Lanka. The facility in New Zealand, on the roof of Auckland University's Physics Department, has a slightly smaller budget but has led to modest but useful research and teaching. Lack of financial backing and expertise are often blamed for weak science in developing countries, but examination shows most of these countries have adequately skilled people, and plenty of resources for religion and military. General lack of motivation for science appears to be the principal reason. This lack of interest and highly inefficient bureaucracies are common to scientifically unproductive countries. They mostly lack the cultural and philosphical base of the European Renaissance that motivate the pursuit of modern science, an activity that violates human preferences. There are excellent facilities (ESO, SAAO, Cerro Tololo, and GONG) in some of these same countries, when administered from the West.

1. Some Personal Experiences in Sri Lanka

In 1993 October, I received a phone call from my long-time friend, Sir Arthur Clarke. He asked me if I would attend a Sri Lanka government meeting as his representative, because he himself was unable to attend. The meeting was chaired by K. Austin Perera, the Secretary of the Ministry of Industries, Science and Technology, and was concerned with locating an observing site for a modern 45-cm reflecting telescope and with setting up a related administration. The telescope was being donated by JICA, a government department in Japan that administers aid to Third-World countries. As I am an observational astronomer with some experience in the early development of observatories, and as a quasi-resident in Sri Lanka for several years, he felt that I might help the committee. I happily agreed and requested that Mr. Simon Tulloch, a visiting astronomer from England, also attend.

The committee was composed of a panel of technical experts in meteorology, computer science, and engineering who were particularly familiar with the practical realities associated with government bureaucracies in Sri Lanka. It was only a matter of minutes before Mr. Tulloch and I saw the same conflicts arising in this meeting that Americans and Europeans had debated and resolved 30 years earlier about locating observatories (Irwin, 1952, 1966, and 1966). The issues at the meeting were mostly ones dealing with convenience of the location versus an effective environment for a telescope, but also ones dealing with administering a modern observatory with a zero-rupee budget.

Somewhat nonplussed, Mr. Tulloch and I heard a quotation by a notable Japanese astronomer, "With modern techniques and detectors you can compensate for a bad location", and the committee added a similar one of its own, "security at a remote location in Sri Lanka would be impossible." In that meeting only Mr. Tulloch and I were motivated to look for a location with clear sky conditions, while the others wanted it located "downtown" for convenience and security as a sort of end in itself. Considering the serious mood, I refrained from quoting for the committee Sir Arthur's celebrated quip about observatory sites, "Big telescopes should be built at bad sites to compensate for the poor conditions." It might have been taken seriously.

On the question of administering the facility, the committee was unanimously skeptical. Not a single rupee had been budgeted for installation or operation, and any institution accepting the facility could be inviting a white elephant.

Mr. Tulloch and I saw two pressing concerns. No one actually knew the sky conditions for astronomical purposes in Sri Lanka, and without some sort of immediate and basic site survey there was a certain disaster in the making. As there were no more than a few old typed comments on musty old paper by Dr. Tom Gehrels in the 1950s about potential sites and some brief suggestions by a local meteorologist, we quickly volunteered for a basic site survey. In retrospect, I know today that our collecting the facts was the last thing that that committee wanted.

For two weeks in 1993 October, Mr. Tulloch and I visited the central highlands of Sri Lanka from Anuradhapura to Samanalawewa. Upon conclusion, we submitted a formal report to the committee, and after my departure, Mr. Tulloch continued the survey alone. He identified several possible and practical sites for situating a small astronomical observatory (Tulloch, 1996). In his recommendations, he noted the particularly useless location being considered on top of the Arthur Clarke Institute for Modern Technologies (ACCIMT) in Moratuwa, where he was employed.

Fortuitously, I encountered a particularly ironic moment during the survey. One evening at Kandy, a substantial Sri Lankan cultural centre, I enjoyed a chance encounter with a prominent senior civil engineer, Dr. Ray Wijewardena, who was actively involved with a government water-management project and who took interest in our site-surveying activities. Upon hearing the full story,

Figure 1.   The 45-cm GOTO telescope at the Arthur Clarke Institute (courtesy ACCIMT).

Dr. Wijewardena asserted that an astronomical research observatory was "the last thing this country needs", and he went on to assure me that he would personally see to it that the project would be stopped. He did not. While not quite the response I appreciated at the time, I later came to appreciate his considerable wisdom, which was based on far more Sri Lankan experience than I had had.

All of our suggestions and reports were ignored, and early in 1996 the GOTO Japanese 45-cm telescope was installed, during the UN/ESA "Workshop on Basic Space Science: From Small Telescopes to Space Missions", at the Arthur Clarke Institute, where it is currently located as a white elephant among bright urban lights and in the remains of a tropical coastal rain forest (Figure 1).

The theme of the conference was concerned with the application of small telescopes to current research problems, and experts gave interesting papers in fields germane to this focus. Informally during the meeting, the chairman, Prof. Hans J. Haubold, urged individuals to follow the conference with on-going support for the astronomy programme in Sri Lanka.

The initial cost of the facility was primarily in the staggering $500,000 cost of the GOTO telescope plus a rolling roof atop the Arthur Clarke Institute for Modern Technologies. Today, it is maintained there at a cost to the Sri Lankan government of $5000 per year, which is used for the maintenance of the telescope and building, the salary of one trainee, and a promotional programme.

Although students at the University of Moratuwa have told me that they have not benefitted in any way by the presence of the telescope, and although there is still no instrumentation for the telescope, the facility has enjoyed some significant success in impressing schoolchildren. Japan continues to help the facility occasionally, but as a research tool, the telescope is only draining a painfully poor country of limited resources, just as Dr. Wijewardena predicted. So, there sits a high-quality and unusable 45-cm telescope, which to date has not produced a single datum of meaningful astronomical data, and no research student has benefited by its being there.

2. A Comparison with the University of Auckland

In a different corner of the former British Empire, there is a country which has followed twentieth-century development with a different philosophical base from that of Sri Lanka. In New Zealand there is a similar project in progress, and it is one for which a comparative analysis is revealing. At the Physics Department at the University of Auckland in New Zealand, there is a modest astronomical pursuit on the roof top of a building similar to the Arthur Clarke Institute (Figure 2). There, a 35-cm telescope is currently in use every clear night at a dreadful observatory location, which is surrounded by a large city and extensive lighting, but that is the limit of the similarities.

The telescope is used to train about 150 first-year university laboratory students and about 30 stage-three students. The 400 hours/year of observing at Auckland over three years of activity has resulted in data collected for five to ten cataclysmic binaries and about three publications. Standard deviations for the differential time-series CCD photometry is typically 5-9 mmag at $V = 13.0$.

In short, a shoestring budget in New Zealand has produced a modest but real scientific activity which is producing educational and research benefits commensurate with their investment. The success in New Zealand can be attributed to some clear skies compared to almost none at the Sri Lanka site and to a programme in Auckland that includes a good understanding and precise calibrations of their instruments. Most of all, it lies on a foundation which is a long British tradition of science and education. In Sri Lanka an extremely expensive donation and investment has produced almost nothing. With such a striking comparison we are compelled to ask the question in the heading of the next section.

3. What went Wrong in Sri Lanka?

Dig deep, and ye shall find! For any institution in a First-World country that is interested in helping an observatory in the Third World, this is an important question, but I have been repeatedly impressed that the answers I have heard are superficial and have missed the point. Whatever the correct answer it does indeed lie deep.

In attempting to install modern pure science into Third-World countries, we are behaving to a considerable extent like Christian missionaries, for better or worse. Should we be doing missionary work? On the other hand, what happens

Figure 2.   The 35-cm telescope at the Physics Department of Auckland University.

when those in developing countries are left to their own devices in pursuing modern science?

The philosophical base for modern science, i.e., the paradigm currently in practice with the combination of empirical and deductive elements, is inexorably linked to the Renaissance of Western Europe (Jones, 1952). Modern science emerged there as a search for relief from desperately bad times, in particular as a reaction to a cruel and repressive religion. The scene was set in Europe with all the players on stage. Christianity had failed them badly with the Black Death, and external elements of science philosophy from China, India, and Arabia played an important role as a stimulus. Today, there are many places where there has never been a Renaissance, i.e., countries with totalitarian political or religious-based power structures. Certainly, none of the Third-World countries have made such a transformation.

We are compelled to pause and ask why so many people are pursuing pure science today. Worldwide, people basically seem to dislike it, but in some places it is grudgingly tolerated. Again, why? Many of us scientists are pretty naïve about this, because we love it so much. More succinctly, the celebrated biologist, Harvard's E. O. Wilson, has said, "Man would rather believe than know" (Leakey and Lewin, 1993). Clearly science violates this preference. It has repeatedly exposed facts that most people do not want to see. To this day, the United States has trouble coping with Darwinian evolution.

Of course, the motives for doing science that you hear depend on whom you ask. As examples, we scientists do it because we have an obsessive delight in pursuing our hobbies. America has done it to produce technology to maintain dominance with their military machine and commercial interests. Singapore and Japan do it to strengthen the bases of their industrial machines. The Third World pursues it occasionally in believing it will produce more material wealth, as in First-World countries, and they are sadly mistaken! Regardless of the motives, the current spectacular success of astronomical research in producing scientific knowledge can be neither denied nor ignored.

There are 67 Member Countries in the IAU, and we might benefit by asking which of their observatories have been productive in astronomical research and education, which have not, and why. Certainly, there are wonderful, productive, big, and small observatories today: Cerro Tololo, ESO, AAO, SAAO, Mauna Kea, HST, NASA and ESA probes, and GONG to name a few. Do they have something in common? Are they obviously different from astronomically unproductive countries in Africa, South America, and nearly all of Asia? I believe the successful ones are almost entirely found in cultures which are products of the European Renaissance.

Although the Renaissance was the cradle for the philosophy of modern science, our current paradigm, there have been other productive times and places. During the collapse of intellectual pursuits in Christian Europe in the Dark Ages, astronomy was pursued mostly for religious applications in Muslim societies, countries that are today particularly unproductive scientifically. There are several countries currently making significant investments in installing astronomical observatories primarily on their own - China, South Korea, Egypt, Iran, and especially India (Jayawardhana, 1994) to name a few. Personally, I am pessimistic about these efforts, because I believe that primarily it is the Western countries that have developed societies with the necessary brash attitudes, free thought, and tolerance which are conducive to creative pure science, but we can all hope that with the restructuring of our Global Village at the end of this century and the consequential integration of cultures, they will nevertheless enjoy a degree of success. I believe they would be far more productive if they would at least develop strong partnerships with individual Western observatories.

In debating these issues, one often hears the view that science in developing countries has not proceeded more rapidly because of poor financial backing and a lack of expertise. I disagree with both of these excuses. In fact the lacking of both of these things is a direct reflection of priorities, and except in extreme cases is not a consequence of poor economies. Were the resources that we see devoted in some countries to the military, or to corrupt politicians, or to religious fanaticism, diverted into pursuing astronomy, those countries could afford Keck telescopes. I am not suggesting that such places would be better off pursuing astronomy than their current military, political, and religious hobbies, but it is clear that the pursuit of astronomy and other pure sciences is not exactly their preoccupation.

As for creativity and expertise, it would seem that the country that has given Ramanujan and Chandrasekhar to the world is not exactly short on it. From my own personal experiences I have often been impressed by the remarkable amount of local skill one finds in even the most remote and simple villages in undeveloped countries. Nearly always, such people are highly receptive to training and opportunities. That is certainly true of Sri Lanka for example. Rajesh Kochhar of Bangalore, India, has suggested three points in the development of Third-World astronomy with foreign help (Kochhar, 1996):

- training local manpower with expert help from countries with established programmes,

- maintaining and upgrading equipment with foreign help,

- providing a stimulus to sustain the interest.

While Dr. Kochhar's outline succinctly covers the essential points for such developments, I wish he had mentioned the barriers posed by the formidable government bureaucracies invariably found in developing countries, for here only the local nationals are capable of negotiating the nightmares these governments create. Foreigners are helpless to do so. Most of all, I believe he missed the key point: a modern research observatory in the Third World will be far more successful if it has a strong association and partnership with a foreign observatory, in particular one in the West!

Today, in some Third-World countries, highly productive observatories are producing copious data where they are administered from Western countries. In Chile, there is Cerro Tololo, which is headquartered in Arizona, the European Southern Observatory, which is administered by many European countries with headquarters in Germany, and the Las Campanas Observatory, which is administered from Pasadena, California. While the particularly productive South African Astronomical Observatory is currently owned and operated from within that developing country, its international staff is largely composed of British Commonwealth and American imports.

In Sri Lanka there is an entrenched and powerful bureaucracy and attempts to invest there, made by foreign companies, have often met with such inefficiency and corruption that those companies have abandoned their efforts for other shores. At the Arthur Clarke Institute some staff feel discouraged, because they believe an effective astronomical staff cannot be hired within the restrictive hiring criteria imposed upon them. This sad condition will not be corrected any time soon, and it is probably the principal barrier to a partnership with a foreign country, even for a non-commercial and educational facility such as an astronomical observatory.

4. Conclusions and Summary

In concluding, I would like to answer "yes" to the interrogative title of this talk. A modern telescope in Sri Lanka and other Third-World countries, could benefit those countries and international astronomy significantly when they have a close scientific and administrative relationship with a foreign observatory, hopefully a Western one which provides the scientific leadership. For local young people who want to study science, such observatories offer educational opportunities. For the foreign institutions, it means fascinating data, often obtained from locations with desirable combinations of latitude, longitude, and climate. The resulting science education means an enrichment for a type of education that offers experience in problem solving and deductive reasoning.

It would be a serious mistake to confuse the past and traditional concepts of colonialism with the highly successful Western leadership found operating currently at astronomical observatories in developing countries. We do not find at them patronising attitudes or innuendoes of domination, subjugation, or weak cultures. Although Western leadership at them is a necessary ingredient, we are talking here about collaborations based on mutual respect and benefit. Sir Arthur Clarke recently expressed to me the view that although the well-intended gift of the GOTO telescope has produced negligible results, astronomical telescopes are the most long lived of scientific instruments and that, eventually, when Sri Lanka is again politically stable, this telescope's time will come. Unfortunately, I am unable to share Sir Arthur's prognosis on this point.

Certainly, there are many notable productive old telescopes in use today, just as he has said, but there are also counter examples. One recalls the Jewitt Schmidt telescope at the Harvard College Observatory, which was built on a weakly conceived programme. In the entire lifetime of that instrument, it did not produce a single datum of meaningful astronomical data. The expensive Damon cameras there are another example.

I believe the sad political instability of Sri Lanka is also not relevant. Only when the telescope is associated with a partnership to include leadership from a Western observatory will the project at the ACCIMT stand a chance of developing a meaningful programme. Just as Dr. Wijewardena predicted seven years ago, any attempts to prop up the project with further developments will waste limited resources, until and unless such an administrative restructuring is established.

For Sri Lanka, a continuation by a Western astronomer of the initial site survey by Simon Tulloch would be an essential next step in setting up a productive telescope. Educational centres like the University of Moratuwa have good faculties and talented students who would love nothing more than the challenge of setting up a simple but serious research facility with the guidance of Western astronomers. A modest investment in such a site survey would stimulate the enthusiastic students, who could be employed for many of the activities. If successful, the survey could be followed by using impressive local talent to design and build a small automated telescope, perhaps with further guidance from an institute like the Global Network of Automated Telescopes (GNAT).

I hope my critical remarks today are not discouraging for a country like Sri Lanka, and let me mention a truly striking impression I've had there. In the past few years, it has been my pleasure to participate in seminars on astronomical topics with high-school and university students throughout Sri Lanka, sometimes in rather remote locations. I have been impressed consistently by the enthusiastic response I have received from these young students, many of whom would love nothing more than to pursue serious careers in science or engineering.

Finally, let me close with an interesting experience I have recently enjoyed in Sri Lanka. Just a few months ago, I was introduced by Sir Arthur Clarke to an established lawyer in Sri Lanka, Mr. Ranjan Sriskantha, who approached the two of us with a particularly intriguing idea. Mr. Sriskantha has proposed a donation of his property in the highlands near Kandy as a site for an astronomical observatory. Mr. Sriskantha's site may indeed offer a dark-sky location at a moderate altitude, easy access, adequate security, and proximity to a major university, all the things described as impossible at the government meeting seven years ago. Site testing would quickly reveal the realities, but from conversations with Mr. Sriskantha I was struck by something subtle and perhaps more important.

Mr. Sriskantha's enthusiasm for astronomy comes from the private sector in that country. While I do not want to overinterpret this significance, the private sector of any country usually reflects more accurately the intellectual mood than government-administered programmes, and for a country in which innovation is systematically strangled by government bureaucrats, perhaps the private sector may offer efficiency and more responsive activities.

Acknowledgments.   I gratefully acknowledge helpful comments from Drs. Philip Catton, Dennis Dutton, Owen Gingerich, Robert Latzer, William Moreau, and Sir Arthur Clarke.

References

Irwin, John 1952, *Science* **115**, 223.
Irwin, John 1966, *Science* **152**, 1597.
Irwin, John 1966, *Science* **154**, 1275.
Jayawardhana, Ray 1994 *Science* **264**, 502.
Jones, W. T. 1952, *Hobbes to Hume, A History of Western Philosophy.*
Kochhar, Rajesh 1996, *Small Telescopes in Research and Education.*
Leakey, Richard and Lewin, Roger 1993, *Origins Reconsidered.*
Tulloch, S.M. 1996, *MNASSA* bf 55, 10.

Discussion

Wentzel commented that the TAD IAU project discussed with the Physics Department of the University of Colombo the possible introduction of some astronomy courses (from visiting professors) in connection with the nearby new telescope. They insisted that physicists active in school astronomy should be trained abroad – impossible for TAD to finance. This suggested to him that there was no real administrative interest at the University.

Narlikar remarked that there has to be a perceived need for a facility before it is donated or created. At IUCAA they began with 35-cm and 40-cm telescopes for Indian universities and these created an awareness of the need for observational astronomy in the universities, leading to a proposal for a 2-m telescope, which should be ready in 2001 and will generate considerable research activity.

Editor's note: See the clarification of Japanese procedure for donating telescopes in Kitamura's poster paper, pp. 312-3. For another view of the relative importance of security in decisions concerning sites for an observatory, see the paper by Onuora, p. 329-332.)

Astronomy for Developing Countries
IAU Special Session at the 24th General Assembly, 2001
Alan H. Batten, ed.

Internet Resources for Astronomers Worldwide

George Helou

IPAC, Mail Code 100-22, California Institute of Technology,
1200 E. California Blvd, Pasadena, California 91125, U.S.A. e-mail:
gxh@ipac.caltech.edu

Abstract. On today's Internet, resources for astronomical research abound, and are available from wherever a connection is available. These are mostly in the form of freely accessible databases, ranging from literature services to space-mission archives. They offer instant access to the latest published data and papers, and unique opportunities for archival and innovative research. A sampling of such services will be listed, with notes on accessibility from far-flung points on the planet.

1. Overview

Internet-based resources are now an integral part of the research tools of astronomers, just as they are an essential element of international business and information flow. They can also be an invaluable tool for instruction in the sciences at all levels from grade school to college. These resources are unique in that they are readily accessible from wherever an internet connection is possible, and are free of charge in the overwhelming majority of cases.

There is quite a proliferation of these resources today, covering all wavelength ranges from γ-rays to the radio, and all areas of research from planetary science to cosmology. They can be generally assigned to one of three main categories.

- Databases and data services, offering primarily catalogs and sky images from surveys covering part or all of the sky, at any number of wavelengths across the electromagnetic spectrum. Prominent examples are the Digital Palomar Sky Survey data server at NASA's Space Telescope Science Institute (STScI) in the U.S.A. at http://archive.stsci.edu/dss, the Two Micron All-Sky Survey (2MASS) data server at NASA's Infrared Processing and Analysis Center (IPAC), U.S.A. at http://www.ipac.caltech.edu/2mass/overview/access.html, or the variety of catalogs accessible at the Centre de Données de Strasbourg (CDS) in France at http://cdsweb.u-strasbg.fr.

- Data archives from space missions or ground-based observatories, distinguished from the above by their origin as a series of targeted observations rather than an unbiased survey. Examples include the Infrared Space Observatory (ISO) archive at ESA's ISO Data Center in Spain at http:/www.iso.vilspa.esa.es, and the collection of mission archives at NASA's High Energy Astrophysics Science Archive Research Center (HEASARC) at http://legacy.gsfc.nasa.gov

- Information services with extensive links to the literature for instance, or to other archives and databases. Examples include the NASA-funded Astrophysics Data System (ADS), for biliographies, at http://adswww.harvard.edu, the NASA/IPAC Extragalactic Database (NED) at IPAC at http://nedwww.ipac.caltech.edu, and the SIMBAD service in Strasbourg, France at http://cdsweb.u-strasbg.fr/Simbad.html.

Most of these services are straightforward to use, and reasonably well documented, so that practising astronomers or physicists ought to be able to obtain what data they seek within a reasonable level of effort. There are a few websites from which one can branch out to a large number of these services, such as the Space Science Data System (SSDS) page at http://ssds.nasa.gov. One example the author is particularly familiar with is the IPAC main page http://www.ipac.caltech.edu, from which one can access services of all three types above, as well as many other similar services around the world. In particular, 2MASS offers a data set with tremendous untapped potential for research and discovery.

2.   Access and Exploitation

Access to the Internet and its resources offers instant access to the latest published literature and data. The new paradigm of astronomical research stresses electronic journals and data sets, with the internet itself serving as the library. Journal subscriptions are increasingly paperless, meaning they consist of data access rights to current issues rather than paper copies sent by mail. Back issues of most journals can even be accessed free of charge, including the full detail of their articles. Archives of space and ground-based observatories are placed on-line, and the data are retrievable without need for shipping tapes. The only serious limitation is bandwidth, and this is constantly improving.

Access to these resources presents unique opportunities for archival and creative research, and for training and education. On the research side, it is well known that only a small fraction of the useful science content in data archives gets published in the first two to three years of the data's existence, leaving room for much innovative data analysis and interpretation. In addition, cross-archive comparisons can yield truly exciting results, and even a smaller fraction of these potential results are ever pursued or published. On the training side, exposure of undergraduate students to data sets and archives is an excellent preparation, as well as a motivating force for further education and research. Most of the websites mentioned above include an "education and outreach" section, which can be very valuable as a teaching resource in schools at various levels.

Apart from the direct data and information resources it represents, the Internet is a powerful medium for collaborations in research and education. Unhampered by borders, it can bring together teams of researchers in multiple developing countries, or establish bridges between these and their colleagues in industrialized countries.

Some of the data services listed above provide interfaces that are less demanding in terms of technical requirements at the user end, and in terms of required bandwidth. For instance, one can use NED in e-mail mode, or use an interactive interface based on ASCII characters and VT-100 displays, accessible by invoking telnet ned.ipac.caltech.edu. The lower bandwidth requirement is a particularly valuable aspect of such interfaces, and it is a consideration for all users, not just those in developing countries. Feedback to the service providers from all users, regardless of their geographic location, on the value of these less demanding interfaces is important to ensure that the teams providing the services do continue to support them.

3. Conclusions

The Internet has revolutionized many areas of human activity, from business to politics. It has changed dramatically the practice and opportunities of science in industrialized nations, touching equally all aspects from education to research. The Internet has the potential to bring similar changes to science, and more especially astronomy, in developing countries. The only significant threshold is the availability of access to the Internet. Scientists sometimes prefer not to make on-line resources their primary source of research material. Even then, they could ill afford to ignore the Internet as a way to access current literature, data and information in order to place their research in the context of existing knowledge. Just like the sky is there to be observed, the electronic Universe is there to be studied or simply consulted, but definitely not to be ignored.

Acknowledgments.   This work was supported by funding from the U.S. National Aeronautics and Space Administration, and carried out at the Infrared Processing and Analysis Center and the Jet Propulsion Laboratory of the California Institute of Technology.

Discussion
Corbin remarked that "free library access" for the current year and the two immediately preceding years is usually only available to those whose library has paid the full electronic subscription – which is the same or only a little less than the paper subscription. This is unfortunate for astronomers in most developing countries since it is difficult for them to have access to the current literature, except for the abstracts. Helou agreed that it is still expensive to access the full texts of current articles, because journals operate on a business model. However, the Internet can speed up access, once the subscription is paid and circumvent the delays and hazards to which paper copies shipped by mail are subject.

Astronomy for Developing Countries
IAU Special Session at the 24th General Assembly, 2001
Alan H. Batten, ed.

Astronomy Research via the Internet

Kavan U. Ratnatunga

Dept. of Physics, Carnegie Mellon University, Pittsburgh, PA 15213, U.S.A. e-mail: kavan@astro.phys.cmu.edu

Abstract.
Small developing countries may not have a dark site with good seeing for an astronomical observatory or be able to afford the financial commitment to set up and support such a facility. Much of astronomical research today is however done with remote observations, such as from telescopes in space, or obtained by service observing at large facilities on the ground. Cutting-edge astronomical research can now be done with low-cost computers, with a good Internet connection to get on-line access to astronomical observations, journals and most recent preprints. E-mail allows fast easy collaboration between research scientists around the world. An international program with some short-term collaborative visits, could mine data and publish results from available astronomical observations for a fraction of the investment and cost of running even a small local observatory. Students who have been trained in the use of computers and software by such a program would also be more employable in the current job market. The Internet can reach you wherever you like to be and give you direct access to whatever you need for astronomical research.

1. Introduction

The exponential growth of astronomical research in the twentieth century, has left the developing nations far behind in contributions to the field. Pure science research in all non-theoretical fields has required very expensive technology which cannot be justified in the absence of immediate practical benefit to the developing nation.

There is a tendency to assume that astronomers in developing countries need to limit themselves to do astronomy research with less expensive small telescopes. Many talks at this Special Session discuss many of the basic but useful astronomical observations being done at small observatories around the world.

Let me first state clearly to avoid being misunderstood that I am not in anyway doubting the role that small telescopes can play in astronomical research or in inspiring students to study astronomy. Continuous monitoring of planets, comets, variable stars and other transient astronomical events benefit significantly from a global network of telescopes. Many developing nations could fill important gaps in such coverage. For example my home nation of Sri Lanka could fill the southernmost latitude between the longitudes of South Africa and Western Australia, north of Antarctica.

The astronomical research environment has however changed over the last five years in many significant ways. The price of computers needed for data analysis and cost of on-line data storage has dropped tenfold. The amount of astronomical data available in on-line archives from both ground-based and spaced-based observations is growing exponentially. Submission of papers to most leading journals is done on-line. Practically all leading astronomy journals back to their first volumes are now archived at Astrophysics Data System (ADS http://adsabs.harvard.edu/). The latest preprints are also archived at astro-ph (http://xxx.lanl.gov/) and updated daily.

The point I wish to make is that an on-site local telescope is now not need to start active research and education in astronomy. For example, Hubble Space-Telescope observations (HST http://archive.stsci.edu/) are released after one year and images taken in parallel mode become freely available on the Internet a day after the observations are made. Everyone on the Internet has equal opportunity to make and publish discoveries from them.

Cutting-edge astronomical research can therefore be done from any place in the world. All that is needed are low-cost computers with a good Internet connection to get on-line access to astronomical, observations, journals and most recent preprints. Since the investment in Internet connectivity is now justified, not only by research, but by e-commerce it is simpler for developing countries to obtain financial aid to establish such network connectivity, which then helps scientific research and the economy in many ways.

I will discuss below, in order of increasing importance, what I consider to be the main advantages of doing astronomy research over the Internet, pointing out the opportunities and limitations. Finally I will discuss as an example the situation in my home nation Sri Lanka, where enthusiasm and spurts of investment in conventional astronomy, have as yet, unfortunately, not yielded any significant research progress.

1.1. Independent of site

Astronomical observational sites with good seeing are rare. Urban growth and accompanying light pollution make it increasingly difficult to find a reasonable dark site. A remote site, even if one exists, leads to practical logistical problems of transport and accommodation. Access roads and guest-observer support are substantial additions to the cost. This has led to many small telescopes in developing countries being installed in bright urban areas and remaining practically useless for astronomical research.

However, the Internet can reach you wherever you like to be and give you direct access to whatever you need for astronomical research. For example, Sir Arthur C. Clarke, who predicted global communications based on geo-stationary satellites, has chosen to make Sri Lanka his home. The Internet connects him to the rest of the global village. How many academics, particularly expatriates, in all fields could be encouraged to return or to spend a productive sabbatical in their developing home country if network access needed for efficient communications and research were to be set up! In the presence of truly global communications there is no need to continue to exile oneself in a foreign land.

1.2. Not dependent on weather

Bad weather seems always to follow you even at good sites. The beauty of the solar corona watched with the naked-eye during a total eclipse of the Sun is not describable by words or recorded on film. It has to be seen to be appreciated. I was at the driest spot in the world, the Atacama desert in Bolivia, to observe my first total eclipse of the Sun in November 1994. It rained the night before!

I agree there is clearly a vast difference in experiencing personal observations compared with seeing images in a book or getting digital data over the Internet. Even relatively bright Messier objects through a small telescope may-be a disappointment in comparison to seeing all those beautiful long exposure image processed photographs of them on the Internet (http://www.seds.org/messier/).

Cloud cover often dampens one's enthusiasm to continue regular observations, particularly of transient events. The consequent lack of observations could lead to a complete loss of productivity when one tries to start an active research program.

However, serious quantitative astronomy research is not now done with observations made at an eyepiece. Is there any difference if the observations come from the instruments on a local telescope or come to your computer over the Internet from a telescope far away?

An astronomy research program set-up which includes the analysis of remote observations is obviously not dependent on the local weather and need not be interrupted even if one is temporarily disconnected from the Internet. No program needs to be data starved. There are far more astronomical data with free on-line access than it is humanly possible for the original observers to look at.

1.3. No need for expensive instrumentation

Telescopes need to be equipped with more than a set of eyepieces for any observations to be used for quantitative astronomical research. Telescopes need special instrumentation such as image detectors and spectrographs. Some spectral regions (e.g., infrared) require expensive cryogenics. Investing in and maintaining an expensive observatory which is useful only for astronomers can be unaffordable for a developing country.

All you need to get started on astronomy research over the Internet is a low-cost personal computer and a link to the Internet. Most of the major software analysis packages are freely available. Unlike costly instrumentation, the development of software requires only skill, a compiler, and a time commitment.

Astronomy resources on the Internet change over time and any detailed list would at most be a snapshot. In the appendix I list some of the major sites probably biased by my personal preference.

AstroWeb (http://cdsweb.u-strasbg.fr/astroweb.html) is an astronomical Internet Resources database which has been maintained since 1994 by a small world-wide consortium of institutions and contains pointers to potentially relevant resources available via Internet. There are a number of mirrors of this site around the globe. It is a good starting place to find anything related with astronomy. It provides searchable links to find telescope-time applications, data centers with astronomy archives, on-line astronomy journals, observation-analysis software, astronomy-conference information etc. It is not based on indexing automated downloads of web-sites. It is a moderated data-base which ensures that you don't need to search through a lot of unimportant URL's which happened to match search keywords to find the most useful sites which match your query. It depends on authors to submit to the site and, therefore, if you have set up an Internet web site associated with astronomy, then you should add it on-line to this database. Another advantage of this site is that an automated Tcl script checks the validity of the links every day to warn users of any that are broken.

1.4. Remote observing

Service observations are now common since they are a more efficient way to share time on a large-telescope facility. With space-based telescopes there is no choice. As instrumentation becomes more complex a local expert is in any case needed to set it up. Most often the required observations can easily be predefined precisely with exposure times or signal required. Telescope-floor decisions are typically made to use available time slots, often degraded by weather, to optimize which observations are taken. This is not the case for service observations which are aranged in a queue and made whenever the weather, seeing and photometric conditions required for each program are satisfied. Service observing obivates the need for costly and time-consuming travel to an observatory, from which one often returns with no data because of bad weather. If active interaction is important you can even observe remotely over the Internet. This could even be during the day, if you are using a telescope half-way across the world.

For example, scientists of any nationality or affiliation may submit to NASA Hubble-Space-Telescope (HST http://www.stsci.edu/) proposals. Restrictions are only for funding requests. All you need is to submit excellent observational proposals over the Internet to win the telescope time required. For other facilities all one needs is at most a collaboration with an observer with access to that telescope. The observations are made as requested and the data are sent by tape or can be obtained over the Internet.

1.5. Rapid decrease in cost of computers

Twenty years ago astronomical data-analysis was done on large VAX/VMS computers which cost over U.S. $100,000. Only large departments or universities as a whole could afford them. Users shared time on this central computer via terminals. Most developing nations could not afford them without significant grants. Ten years ago the same power was available on SUN/UNIX workstations which cost over U.S. $10,000 but started to become personal desk-top units. Today personal computers are more than 10 times powerful and cost under U.S. $1000. Computer hard-disk space is also important for astronomy data-analysis. Ten years ago it cost over U.S. $2000 per GB. Today disk storage costs under U.S. $10 per GB.

Increasingly, in many academic environments, the free shareware UNIX operating system known as LINUX has enabled home-market personal computers to replace the more costly workstations without sacrificing software portability, numerical computing power or operational stability.

Computers depreciate in value very fast. So investment in the needed computer and data storage should be made only as the research program grows. A computer laboratory has a half-life of under 2-years. A computer laboratory set up five years ago would probably be of no residual value. In contrast a telescope will, in principle, have an operational life-time of probably over 50 years, if it is set up and used in a good site, and still retain its value.

However, computer costs have dropped so much that they are now small compared to operational costs such as salary, and travel. The cost of computers and disk storage are clearly not limiting factors in setting up a computer laboratory for astronomical research.

Good Internet connectivity is also becoming affordable although it is still a non-negligible cost for a developing nation. I read on the web that connectivity with T1 bandwidth (1.5Mbits/sec) costs about $1000 per month in the U.S. and is sufficient for only about 75 users. Although telecoms may charge ten times this figure, in reality it is no more expensive to a government of a developing nation. A T3 (28Mbits/sec) line would be needed to download files of order 100MB or more in reasonable time.

1.6. Common research facilities

The set-up of computer facilities and Internet connectivity are common to most fields of scientific research. It is now also used by e-commerce and the public at large. This is a very important consideration when it is proposed to start any fundamental scienctific research like astronomy in developing countries, where everything needs to be related to practical needs of the country. Astronomy is clearly useful for the development of the intellectual environment but we must recognize that it is a luxury for which it is probably difficult to justify independent funding.

Computers serve both the analysis of observations and theoretical studies. E-mail allows easy collaboration with research scientists around the world. Voice and even video-phone calls over the Internet bring added possibilities for fast and inexpensive communication. However for active research, one needs to be able to get fast response during working hours for remote interactive log-in and to have the ability to transfer of the order of 100MB or more on demand, without unreasonable time delays. Developing nations need to negotiate with funding agencies to obtain seed money to set up collaborative efforts for analysis of observations and data mining.

1.7. Job security for students

A Ph.D. in astronomy does not ensure a research or teaching position. In reality many good post-doctoral fellows, even in the West, need to switch to non-astronomy jobs. Other than as an indicator of a level of intellectual maturity, expertise in pure astronomy has no practical applications in commercial industry. Students familiar with computer data-analysis and the Internet, however, find it much simpler to find gainful employment. The growth of the Internet is so rapid that for a long time there will ample Internet-related jobs. The type of expertise one gains while doing research is important. A person with a bachelor's degree and good computer skills is probably more employable than one with a Ph.D. in astronomy but without advanced computer capabilities.

1.8.   Extensive on-line archives of journals.

The Internet has more astronomy journals on-line than most major university libraries including Carnegie Mellon have on their shelves. Astronomy journals back to their first issues are now archived digitally at Astrophysics Data System (ADS) at http://adsabs.harvard.edu with mirror sites around the globe.

No longer do you need to suffer months of delay waiting for journals to arrive, posted by sea-mail because of excessive air-mail costs. Everyone on the Internet can see publications at the same time. Electronic submission of papers to practically all journals reduces delays and postage costs in the referee and publication process. It has also minimized the effort needed for proof corrections since the submitted manuscripts in LaTeX or other standard word-processor format are directly translated to the format needed for publication.

The latest preprints are also archived at astro-ph (http://xxx.lanl.gov) and updated daily. New abstracts are sent out by e-mail. Since most astronomers now read the latest papers from this free service it has almost replaced the costly distribution of preprints. Submission to the astro-ph server is now considered an essential part of publication, to ensure that a paper will come to the attention of the astronomy research community.

However, it is important to negotiate with the on-line journal publishers to allow access to recent on-line editions for developing nations without institutions being required to subscribe to the expensive printed copy. Institutions should also have access to commercial products such as the ISI's *Web of Science*, a journal-article data-base that covers more than 5300 major journals in the natural sciences, mathematics, engineering, technology, and medicine.

1.9.   Leading instruments and telescopes

Observations taken with the leading instruments like the Hubble Space Telescope and large ground-based national facilities can be down-loaded on the Internet. A recent National Academy of Science report, *Astronomy and Astrophysics in the New Millennium*, discusses a virtual observatory as a very important initiative in astronomy over the next decade; a digital sky in all wavelengths based on the massive data sets being created with tools to explore the data base.

All space-based missions and most large ground-based observatories maintain a complete data archive of observations. Almost all HST observations are put in the public domain after one year. However you are not limited to astronomical observations which have been milked for the best science. Pure-Parallel observations made with the HST are put on-line the day after observation, free for an astronomer from anywhere in cyberspace to analyze.

1.10.   Archived data available freely

Large ground-based and space-based surveys and individual observations are generating more data than most observers can use, except for the very specific application for which the observations were made. There is a lot more invaluable astronomical information in most of these observations, for different applications, if one has the time to look more carefully.

There is a lot of archived data on-line with free Internet access, much of which has not been analyzed in detail, nor even been looked at in detail, because the original observers had insufficient funding.

For example, the first gravitational lens to be discovered by the Hubble Space telescope was found by us as a part of the Medium Deep Survey (MDS) on archival data (Ratnatunga et al. 1995), after the observations were released to the public one year after they had been made. The only advantage that the MDS group had was the software we had developed to handle automatically some aspects of the data reduction (Ratnatunga et al. 1996). Software like this could be easily obtained in collaboration or even developed without large investment. However much of our discovery was due to the fact that, in the pipeline of automated analysis, we included in addition a careful visual inspection of the observations. The trained human eye is still far better than any software in picking out the interesting and unusual objects, and far less visual inspection is done now than should be – a lapse that any developing country properly connected to the Internet can exploit.

There are many opportunities to develop programs for data mining. It is much less expensive than data acquisition and the credit of discovery is with the person who publishes the results not with the person who made the observation. There are probably lots of hidden serendipitous discoveries to be made in space-based observational data obtained at astronomical cost.

2. An example: Sri Lanka

Astronomical research in Sri Lanka is probably typical of that in many developing countries. I use Sri Lanka as an example to illustrate how, like many other developing nations, it could develop more by the Internet than with a small telescope. I have observed first hand its attempts to develop in astronomy for over 30 years.

The absence of astronomical research in Lanka is *not* as Gehrels (1988) suggested in his review of Sri Lanka's Telescope (I quote) "Could it be that inquiry into our origins has little appeal to Buddhists ?". Questioning authority and the status quo is a cornerstone of Buddhism as the *Kalama Sutta* (http://www.accesstoinsight.org/lib/bps/wheels/wheel008.html) expresses very eloquently. This was the Eastern experience of the Renaissance and the cultural attitudes, such as a willingness to confront authority and reject religion which Snowden (2001) feels is now lacking in Sri Lanka. In any case, only a minute fraction of the population in the West or the East are associated with research in fundamental science such as astronomy, the mind-set of the current population at large is irrelevant to the establishment of astronomy research programs. For example, 47% of Americans– and a quarter of college graduates –believe humans did not evolve, but were created by God a few thousand years ago, and yet the U.S.A. is the world's leading scientific nation. (MacKenzie 2000).

Eastern civilization in particular has had an interest in the cosmos from ancient times. They accepted the vastness of space and time while religious belief in the West adopted a geocentric universe created 6000 years ago. Hindus for example believed in a cyclic universe with a period of 7.3 billion years. I quote from the 1822 book *Hindoostan* edited by Frederic Shoberl "A learned Bramin laughed, on being told that we Europeans reckon only about six thousand years since the creation of the world, and pointing to an old man with a long beard, asked if it was possible to believe that he was born but the preceding day."

Astronomical research in Sri Lanka dates back to the early 1900s when Major P. B. Molesworth (1867-1906) ordered himself a 32-cm photographically equipped Newtonian reflector from George Calver. He housed it in a observatory in Trincomalee and did significant research on Jupiter which he published in the *Monthly Notices of the Royal Astronomical Society* (Molesworth 1905). The telescope was moved after his death to Colombo. I had the opportunity to use this telescope when attending the University of Ceylon. A small 10-cm telescope gifted to Royal College which I attended had got me interested in astronomy. I was also encouraged by regular meetings of the Ceylon Astronomical Association founded in 1959 with Arthur Clarke as Patron. He has contributed very positively to encouraging an interest in astronomy in Sri Lanka. A 25-cm telescope was set up by Herschel Gunawardena at the Colombo Observatory. There was clearly sufficient interest in astronomy in Sri Lanka to initiate serious amateur observations and to motivate a few of us to take up astronomy as a career (http://lakdiva.com/astronomers/), even 25 years ago.

I also know of many astronomers with special interest in Sri Lanka. In 1975, Prof. Tom Gehrels of the University of Arizona, who had been enchanted by Ceylon in 1945, offered a gift of a 1.78-m telescope mirror (Gehrels 1984). Sufficient local interest in the project was very wisely required to build the rest. With no good site to ensure a useful return on the required investment, the telescope was never built. Sites which were in the dry zone were politically unsafe at that time and are even more so today, since it is that part of the country which is currently in a state of civil war. Sites of high altitude in the wet zone had over 75% of the days cloudy.

In the mid-1980s, the growth of instant electronic communication started opening the way for doing research over the Internet as we know it today. Interest of Lankans abroad in this new medium of communication was simulated by a need for news about Sri Lanka which was hardly covered in the foreign press. An informal e-mail group SLnet was formed in 1988 to exchange news, and Lankans were among the first dozen countries to form a Usernet group soc.culture.sri-lanka in 1989. A non-profit organization LAcNet was formed in 1991 which maintained a dial-up e-mail service till Lanka was connected directly to the Internet in 1995. One of the primary aims of this organization is to promote academic collaboration over the Internet between students in Sri Lanka and Lankan scientists overseas. In 1991/92, I spent a sabbatical year at the Institute for Fundamental Studies (IFS) in Kandy, Sri Lanka, exploring the possibility of doing astronomical research from Sri Lanka aided by the Internet which still needs to be funded properly.

Around the same time, IFS was offered a 45-cm telescope by Japan. Being the only professional astronomer resident in the island at that time, I tried in vain to negotiate that most of the aid should be used to set up an Internet connection into Lanka and a computer laboratory which could be used for active astronomical research. A small mobile telescope (even up to 36-cm aperture can be an off-the-shelf item and relatively inexpensive) would have been more useful to serve the purpose of inspiring students to study astronomy. Although I stalled the telescope project, it restarted after I left Sri Lanka. A 45-cm telescope gifted by Japan was inaugurated at the UN/ESA Workshop on Basic Space Science, held in Colombo in Jan 1996. It was equipped with a photometer which is useless at the brightly lit urban site. More recently a CCD was gifted, the images from which will probably suffer from the instability of the telescope (mounted on the fourth floor of a building) caused by trucks on the nearby highway.

It remains as I predicted practically unused and, as far as I know, used only as an exhibit for visiting schoolchildren. For maintenance and supervision it costs about U.S. $5000 per year of the very limited resources available for astronomy. For more details see Snowden (2001). Back in 1964 the Sri Lankan government, using a gift from East Germany, set up a Zeiss Planetarium as a part of an Industrial Exhibition. For a long time that was considered by many politicians as funding the needs of astronomical research. The new "white elephant" will similarly hurt more than help any future effort to invest in needs of astronomical research in Sri Lanka.

That the 45-cm telescope has served no purpose for astronomical research or education in Lanka is not entirely due to a lack of interest. Bureaucrats in the host institution, who have no motivation to use the instrument themselves, do not even allow access to it by students from the nearby university or to local amateurs since the telescope is listed as costing $300,000 of aid to Sri Lanka. An active amateur group located nearby at the Institute for Integral Education run by Father Mervyn Fernando could be learning on this instrument. It is sadly ironic that, instead, they received a gift of a 20-cm telescope from George Coyne, the Director of the Vatican Observatory. That this smaller telescope can be moved to a site reasonable for observation helps the group to use it more frequently.

Personal computers are now freely available in Sri Lanka at reasonable cost. Internet bandwidth is what is most lacking. Future plans are for a 0.5Mbit line which seems clearly inadequate for supporting a user community of very much more than 25 users. The academic bandwidth into the whole country is less than what would be considered acceptable even for a small university in the U.S.A.

3. Conclusion

A small telescope installed in a bright urban setting may not be the best way to develop a small astronomy research program in a developing nation. A decision to set up such an instrument needs to be based on the availability of a reasonable site, and the recommendations of local astronomers with expertise in both astronomy and the aspirations of the local user community.

In the global village we now live in, international collaborations are practical if the developing nation is properly networked. The Internet also gives you access to enormous data archives with the latest observations for quantitative analysis with computer systems which are now not very expensive. It is truly an optimum time for any developing nation interested in astronomy research to seriously consider doing it via the Internet.

Astronomical research in Sri Lanka is probably typical of many developing countries. Although there is a strong amateur astronomy interest in the country there are less than a dozen Lankans who are professional astronomers; all of them are doing research outside Sri Lanka. Could they or more recent graduates be attracted back? I think this is a clear possibility if the infrastructure needed for research is set up in Sri Lanka. Linking Sri Lanka to the Internet with an electronic superhighway is not the only need, but it is clearly the minimum before one can consider the possibility of active research seriously.

Acknowledgments. I wish thank the organizing committee for inviting me to give this contribution and for the financial support in attending the 24th IAU general assembly in Manchester, U.K.

References

Gehrels, T. 1984 in *Fundamental Studies and the Future of Science* ed. C. Wickramasinghe (Cardiff:University College) p.377.

Gehrels, T. 1988 *On the Glassy Sea: An Astronomer's Jouney*, AIP, New York, p. 236.

Molesworth, P. B. 1905 MNRAS65 691

Ratnatunga, K. U., Ostrander, E. J., Griffiths, R. E., & Im, M. 1995 ApJ453 L5

Ratnatunga, K. U., Griffiths, R. E. & Ostrander, E. J. 1999, AJ118 86

Snowden, M. S., 2001, this volume, pp. 266-275.

MacKenzie, D. 2000, *New Scientist* 22-April 2000.

Appendix

I give below a sampling of the major Internet sites with URL's of astronomical databases. A more complete and current list could be obtained from **AstroWeb** (http://cdsweb.u-strasbg.fr/astroweb.html).

Centre de Donnes astronomiques de Strasbourg (**CDS** http://cdsweb.u-strasbg.fr/) which hosts *Set of Identifications, Measurements, and Bibliography for Astronomical Data* (**SIMBAD** http://simbad.u-strasbg.fr/Simbad/) an astronomical database provides basic data, cross-identifications and bibliography for currently about 3 million astronomical objects outside the solar system.

VizieR (**CDS** http://vizier.u-strasbg.fr/)provides on line, a library of computer-readable astronomical catalogues and data tables with documentation. It is mirrored in Astronomy Data Centers in USA (**ADC** http://adc.gsfc.nasa.gov/), Japan (**ADAC** http://adac.mtk.nao.ac.jp/), and India (**IUCAA** http://www.iucaa.ernet.in/)

High Energy Astrophysics Science Archive Research Center (**HEASARC** http://heasarc.gsfc.nasa.gov/). A source of γ-ray, X-ray, and extreme ultraviolet observations of cosmic sources with direct links associated support facilities. It hosts a Virtual Observatory (**SkyView** http://skyview.gsfc.nasa.gov/) for generating images of any part of the sky at wavelengths in all regimes from Radio to Gamma-Rays.

The Hubble Data Archive (**HDA** http://archive.stsci.edu/) which also supports a Multi-mission Archive (**MAST** http://archive.stsci.edu/mast.html) with a variety of astronomical data archives, with focus in the optical, ultraviolet, and near-infrared parts of the spectrum.

Infrared Processing and Analysis Center (**IPAC** http://www.ipac.caltech.edu/) maintain infrared data archives and access tools and hosts the NASA Extra-galactic Database (**NED** http://nedwww.ipac.caltech.edu/) with over 3 million objects and 150 thousand red-shifts.

Planetary Data System (**PDS** http://pds.jpl.nasa.gov/) archives digital data from past and present NASA planetary missions, astronomical observations, and laboratory measurements. All data is classified Technology and Software Publicly Available (TSPA) to be exported outside the United States.

Canadian Astronomy Data Center (**CADC** http://cadcwww.dao.nrc.ca/) which host the archive observations since 1990 from the Canada-France-Hawaii 4-meter Telescope on Mauna Kea, Hawaii, (**CFHT** http://cadcwww.dao.nrc.ca/cfht/). Links are provided to similar archives setup at many major ground-based observatories worldwide using CADC software and expertise.

Sloan Digital Sky Survey (**SDSS** http://www.sdss.org/) which will systematically map one fourth of the sky to probe the large-scale structure of the universe. The survey will produce a catalog of roughly 100 million objects, with red-shifts to more than a million galaxies and quasars.

Astrophysics Data System (**ADS** http://adswww.harvard.edu/) A NASA-funded Abstract Service which provides access to over two million astronomy related abstracts which can be searched by author, title, object name, or keywords. It has also made agreements with publishes of astronomical journals to provide links to scanned images of over 40,000 articles appearing in most of the major astronomical journals and over 3 years old.

Physics e-Print archive (**XXX** http://xxx.lanl.gov/) which includes astrophysics (astro-ph). Mirror sites are being established all over the world. Replacing postal distribution of preprints.

AAS job Register (**AAS** http://www.aas.org/JobRegister/aasjobs.html) A monthly posting of worldwide astronomy related jobs maintained by the American Astronomical Society.

Digitized Sky Survey (**DSS** http://stdatu.stsci.edu/dss/) This comprises a set of all-sky photographic surveys conducted with the Palomar and UK 48-inch Schmidt telescopes and scanned using a PDS microdensitometer to scale of about 1.0 to 1.7 arcseconds per pixel.

Global Network of Astronomical Telescopes (**GNAT** http://www.gnat.org/) A non-profit organization dedicated as an information source for all those interested in research and education using relatively small astronomical telescopes.

International Dark-Sky Association (**IDA** http://www.darksky.org/) A non-profit organization with a goal to stopping the adverse environmental impact of light pollution and space debris through education about the value and effectiveness of quality nighttime lighting.

International Astronomical Union (**IAU** http://www.iau.org/) Announcements, Information bulletin and services.

Discussion Martinez agreed that a good Internet connection may be preferable to the building of a "research-grade" telescope but thought that the promotion of linkages between scientists in developed and developing countries is also essential. Without such links, scientists in developing countries will not be in a position to conduct competitive research and the gap in research capacity will get wider. Providing access to our data is not sufficient; we have to provide our colleagues with our time, expertise and research programmes as well. Ratnatunga agreed but thought that short visits would create the links which could be maintained over the Internet by e-mail.

Al-Sabti said that reserarch based on the Internet was useful but should not be a substitute for building telescopes and observatories. Doing astronomy through the Internet will create dependency on advanced countries similar to that of the developing world on the technology of the West. Ratnatunga replied that technology of instrumentation was only one aspect of the problem. Analysis software has to be developed and this can be done with only a PC. Percy agreed about the Internet but pointed out that good research can be carried out with a telescope costing $ 10,000 on an urban site (as Snowden, p. 269, had pointed out). Ratnatunga replied that if he had $300,000 to spend, he would spend $ 10,000 on the telescope and the rest to establish good Internet connections!

Kochhar said that, by extending Ratnatunga's argument, one need not cook food at home because cooked food can be bought from the market. Dworetsky suggested that the analogy should be that a fine picnic has been set out; the food is free, but bring your own utensils. In every hundredth cake there is a gold nugget hidden. Ratnatunga thought this much the better analogy. Schreuder suggested that a summary might be: telescopes are better for education, the Internet for research. Ratnatunga said that the Internet can be used for education as well. Developing software for analysis is educational and data from the Internet are no different from those from a telescope.

Astronomy for Developing Countries
IAU Special Session at the 24th General Assembly, 2001
Alan H. Batten, ed.

The Role of Astronomical Catalogues in Modern Theory and Observation

Oleg Yu. Malkov, Alexander V. Tutukov, Dana A. Kovaleva

Institute of Astronomy, 48 Pyatnitskaya St., Moscow 109017, Russia.
e-mail: malkov@inasan.rssi.ru

Abstract. Astronomical catalogues are powerful tools for carrying out modern theoretical and observational studies. Analysis of catalogue data enables important information to be obtained and does not require modern expensive observational techniques to reach reliable scientific results. In this paper we give some examples of such approaches and present some results of successful work with catalogues.

1. Introduction

Astronomical catalogues represent an important branch of modern astronomy. Besides their obvious use as lists of positions, photometry and other data for celestial objects, the great feature of astronomical catalogues is their effectiveness as powerful tools to carry out modern theoretical and observational studies. About 1000 modern astronomical catalogues accumulate accessible information on the principal astronomical objects: stars, galaxies and now extrasolar planets. Many astronomical centres in the world are involved in compilation and accumulation of catalogues; and maintain sets of astronomical catalogues in archives that are publicly accessible (by means of standard Internet tools).

Any stellar catalogue can be considered as a product of the history of star formation, the consequent stellar evolution, and observational selection effects. Therefore, astronomical catalogues offer an exceptional opportunity for highly effective investigations of stellar evolution, provided that we can correctly take into account the effects of observational selection and present reliable ideas about history of star formation. Analysis of catalogue data enables important information about the evolution of stars and other astronomical objects to be obtained. Such analysis does not require modern expensive observational techniques to reach reliable scientific results. In this paper we give some examples of such approaches and present some results of successful work with catalogues in the Institute of Astronomy of the Russian Academy of Sciences.

2. Study of Physical Properties of Binary Stars

The data for about 1000 spectroscopic and about 3000 visual binaries provide an opportunity to derive the birth-function of binary stars in our Galaxy. This birth-function gives the formation rate of binary stars as a function of the primary mass, of the components' mass-ratio and of the major semi-axis. Analysis

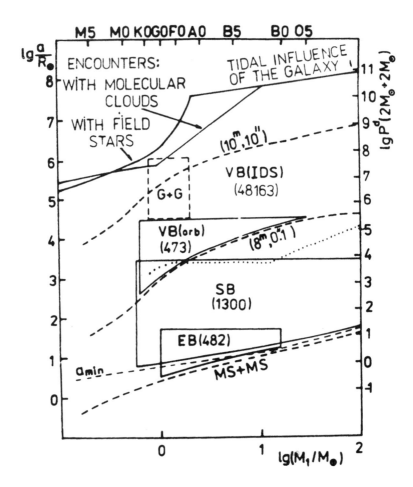

Figure 1.   Position of binary stars of different types in log(primary mass) — log(linear separation) diagram. EB: eclipsing binaries, SB: spectroscopic binaries, VB: visual binaries, VB(orb): VB with known orbits. The MS+MS line indicates the lowest lomit of separations set in by the contact of two homogeneous stars. The line a_{min} shows the minimal separation for new-born stars. Thick lines in the upper part of the diagram indicate the maximum major semi-axes of binary orbits. The region G+G is occupied by VB with giant components. The dotted line limits the region of evolutionary close binaries.

of the distribution over orbital angular momentum made it possible to estimate the number of planetary systems in our Galaxy. The birth-function and current theory of evolution of binary stars helped to compile the numerical scenario program for study of evolution of binary stars. This program reproduces the current population of binaries and provides the possibility of comparing the main properties of the model and observations. Reasonable agreement with the observations gives good grounds for believing in all the included ingredients. The complex of investigations of binary stars provides a good example of current evolutionary population approach to astronomical catalogues.

This work is based on the data from the following catalogues: *Index Catalogue of Visual Double Stars* (Jeffers, van den Bos, & Greeby 1963), *Catalogue of Physical Parameters of Spectroscopic Binary Stars* (Kraicheva et al. 1980), *The Fourth Catalog of Orbits of Visual Binary Stars* (Worley & Heintz 1983), *Eighth Catalogue of the Orbital Elements of Spectroscopic Binary Systems* (Batten, Fletcher, & McCarthy 1989), *The Washington Visual Double Star Catalog* (Worley & Douglass 1996). The position of binary stars of different types in $\log m_1 - \log a$ plane is presented in Figure 1.

3. Analysis of Properties and Evolution of Extra-Solar Planets

Another example of the application of astronomical data to the analysis of the physical properties and evolution of astronomical objects is the quickly growing catalogue of extra-solar planets (Tutukov, unpublished). Several dozens of them were discovered during last years. Figure 2 shows these planets, together with planets of the solar system: the major semi-axes of the orbits are plotted against the mass of the respective central star. There are several important lines plotted as well. "The edge" shows the external border of planetary systems with external planets forming through collision in the lifetime of the central star for stars with masses above the solar mass and during the Hubble time for systems with with central stars of a lower mass. The lines "$T_d = 1500K$" and "$T_d = 180K$" indicate isotherms. Dust in the region with $T_d > 1500K$ will be evaporated, thus excluding planet formation. Dust in $T_d \leq 180K$ can have ice envelopes, and planets formed of such dust will have powerful gas envelopes.

It is evident now that planets can exist only in the area between the lines "the edge" and "$T_d = 1500K$". Observations support this evident prediction. The most part of known planets are concentrated around stars of solar mass. This result is a simple consequence of an evident effect of observational selection. Stars with masses above $1.5m_\odot$ are quick rotators with wide spectral lines. This practically excludes the search for velocity variability of very low amplitudes. MS stars with masses much below the solar mass are faint because of selection effects, and their variability is also undetectable. Therefore solar-mass stars are the best candidates to search for radial-velocity variability since they are bright enough and have very narrow lines in their spectra.

4. Large Catalogues Data-Retrieval Software

Large astronomical catalogues, containing astrometric and photometric information for millions of objects are widely used by the astronomical community

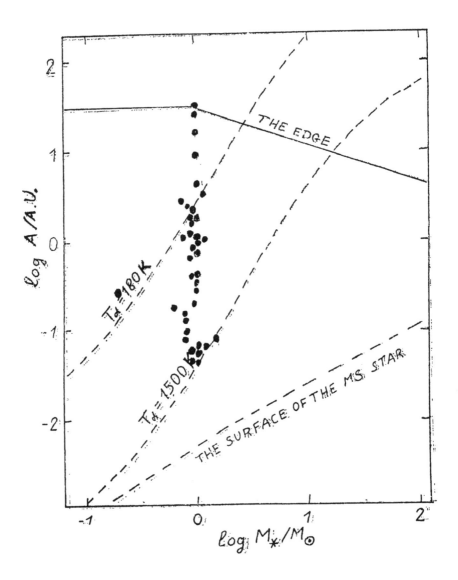

Figure 2.   Position of planets in the plane (mass of the central-star –
major semi-axis of the orbit). The meaning of lines is explained in the
text.

for all sorts of applications, particularly, for preparing and carrying out observations with ground-based and space-based telescopes, for quick identification of objects and first interpretation of results. Among these catalogues are the *Guide Star Catalog (ver 1.1)*, hereafter GSC (Lasker et al. 1990, Russell et al. 1990, Jenkner et al. 1990) and *USNO Catalog of Astrometric Standards* family, hereafter USNO-A (Monet 1996). However, getting at the actual catalogue data is not quite straightforward, due to the huge size of the catalogues and a somewhat complicated internal format. To facilitate data retrieval, user-friendly programs have been created for GSC (Malkov & Smirnov 1995) and for USNO-A (Malkov & Smirnov 1999) that let one look directly at the data in the large catalogues, either as a graphical sky map, a plot, or a simple text table. The programs can read a sampling of the catalogue data for a given sky region, store this sampling in a text file, and display a graphical map of the sampled region.

We have also designed XSKYMAP — a widget-based IDL application for visualization of astronomical catalogues. XSKYMAP (Smirnov & Malkov 1999) supports the ZGSC — a compressed version of the GSC that was constructed by Smirnov & Malkov (1997). ZGSC employs a custom binary format and an adaptive compression algorithm to achieve 6:1 lossless compression of GSC — down to about 200 Mb (from 1.2 Gb). XSKYMAP also supports the *PPM* family of astrometric catalogues, namely the *Catalog of Positions and Proper Motions*, the *Catalog of Positions and Proper Motions – South*, the *Bright Stars Supplement* to the *PPM* and *PPM South Catalog*, (Revised Edition), and the *90000 Stars Supplement* to the *PPM Star Catalog* (Roeser & Bastian 1988; Roeser & Bastian, 1993; Roeser, Bastian, & Kuzmin 1993). GSC's inherent depth of field is supplemented by extremely precise positions of relatively brighter stars from the *PPM*.

XSKYMAP provides a wide range of visualization tools for various applications. The current version has been integrated with the control software for the Galileo Italian National Telescope as an observational support tool (Pasian et al. 1998); the primary applications being generation of finder charts and preliminary telescope positioning.

5. Testing the Galaxy Model with the Guide Star Catalog

In another project we planned to develop methods of using the *GSC* as a source of statistical data, and apply them to the Bahcall-Soneira Galaxy model (Bahcall & Soneira 1980), with the aim of testing and extending the latter into lower galactic latitudes. A secondary goal is a detailed investigation of the photometric and statistical properties and irregularities of the *GSC*, as well as development of methods and software to deal with them. We modified the Bahcall-Soneira system to produce results in the native photometric bands of the *GSC*. We used our programs to extract *GSC* data sets from 40 small regions evenly distributed across the sky. To avoid dealing with blurred plate edges and plate overlaps at this stage, each area was shifted to the nearest plate centre, and the effective radius decreased to leave out any irregularities. To compare the *GSC* star counts with the theoretical model distributions, we developed a suite of IDL programs, and performed standard statistical tests (χ^2, Kolmogorov-Smirnov).

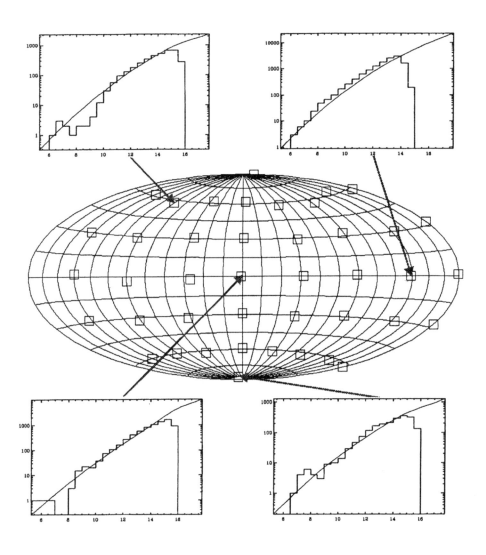

Figure 3.   Comparison of the Bahcall-Soneira Galaxy model (curves) with the GSC data (histograms) for some areas from a set uniformly distributed across the sky

Initial statistical comparisons between theoretical star counts and those obtained from the *GSC* (see Figure 3) showed interesting and sometimes significant discrepancies, namely: a deficit of bright stars ($11^m - 12^m$) in the *GSC* relative to the model exists at high galactic latitudes; a significant deficit of faint stars in the *GSC* is present at the prime meridian; a surplus of faint stars in the *GSC* can be seen at most of the other areas. Several explanations for the trends were suggested in (Malkov & Smirnov 1994). However, we have found that the *GSC* does not easily lend itself to statistical studies, due to various irregularities in the stellar classifications. Therefore, we have had to design methods of reclassifying *GSC* data, using both statistical and artificial intelligence methods.

6. Modelling of the Quasi-Empirical Mass-Luminosity and Mass-Radius Relations for the Region of Low and Intermediate Mass Stars

The mass-luminosity relation (MLR) plays a prominent role in many astrophysical fields. First of all it helps to estimate stellar mass - an all-important stellar parameter, mostly only indirectly known from observations. The MLR is also especially valuable for transformation of the observed luminosity function to the initial mass function, which is of key importance in star-formation and galactic-evolution studies. As the transformation involves the MLR derivative, knowledge of the precise shape of the MLR is highly desirable. The MLR fine structure in the domains of low-mass ($0.08m_\odot - 0.8m_\odot$) and of intermediate-mass stars ($0.8m_\odot - 6m_\odot$) is a subject of a constant interest to investigators at the present time for several reasons.

The number of accurate observational data (masses, luminosities, photometry etc.) rapidly increases for moderate-mass stars. The detached double-lined eclipsing binaries give us a set of most accurate, fundamental determinations of physical parameters of the stars (mass, radius, effective temperatures, luminosity, etc.). Present typical accuracy for this kind of star enables us to hope for deeper astrophysical insight then merely "improving" "mean" relations. But when uncertainties in masses become less then 5%, the deviations of the individual systems from a mean relation are due not to observational errors, but to real differences in evolution and composition. That is why averaging any number of accurate binary masses and luminosities will not improve one-dimensional calibrations decisively.

We constructed a set of the main-sequence double-line detached eclipsing binaries by selection from the *Catalogue of Astrophysical Parameters of Binary Systems* (Malkov, 1993). As only three eclipsing systems have low-mass components, we completed the investigated set in the low-mass domain with high-quality data for dynamical masses and luminosities of the components of visual and spectroscopic low-mass binaries. Thus, the final set contains data for 43 intermediate-mass eclipsing binaries, three low-mass eclipsing binaries and 22 low-mass double and triple visual and spectroscopic systems (48 components). Data for masses, radii, effective temperatures and luminosities of the eclipsing binaries were selected from the the *Catalogue of Astrophysical Parameters*; the set of low-mass systems was mainly extracted from the *Catalogue of Nearby Stars* (Gliese & Jahreiss 1991). Dynamical masses for these systems were se-

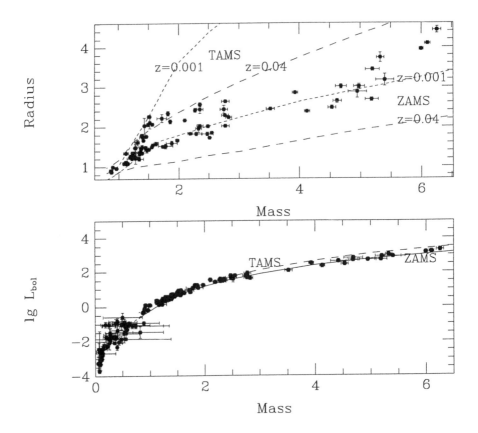

Figure 4.   The observational data are presented in the mass-radius and mass-luminosity planes. The data for components of eclipsing binaries (all stars with $m > 0.8m_{\odot}$) are selected from Malkov's (1993) catalogue; the data for the low-nmass stars are mainly extracted from the original publications, the parallaxes needed for calculation of the bolometric luminosities from the photometry, and for verification of the mass data, were taken from the catalogues of van Altena et al. (1991) and *HIPPARCOS Output Catalogue (ESA 1997)*. ZAMS and TAMS lines shown in the mass-luminosity plane have been calculated for solar abundance, according to the grids of stellar models published by the Geneva group (Schaller et al. 1992, Shaerer et al. 1993a,b, Charbonnel et al. 1993, 1996 and 1999, Meynet et al. 1994, Mowlavi et al. 1998); ZAMS for the low-mass stars is from Baraffe et al. (1998). In the mass-radius plane both ZAMS and TAMS are shown for two limiting abundances for Population I stars. It can easily be seen that the scatter of the observational data (especially in the mass-radius plane) is much greater than the estimated observational errors.

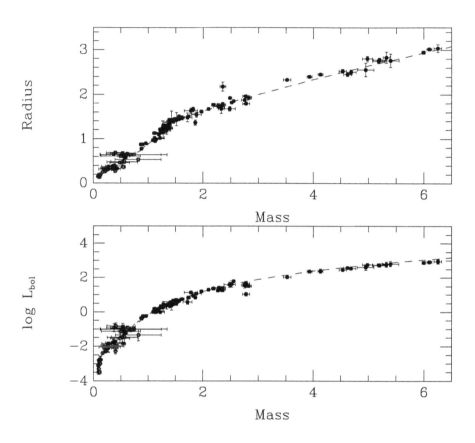

Figure 5.   The same set of stars with the data for radii and luminosities "corrected" for the effects of evolution and abundance variations. The scatter shown around the plotted solar-abundance ZAMS lines is caused mainly by observational error.

lected from the original publications, as well as the multicolor photometry. The parallaxes necessary for calculation of the bolometric luminosities from the photometry and for verification of the mass data were taken from the catalogues of van Altena et al. (1991) and *HIPPARCOS Output Catalogue (ESA 1997)*. The data are presented in Figure 5.

We tried to reveal existence of fine structure of the MLR by looking for systematic differences between the present models and modern data on luminosities and dynamical masses of MS stars of spectra from late B to M. To remove the influence of the effects of evolution within the main sequence and of the chemical-composition dispersion, we calculated for the investigated set of stars most probable ages and metallicities derived from recent stellar models using constraints arising from the common origin of components (see Figure 5). Analysis of the empirical MLR, considering the influence of these effects, shows in general perfect agreement of observations with present-day models. We also note an existence of systematic deviations of theoretical MLR between $1m_\odot$ and $3m_\odot$.

7. Conclusions

One can see that astronomical catalogues (and, more generally, astronomical data sources) are good examples of how

- modern data can be made operatively available to astronomers in developing countries by means of standard Internet tools;

- modern theoretical studies can be carried out without involving large telescopes;

- statistical properties of astronomical objects provide us with the most common parameters of their families: stars, galaxies, clusters, planets.

Acknowledgments. OM thanks the SOC of the conference for financial support that made this presentation possible. OM and DK also asknowledge the Russian Academy of Sciences for the aid provided in the frame of the "Young Scientists Support Program".

References

Bahcall, J.N. & Soneira, R.M. 1980, ApJS, 44, 73

Baraffe, I., Chabrier, G., Allard, F., & Hautschildt, P. H. A&A, 337, 403, 1998

Batten, A.H., Fletcher, J.M., & McCarthy, D.G. 1989 Publ. Dominion Astrophys. Obs. 17

Charbonnel, C., Dappen, W., Shaerer, G., Bernasoni, P. A., Meader A., Meynet, G., & Mowlavi, N., 1999, A&AS, 135, 405

Charbonnel, C., Meynet, G., Meader, A., & Shaerer, G. 1996, A&AS, 115, 339

Charbonnel, C., Meynet, G., Meader, A., Shaller, D., & Shaerer, G. 1993, A&AS, 101, 415

ESA 1997, *The HIPPARCHOS and Tycho Catalogues*, ESA SP-1200

Gliese, W. & Jahreiss, H. 1991, *Nearby Stars*, Preliminary 3rd Version, Astron. Rechen-Institut, Heidelberg

Jeffers, H.M., van den Bos, W.H., & Greeby F.M. 1963, Lick Obs. Publ. 21.

Jenkner, H., Lasker, B. M., Sturch, C.R., McLean, B.J., Shara, M. M. and Russell, J. L., 1990, AJ, 99, 2081

Kraicheva, Z., Popova, E., Tutukov, A., & Yungelson, L. 1980 Bull. Inform. CDS 19, 71

Lasker, B.M., Sturch, C.S., McLean, B.J., Russell, J.L., Jenkner, H., and Shara, M.M. 1990, AJ, 99, 1019

Malkov, O.Yu. 1993, Bull. Inf. CDS, 42, 27

Malkov, O.Yu. & Smirnov, O.M. 1994, in ASP Conf. Ser. Vol. 61, Astronomical Data Analysis Software and Systems III, ed. D.R. Crabtree, R.J. Hanisch, & J. Barnes (San Francisco: ASP), 187

Malkov, O.Yu. & Smirnov, O.M. 1995, in ASP Conf. Ser. Vol. 77, Astronomical Data Analysis Software and Systems IV, ed. R.A. Shaw, H.E. Payne, & J.J.E.Hayes (San Francisco: ASP), 182

Malkov, O.Yu. & Smirnov, O.M. 1999, in ASP Conf. Ser. Vol. 172, *Astronomical Data Analysis Software and Systems VIII*, ed. D.M. Mehringer, R.L. Plante, & D.A. Roberts (San Francisco: ASP), 407

Meynet, G., Meader, A., Shaller, D., Shaerer, G., & Charbonnel, C. 1994, A&AS, 103, 97

Monet, D. 1996, USNO-A: *A Catalog of Astrometric Standards*, (Washington: USNO)

Mowlavi, N., Shaerer, G., Meynet, G., Bernasconi, P. A., Charbonnel, C., & Meader, A. 1998, A&AS, 128, 471

Pasian, F., Marcucci, P., Pucillo, M., Vuerli, C., Malkov, O.Yu., Smirnov, O.M., Monai, S., Conconi, P., & Molinari, E. 1998, in ASP Conf. Ser., Vol. 145, *Astronomical Data Analysis Software and Systems VII*, ed. R. Albrecht, R.N. Hook, & H.A. Bushouse (San Francisco: ASP), 433

Roeser, S. & Bastian, U. 1988, A&AS, 74, 449

Roeser, S. & Bastian, U. 1993, Bull. Inform. CDS, 42, 11

Roeser, S., Bastian, U., & Kuzmin, A. 1993, A&AS, 105, 301

Russell, J.L., Lasker, B.L., McLean, B.J., Sturch, C.R., and Jenkner, H. 1990, AJ, 99, 2059

Shaller, D., Shaerer, G., Meynet, G., & Meader, A. 1992, A&AS, 96, 269

Shaerer, G., Meynet, G., Meader, A., & Shaller, D. 1993a, A&AS, 98, 523

Shaerer, G., Charbonnel, C., Meynet, G., Meader, A., & Shaller, D. 1993b, A&AS, 102, 339

Smirnov, O.M. & Malkov, O.Yu. 1997, in ASP Conf. Ser. Vol. 125, *Astronomical Data Analysis Software and Systems VIII*, ed. G. Hunt & H.E. Payne (San Francisco: ASP), 426

Smirnov, O.M. & Malkov, O.Yu. 1999, in ASP Conf. Ser. Vol. 172, *Astronomical Data Analysis Software and Systems VIII*, ed. D.M. Mehringer, R.L. Plante, & D.A. Roberts (San Francisco: ASP), 442

van Altena, W.F., Lee, J.T., & Hoffleit, D., 1991, *The General Catalogue of Trigonometric Stellar Parallaxes*, Preliminary Version, Yale University Observatory

Worley, C.E. & Heintz, W.D. 1983, Publ. U.S. Naval Obs. (2) 24, part VII

Worley, C.E. & Douglass, G.G. 1996, US Naval Observatory (unpublished).

Discussion

Chamcham asked if the data of all the catalogues had been reduced. Malkov replied that some had been. In deriving the mass-luminosity relations, they reduced the observed luminosities of the stars "removing" the changes caused by evolutionary effects (thus obtaining the ZAMS relation) and they reduced the differences arising from abundance variations, obtaining the relation for solar chemical composition. Fierro asked how many students in the Moscow Centre for Astronomical Data used these catalogues in their work. Malkov believed that they all did.

Astronomy for Developing Countries
IAU Special Session at the 24th General Assembly, 2001
Alan H. Batten, ed.

Design of a Small Automated Telescope for Indian Universities

Mandayam N. Anandaram, and B.A. Kagali

Department of Physics, Bangalore University, Bangalore 560 056, India.
e-mail:anandaram_mn@yahoo.com

S.P. Bhatnagar

Department of Physics, Bhavnagar University, Bhavnagar, India

Abstract. We have constructed a computer controlled telescope using a 0.36-m *f*/11 Celestron optical tube assembly for teaching and research applications. We have constructed a heavy duty fork-type equatorial mount fitted with precision machined 24 inch drive disks for both axes. These are friction driven by stepper motors through one inch rollers. We have used an open loop control system triggerable by an ST-4 CCD camera to acquire and track any target object. Our telescope can home in on any target within a range of two arc-minutes. We have employed a commercial stepper motor controller card for which we have written a user friendly pc based telescope control software in C. Photometry using a solid state photometer, and imaging by an ST-6 CCD camera are possible

We consider that this project is suitable for those wishing to construct some parts of a telescope and understand the principles of operation. A simpler model of this telescope could use DC motors instead of stepper motors. We shall be happy to send our design diagrams and details to those interested. This project was funded by the DST, and was assisted by IUCAA, Pune.

1. Introduction

We have undertaken the construction of the 0.36-m automated photoelectric telescope (APT) having been inspired by similar developments abroad, and also facing the need of a small modern telescope for teaching of stellar photometry and carrying out minor research in our universities. At the suggestion of the instrumentation laboratory of IUCAA in Pune, we purchased the 0.36-m Schmidt-Cassegrain optical tube assembly made by Celestron of USA,stepper motors along with their digital driver units made by Aerotech of U.S.A., and the ST-4 and the ST-6 CCD cameras made by SBIG of U.S.A.. The telescope mount and other parts needed for assembling the telescope were locally fabricated with engineering design assistance of IUCAA. We now give a description of the relevant work done by us in India. A photo of the completed telescope mounted at Bhavnagar University sliding-roof observatory is displayed below.

Figure 1.   The 0.36-m APT Telescope

The telescope is illustrated in Figure 1 and a functional block diagram is shown in Figure 2.

2.   The Telescope Mount Design

The telescope mount consists of a heavy-duty fork-type equatorial mount of rectangular cross-section. Two 0.6-m drive disks with hardened chrome plated rims are attached to this fork. These disks are friction driven by one inch diameter rollers to produce RA and DEC axes rotations. The rollers are driven by the stepper motors which require 25,000 steps provided by their digital microstepper translator units. All the mechanical parts were fabricated in local workshops accoding to specifications listed in their detailed engineering diagrams. These parts were then carefully assembled together and tested. In our tests to measure the wobble of the rotating disks with a dial gauge, we found that the wobble was less than 0.1 mm for a complete rotation. This was considered adequate.

The RA base housing was first assembled with the 7.6-cm diameter RA shaft mounted on its bearings. Then the 0.6-m RA disk and the base of the fork were then mounted. One arm of the fork was then fitted with the DEC disk as well as the OTA mounting brackets. The 2.5-cm rollers were then mounted under both disks such that they were in full contact with the smooth rims of the disks with adjustable pressure. The RA drive stepper motor was mounted behind the RA roller inside the RA axis cage and was dynamically coupled to it. Similarly the DEC drive motor was mounted on a fork arm and then coupled to its roller. Finally the OTA was mounted into the fork and weights on both fork arms were counter balanced.

The drive disks have been provided with rotation limiting stops inboth CW and CCW directions. The stoppers are so positioned that when the motion in RA axis is stopped the fork arms are horizontal and inclined at polar lattitude angle. Similarly when the motion in the DEC axis is stopped, the OTA is pointed towards the RA axis shaft. This position is defined as the PARK position of the telescope. To enable computer conrol three opto-coupled electronic position sensors have been provided for each axis. They provide logical high signals to identify the PARK position, the CW rotation limit just before its physical limit, and the CCW rotation limit just before its physical limit on both the RA axis and the DEC axis. When the limit signals are received the control program issues a command to stop the motion and/or reverse the direction of rotation as needed. Our control program utilises the PARK position signal to initialise the RA and DEC disk positions defined as the ZERO position. The DEC disk ZERO position is set to the value of -90 degree declination. The RA ZERO position is set at 180 degrees from the PARK position and from there the telescope can be driven to acquire any target object by software command. The occurance of the CW limit signal allows the disk to commence rotation only in the CCW direction. Similarly an occurrance of the CCW limit signal allows a rotation only in the CW direction. The control program ensures that the telescope always begins its operation by sensing its current position and moves in the required direction if necessary to its PARK position. Then it would move from there to its ZERO position to initialise its position in memory so that the computer can continuously update the current position of both disks by the number of micro step pulses sent to both drive motors. In this way we have achieved an open loop control of the telescope.

3. Control Program Design

The 2.5-cm roller coupled to the motor shaft rotates by 360° on receiving 25,000 stepper pulses. Since the disk diameter is 24 times the roller diameter, we find that the telescope would rotate by one degree for 1666.7 stepper pulses. Thus each pulse causes an angular motion of 2.15 arc-second. This is the design pointing accuracy of our telescope and is used by the computer to determine the number of stepper pulses needed to cause the required angular motion in both axes to acquire the target.

We have employed a high speed programmable stepper motor controller card available in the market to generate pulses at the required rate for slewing the telescope rapidly and also to enable it to track any star after acquiring it. This card is installed in a personal computer and its input/output lines are connected to the digital microstepper controller boxes attached to the motors as well as the outputs of position sensors on the drive disks. Provision exists to add additional control lines from instruments like photometers and CCD cameras.

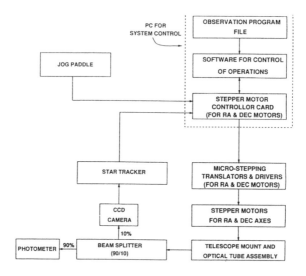

FUNCTIONAL FLOWCHART OF THE APT

Figure 2.   Block Diagram of the Telescope

We have written a user friendly telescope control program in C which offers a menu based choice of operations.

4.   Performance of the Telescope

We have carried out several types of performance checks on the operation of our telescope. In one test for the reproducibility of a specified pointing direction, the telescope was moved in RA or DEC axes by 50° and then moved back to the original direction by program command. We found that the slippage in pointing is 2 arc-minutes in 50°, or,just 2.4 arc-second per degree. This was considered satisfactory. We find that our telescope has a satisfactory target acquisition ability of about 2 arc-minutes in its field of view.

This telescope is presently equipped with a solid state UBV photometer for carrying out photometry, a ST-4 CCD used as a star tracker, and a ST-6 CCD camera for use in imaging applications. We plan to convert the telescope control from open loop to closed-loop control by adding digital position encoders.

In summary, we find that building this telescope has been a challenging and educative experience. We shall be happy to send our detailed design diagrams and helpful advice to any one interested in building a similar telescope.

Acknowledgments.   This project was made possible by DST project grants to both of us (M.N.A and S.P.B), and extensive assistence rendered by IUCAA in Pune, India.

Astronomy for Developing Countries
IAU Special Session at the 24th General Assembly, 2001
Alan H. Batten, ed.

Small Radio Telescopes for Education

Koitiro Maeda

Department of Physics, Hyogo College of Medicine Nishinomiya, Hyogo 663-8501, Japan. e-mail:maeda@hyogomc.kugi.kyoto-u.ac.jp

Noritaka Tokimasa

Nishi-harima Astronomical Observatory Sayo-cho, Hyogo 679-5313, Japan

Abstract. We present small radio telescopes consisting of commercial instruments for satellite TV reception. With these radio telescopes we can observe the 12 GHz emissions from the quiet sun and solar flares. Since microwaves occurring in our environment, e.g., those from a building and a fluorescent lamp, are also detectable, such radio telescopes are useful not only for radio astronomy education but also for physics education.

1. Introduction

At low frequencies (HF and VHF bands), the galactic background radiation, solar bursts, and Jupiter's decametric radiation are strong and detectable with a radio telescope consisting of a simple antenna (e.g., a dipole or a 3-element Yagi) and a commercial communications receiver. Maeda (1990a) reported such simple radio telescopes to observe the galactic background radiation and solar bursts at about 30 MHz. Since the thermal emission from the quiet Sun is relatively weak at low frequencies, some sensitive observations are necessary to detect it. Phase-switched interferometers designed for education were used to observe the quiet Sun, Cas A, Cyg A, and Tau A at 74 MHz (Swenson, 1978) and at 49.5 MHz (Maeda 1990b). The quiet-Sun component becomes stronger with increasing frequency. Radio telescopes for the microwave region have been largely beyond the reach of teachers and amateurs, owing to the high cost of antenna with a large collecting area. Now, however, we can do some interesting astronomy with widely available antennas with sensitive electronics designed for satellite TV reception. The mapping of the Milky Way at a wavelength of 21 cm was made using such an antenna (Schuler, 1994; Rogers, 1996). In this paper we describe the small radio telescopes that are portable and usable for observing the sun at 12 GHz.

2. Small Radio-Telescope at 12 GHz

The small radio telescope that we made consists of a commercial satellite TV antenna with an effective diameter of 35 cm, a commercial IF booster, a home-made detector, and a voltage-measuring device (see Figure 1). The satellite TV

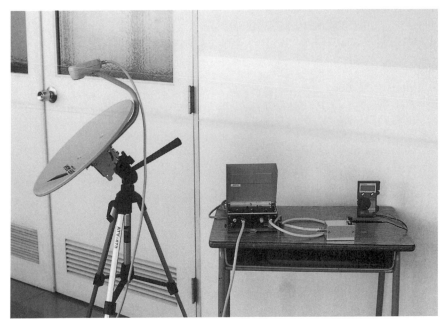

Figure 1.   The simplest radio-telescope we made. The satellite TV antenna of 35 cm effective diameter mounted on a video-camera tripod is at the far left. On the small desk, from left to right an IF booster, a home-made detector, and a digital multi-meter are seen.

antenna we used consists of an offset parabolic reflector and a frequency converter. This antenna receives the right-hand circularly polarized signal. The antenna is mounted on a video-camera tripod using a cylindrical aluminum bar. The antenna beam direction is therefore easily changeable. Since the effective diameter of the antenna is 35 cm, the beamwidth at 12 GHz (a wavelength of 2.5 cm) is about 4 degrees. The nominal antenna gain is about 32 dB. The frequency converter amplifies the received 12-GHz signal by 52 dB and outputs an IF signal of 1 GHz. The noise figure of the converter is 0.9 dB. The 1-GHz signal from the converter is fed into the IF booster and further amplified by about 30 dB. Such a booster is commonly used for satellite TV reception at an apartment and commercially obtainable. The amplified IF signal is detected to get a D.C. voltage. We use a simple homemade detector. A commercial detector is also available. The output D.C. voltage of the detector is about several tens of millivolts and measurable with a digital multi-meter. For obtaining a continuous record, a pen recorder is useful. It is also good for students to use an analog-digital converter to graphically display the data on a personal computer.

3.   Observation of the Sun

The beamwidth of the antenna is about 4°, that is much greater than the angular radius of the Sun, i.e., 0.5°. It is relatively easy to catch the Sun even in a cloudy

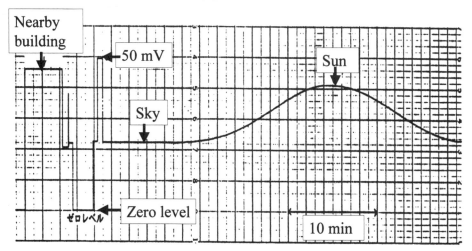

Figure 2. Drift-scan of the Sun recorded on a pen-recorder. The satellite TV antenna having an effective diameter of 45 cm was used. The levels of the sky and a nearby building are also shown.

sky. On a sunny day the sunlight can be used for pointing. A good way is to put a piece of well reflective tape on the parabolic reflector surface. The reflected light from the tape makes a bright spot on the converter input surface. By setting the bright spot at the center of the converter input surface we can easily point the antenna beam at the Sun. The sky background at 12 GHz consists of the cosmic background radiation of 2.7 K and the atmospheric emission (say, about 3 K). A simple calibration method was tested using the sky background radiation and the thermal radiation of the wall of a room. By a linear interpolation the antenna temperature due to the sun was estimated to be about 130 K. The flux density was calculated using this antenna temperature and the effective area of the antenna. We obtained a flux density value of 4.8×10^6 Jy. This value is in good agreement with that given in the textbook of Kruger (1979). Figure 2 shows an example of the drift scan observation of the sun recorded on a pen recorder. The antenna of 45 cm effective diameter was used for this observation.

4. An Advanced Version of the Small Radio-Telescope

Figure 3 shows an advanced version of the small radio telescope. The satellite TV antenna of the effective diameter of 45 cm is used for this radio telescope. The gain and beamwidth of this antenna are about 34 dB and 3.2 degrees, respectively. The satellite TV antenna is set on the equatorial mount that is commonly used for an optical telescope, using a joint as shown in Figure 3. By tracking the Sun and detecting solar bursts we can do solar flare patrol. Since the solar flare affects the terrestrial upper atmosphere in various ways, such an observation is useful for education of the upper atmospheric environment, so called space weather.

Figure 3.   An advanced version of the small radio-telescope. The satellite TV antenna, its effective diameter being 45 cm, is set on an equatorial mount for an optical telescope.

5.   Other Experiments with the Small Radio-Telescope

At 12 GHz everything in our environment radiates radio waves. The human body is emitting the microwaves as thermal radiation at a temperature of about 310 K, which is much greater than that of the noise temperature of the sky. We can easily show the human body is really emitting the microwaves using the small radio telescope. Since it is easy to carry the small radio telescope, we can use it for a lecture in a classroom. In this respect a fluorescent lamp is useful as an artificial radio source. The directivity of the antenna is easily demonstrated using the fluorescent lamp. By blocking the ray path of the radio waves by various materials we can demonstrate what kind of material effectively attenuates the microwaves. An interferometer experiment is also possible using a fluorescent lamp and an aluminum plate as a reflector. We therefore believe that small radio telescopes are widely usable at various levels of physics education.

References

Kruger, A., 1979, *Introduction to Solar Radio Astronomy and Radio Physics*, Reidel, Dordrecht, The Netherlands.

Maeda, K., 1990a, Sky & Telescope 80, (August), 200.

Maeda, K., 1990b, Astronomical Herald, 83, 72. (in Japanese)

Rogers, A.E.E., 1996, Sky & Telescope 92, (August), 75.

Schuler, P.W. III, 1994, Sky & Telescope 87, (March), 91.

Swenson, G.W., 1978, Sky & Telescope 56, (October), 290.

Astronomy for Developing Countries
IAU Special Session at the 24th General Assembly, 2001
Alan H. Batten, ed.

Cultural-Grant Aid in Astronomy for Developing Countries from the Japanese Government

Masatoshi Kitamura

National Astronomical Observatory, Mitaka, Tokyo 181-8588, Japan

Abstract. In order to promote education and research in developing countries, the Japanese Government began in 1982 providing high-grade equipment within the scheme of its ODA (Official Development Assistance). Since then, for astronomical development, twelve planetaria have been donated to eleven countries and seven reflecting telescopes, with accessories, have been installed in Asian and Latin-American countries.

1. Introduction

The programme of providing scientific and educational equipment to developing countries, operated by the Japanese Government, is called the Japanese Cultural Grant Aid and is a part of ODA (Official Development Assistance). The donation of astronomical instruments, such as research telescopes and educational planetaria is part of this Cultural Grant Aid. Application for such Aid must be formally made through the Japanese Embassy in the country concerned, by the submission of an application document. The aid provided is not exclusively for astronomy and, therefore, applications for astronomy must compete with applications from workers in other fields in a country that seeks aid from the Government of Japan.

No special application form exists. In preparing an application document, applicants should give good reasons why the equipment is needed and are advised to include a short history of astronomy education and, if applicable, astronomy research in their country. It is essential to name and to describe a responsible institution within the country, where the equipment can be housed. It is most important to describe any guarantee by the applying country for the provision of the building needed to house the equipment being sought. Finally, the completed application document should be submitted to the Japanese Embassy through the **Host Institute** and the **Ministry of Education** of the country concerned, with signatures by represenatatives of these respective institutions.

Upto the end of the Japanese fiscal year 2000, the following countries and their respective institutes have received the astronomical equipment indicated in the next section.

2. Equipment Donated by the Japanese Government 1986-2001

Reflecting telescopes and accesories have been given to:

- 1987 Singapore, 40-cm reflector (Science Centre)

- 1988 Indonesia, 45-cm reflector (Bosscha Observatory, Institute of Technology, Bandung)

- 1989 Thailand, 45-cm reflector (Chulalongkorn University, Bangkok)

- 1995 Sri Lanka, 45-cm reflector (Arthur C. Clarke Institute for Modern Technologies, near Colombo)

- 1999 Paraguay, 45-cm reflector (Asunción National University)

- 2000 Philippines, 45-cm reflector (PAGASA, Quezon City)

- 2001 Chile, 45-cm reflector (Cerro Calan Astronomical Observatory, University of Chile).

 Planetaria have been given to:

- 1986 Burma (Myanmar) (Pagoda Cultural Centre)

- 1989 Jordan (Haya Cultural Centre)

- 1989 Malaysia (Space Science Education Centre)

- 1990 Philippines (a supplementary projector for the already-existing planetarium in Manila)

- 1993 India (Burdwan University, West Bengal)

- 1993 Argentina (a supplementary projector for the already-existing Planetario de la Ciudad, Buenos Aires)

- 1994 Uruguay (a supplementary projector for the already-existing Planetario de la Ciudad, Montevideo)

- 1998 Vietnam (Cultural Memorial Hall, Vinh City)

- 1998 Thailand (a supplementary projector for the already-existing planetarium in Bangkok)

- 1998 Sri Lanka (a supplementary projector for the already-existing planetarium in Colombo)

- 1999 India (Tamil Nadu Science and Technology Centre, Chennai)

- 2000 Uzbekistan (Tashkent City Park).

Acknowledgments.   The author would like to thank Prof. H. Haubold, Office for Outer Space Affairs of the United Nations and the previous director of the Office, Mr N. Jasentuliyana, for their strong support and assistance in the course of processing the successive applications for Aid. Thanks are also due to Prof. T. Kogure, former Director of Bisei Observatory in Japan, for his kind collaboration.

Abstracts of Poster Papers

Abstracts of poster papers relevant to this section of the Special Session are presented below.

Robotic Telescopes: A Link between Astronomically Developed and Astronomically Developing Countries

Peter Martinez, South African Astronomical Observatory, P.O. Box 9, Observatory 7935, South Africa

Astronomers in developing countries are often confronted with multiple problems of access to good equipment and access to good astronomical sites. While there is no subsititute for personal experience of using qood equipment at a good site, access to robotic telescopes can address these problems of access to some extent. This paper explores the possibilities for at least moderately experienced observational astronomers in developing countries to conduct observing programmes using remote robotic telescopes at good astronomical sites.

The Four College Automated Photoelectric Telescope

Saul J. Adelman et al. Department of Physics, The Citadel, 171 Moultrie St, Charleston, SC, U.S.A.

For the past decade, astronomers from The Citadel, The College of Charleston, the University of Nevada Las Vegas, and Villanova University have operated in Southern Arizona a 0.75-m automated telescope to obtain differential Strömgren uvby, Johnson BV, and Cousins RI photometry of a wide variety of stars. Each school averages the equivalent of about 40 nights/year of high quality photometry at a cost less than using observers. By mixing our programs we obtained observations of certain important stars on nearly every clear night they can be observed. Usually we request a star be observed only once per night. Still continuous coverage can be obtained. The stars are given priorities and scheduled using ATIS. The telescope selects targets from the groups with the highest priority by choosing the one closest to the western edge of the observing window. Some data has been analyzed by undergraduate and graduate students. We use internet to send requests for observations and to retrieve data. We believe our experiences are germane to others interested in automated photometric telescopes. We are open to the possibility of collaborations with other astronomers who are obtaining photometric and spectroscopic data. (Coauthors are: L. Boyd, R.J. Dukes Jr, E.F. Guinan, G.M. McCook and D.M. Pyper, all of the U.S.A..)

Developing an Astronomical Observatory in Paraguay

Alexis E. Troche-Boggino, Universidad Nacional de Asunción, Facultad Politecníca, 01 Agencia Postal Campus U.N.A., Central XI, Paraguay.

Background: Paraguay has some heritage from the astronomy of the Guarani Indians. Buenaventura Suarez S.J. was a pioneer astronomer in the country in the XVIII century. He built various astronomical instruments and imported others from England. He observed eclipses of Jupiter's satellites and of the Sun and Moon. He published his data in a book and through letters. The Japanese O.D.A. has collaborated in obtaining equipment and advised their government to assist Paraguay in building an astronomical observatory, constructing a moving-roof observatory and training astronomers as observatory operators. Future: An astronomical center is on the horizon and some possible fields of research are being considered. Goal: To improve education at all possible levels by not only observing sky wonders, but also showing how instruments work and teaching about data and image processing, saving data and building a data base. Students must learn how a modern scientist works.

Science with Small Observatory Instruments

Ron G. Samec et al., Department of Physics, Bob Jones University, 1700 Wade Hampton Blvd, Greenville, SC, U.S.A.

Good science can be done with small instruments. Very productive research in the area of short period interacting binary systems has been conducted by the authors with PMT's and small CCD cameras attached to 0.28 m to 0.9-m instruments. A summary of important results over the past fourteen years is presented along with current work on the interesting semidetached and contact solar type systems TY UMa, V523 Cas, CN And, VZ Psc and BE Cep. Light curve asymmetries in CN And, BE Cep and possibly others indicate the presence of gas streams. This may signal that they are undergoing evolution into contact, an important but rarely observed stage in binary star evolution. Period studies are presented for each of the systems, documenting interesting orbital histories. V523 Cas and TY UMa display large period changes giving evidence of a light time effect due to a third body. This research has been largely supported at the local level and by small research grants from the American Astronomical Society.

NWG acknowledges participation in the Research Experiences for Undergraduates (REU) program at the Southeastern Association for Research in Astronomy (SARA), sponsored by the National Science Foundation (NSF grant AST9619939). (Co-authors are: M.L. Stoddard, W. van Hamme and N.W. Gothard, D.R. Faulkner and R.L. Walker, all of the U.S.A.)

Section 6: Some Practical Matters

Astronomy for Developing Countries
IAU Special Session at the 24th General Assembly, 2001
Alan H. Batten, ed.

Some Cooperative Activities in East Asia

Norio Kaifu

National Astronomical Observatory of Japan, Mitaka, Osawa 2-21-1, Tokyo, 181-8588 Japan e-mail: kaifu@nao.ac.jp

Abstract. We report the activity of continued and sequential coopera- tion among Asian countries/regions, especially in East Asia. Such efforts started in 1990 from a small-size China-Korea-Japan meeting on star- forming regions. Being aware of the importance of cooperation among those neighboring countries, participants agreed to hold sequential "East Asian Meetings for Astronomy (EAMA)". The 1992 meeting entitled "Millimeter-Wave and Infrared Astronomy" was held in Korea, the 1995 meeting entitled "Ground-Based Astronomy in Asia" was held in Japan, and the 1999 meeting entitled "Observational Astrophysics in Asia and its Future" was held in China. These meetings achieved quite high activity with 100-200 participants, each. An important product of those meet- ings was active exchange between young astronomers, including graduate students. The primary aim of these meetings/activities was to promote small but practical cooperation in the field of astronomical instrumen- tation, as well as to widen the contact among Asian astronomers. An East-Asian co-experiment to search for good sites for a possible "Asian Observatory" was among such efforts. The close cooperation between Japan, China (Peoples' Republic and Taipei) and Korea, on millimeter and sub-millimeter wave technology is another good example of joint de- velopments of new instruments.

1. East-Asian Meetings for Astronomy (EAMA)

In 1990 Chinese and Japanese radio and IR astronomers held a meeting on "Star-Forming Regions" in Huang-Shiang, China, co-chaired by C. P. Liu and N. Kaifu. There were 50 participants from China and 15 from Japan. Three astronomers from Korea were also invited. It was one of the first active meetings for observational astronomy (except the field of solar astronomy) ever held under the cooperation of East-Asian Countries. The important products from this meeting were:

1 Awareness of the importance and usefulness of cooperation between those Asian neighboring countries/regions.

2 From the above point of view, participants agreed to hold sequential "East
Asian Meetings for Astronomy (EAMA)" setting a target for observational
astronomy and instrumentation, and to invite other neighboring countries
as well.

Following the above consensus the second EAMA was organized by Y.K.
Minn and Korean astronomers in 1992. This meeting entitled "Millimeter-Wave
and Infrared Astronomy" was held in Daejon, Korea with 100 astronomers from
Korea, China, Japan, Taiwan, and the U.S.A., and was very fruitful, reflecting the rapidly increasing activities of Korean astronomy. The proceedings were
published by the Korean Astronomical Observatory (edited by Y.K.Minn, 1992).
Importantly, many Korean graduate students attended this meeting, and it was
followed by active exchange among graduate students. Japanese graduate students invited many Chinese, Korean and Taiwanese friends to their "Young Astronomers' Summer School" in 1993, 1994 and 1995. The YASS is a traditional
activity of Japanese graduate students in the field of physics and astronomy; they
organize self-educating summer schools every year by themselves, with limited
financial support from senior scientists. Such exchanges of young astronomers
had a tremendous positive effect in those neighboring countries/regions.

In 1995 the third EAMA, "Ground-Based Astronomy in Asia" was held in
Tokyo, Japan. This was a first attempt in an EAMA to extend the cooperation
to other Asian countries. 215 astronomers from 13 countries/regions attended,
and 210 papers (including posters) were presented. We made an attempt to
list all of the astronomical observatories, institutes, groups and telescopes in
the Asian region, aiming to achieve better information, contacts and cooperation in this region. The proceedings published by the National Astronomical
Observatory of Japan (edited by N. Kaifu) lists all the data collected as well
as the scientific reports presented, and it can be used as a reference book of
the Asian astronomical observatories/institutes though it requires much more
improvement. Some copies of the proceedings are still available from NAOJ on
request.

The latest and 4th EAMA was "Observational Astrophysics in Asia and
its Future" held in Kunming and Lijiang, China. 118 participants from 13
Asian countries/regions discussed future telescopes, instrumentation and cooperation. The proceedings were published by the Chinese Academy of Sciences
(edited by P.S. Chen, 1999). In this meeting participants agreed to hold a young
astronomers/graduate-students seminar in Hawaii in the near future. This has
not yet happened because of the busy situation of the Subaru Telescope (NAOJ),
but such activity will open a lot of good possibilities of scientific cooperation
using the Subaru telescope, and will activate observational astronomy in Asia.
Taiwanese astronomers expressed the intention to invite the next EAMA.

2.   Site-Testing for an Asian Observatory

The primary target of those EAMA activities was to promote small but practical cooperation in the field of astronomical instrumentation as well as to widen
contacts among Asian astronomers. Among those efforts was an East-Asian

collaboration to search for good sites for a possible future "Asian Astronomical Observatory". In 1993 "Site Experiment for Astronomical Observations in the Northwest Region of China" was organized as an East-Asian Regional Cooperation, with support from the Chinese Academy of Sciences and the Japanese Society for Promotion of Science. Twenty-four participants from Mainland China, Korea, Taiwan and Japan visited Qinghai and other dry and high-altitude places in west China and made various site measurements for optical, IR and mm-wave observations. This visit, especially to the Qinghai Station (Purple Mountain Observatory) in Delingha created good relations among the mm-wave astronomers and was followed by the realization of a cooperation program in mm-wave and sub-mm-wave instrumentation between China and Japan, as described later. The report of this experiment was published by NAOJ (edited by N. Kaifu and C.P. Liu, 1995).

3. Cooperation in mm-Wave and sub-mm-Wave Instrumentation

A program of close cooperation program in millimeter-wave and sub-millimeter-wave technology was another by-product of the EAMA, and it is a good example of the small but practical and realistic cooperation that bore scientific fruit.

> 1 *Millimeter-wave cooperation between Purple Mountain Observatory and Nobeyama Radio Observatory, NAOJ.*
> An SIS receiver at 115GHz was developed jointly for the 20-m mm-wave telescope of the Quinghai Station (PMO), and now is in operation.
> POST: A portable sub-mm wave telescope and 230GHz (and 350GHz) SIS receiver were developed jointly by PMO and Nobeyama for the observations in the Delingha area which proved to be a good site for the sub-mm wave observations while above-mentioned 1993 experiments.

> 2 *Millimeter-wave cooperation between Korea Astronomy Observatory and Nobeyama Radio Observatory, NAOJ.*
> 100GHz and 150GHz SIS receivers were jointly developed for the 14-m mm-wave telescope in Daejon, KAO and are producing large amount of observing data.

> 3 *SMA/ASTE cooperation.*
> The Institute of Astronomy and Astrophysics of Academia Sinica, Taiwan (ASIAA), Nobeyama, NAOJ and PMO, China are about to start a new cooperation program for the development of a 650GHz sub-mm wave receiver for the SMA by using Nobeyama SIS mixer.

This last cooperation will also include scientific cooperation by using the SMA which will soon be in operation. In parallel to it, observational cooperation and joint developments will start for the Atacama Sub-mm Telescope Experiments (ASTE). The central instrument of ASTE is a 10-m sub-mm wave telescope which was built by NAOJ and will be transported to the Atacama high plateau (altitude 5000m; near the ALMA site) in 2001. This will be one of the highest performance sub-mm wave telescopes, and will be operated with the cooperation of NAOJ and several groups from Japanese universities. The participation of China (PMO), Taiwan (ASIAA), and possibly Korea in ASTE in the near future will considerably add to the activities both in science and instrumental developments in the field of sub-mm wave astronomy. We foresee more Asian cooperation for the ALMA, a global joint project by North America, Europe and Japan of which we hope to start the construction in 2002.

References

Millimeter-Wave and Infrared Astronomy ed. Minn, Y.K. ed. 1992, Korea Astronomy Observatory

Report on Site Experiments for Astronomical Observations in Northwest Region of China ed. Kaifu, N. and Liu, C.P. eds. 1995, National Astronomical Observatory of Japan

Observational Astrophysics in Asia and its Future ed. Chen, P.S. 1999, Chinese Academy of Sciences

Discussion Anandaram congratulated Kaifu on the successful Japanese efforts made to achieve scientific cooperation through EAMA. He thought that the objectives of EAMA were very similar to those of the Third-World Astronomy Network proposed by Narlikar of (see pp. 324-8). Anandaram hoped that other countries could be brought into this kind of cooperation through suitable arrangements set up under IAU. Kaifu replied that he was aware of Narlikar's proposal and hoped to discuss it with him. Hearnshaw pointed out that the IAU already recognizes the Asian-Pacific region for organizing Asian-Pacific Regional Meetings and the EAMA meetings are outside this structure. He wondered if countries like Australia and New Zealand would be welcome at the EAMA meetings? Kaifu replied that the Asian-Pacific Regional meetings were more general and the EAMA is specifically for observational astronomy and the East-Asian region. Any countries and regions of Asia would be very welcome. There were participants from Australia for ALMA in 1995 and 1999 and from India and even the Middle East. Orchiston commented that Australia already has two professional mm-wave telescopes and there is Australian-Japanese collaboration on the square-kilometer array, but there has not yet been any attempt at Australian-Japanese collaboration in mm radio astronomy. Kaifu was unaware of Australian-Japanese collaboration on the square-kilometer array but agreed that it would be useful to discuss collaboration between Asia and Australia in mm-wave astronomy. Rieu said that he had heard that the Qinghai millimeter telescope was not in active operation because of logistical problems. Qinghai is indeed very far from any radio astronomy observatories in China. He wondered if the problems had been solved. Kaifu replied that the Qinghai station is certainly quite far from the main cities of China but he understood from Chinese friends that the telescope is in active use, especially by Dr. Y. Chi and his group from PMO who took the SIS mm-wave receivers to Qinghai. Wang remarked that the Qinghai telescope is definitely in operation and Delingha is an observing base of the National Astronomical Observatories of the Chinese Academy of Sciences. The young and able group headed by Yang, Shi and Pei is operating the 13.7m mm-telescope and obtaining good results.

Chen commented that the difference between the EAMA consortium and TWAN is that the former deliberately avoids not only the words "third-world" but also the word "countries". He took the opportunity to draw attention to IAU Colloquium 183 "Small-Telescope Astronomy on Global Scales" to be held January 2001 in Taiwan. All were cordially invited.

Third-World Astronomy Network

Jayant V. Narlikar

Inter-University Centre for Astronomy and Astrophysics, Pune 411007, India. e-mail: jvn@iucaa.ernet.in

Abstract. Several developing countries of the Third World have been actively interested in astronomy, as is evidenced by the membership of the IAU. The enthusiasm of individual astronomers from these countries is, however, not matched by the resources available to them to pursue their interest in astronomy, in teaching as well as research, at an above-threshold level. Major problems requiring solutions are (i) isolation from the mainstream work, which leads to research work which is not quite relevant or realistic, and to teaching based on outdated knowledge; (ii) lack of financial resources, leading to shortage of books and journals in the library, insufficient computing power, out-of-date instruments, as well as inability to participate in essential activities like schools, workshops, and major international conferences and symposia; and (iii) lack of hands-on experience with state-of-the-art instrumentation that often leads to good scientists being turned away from astronomical observations towards abstract theories.

Experience of the International Centre for Theoretical Physics at Trieste, Italy and of the inter-university centres in India, like the IUCAA at Pune, has shown that limited resources can be made to go a long way by sharing, networking and intelligent use of communications technology. Based on the above experience, this proposal envisages setting up a *Third World Astronomy Network* (TWAN) under the auspices of the IAU, within the wider ICSU-umbrella with support from the UNESCO as well as participating nations. The TWAN will operate with a few key institutions as local nodal points of a wide network. The objectives of the proposed *TWAN* and the role of the *Nodal Institutions* (NIs) are spelled out in this proposal, along with the budgetary support required.

1. Introduction

The Third-World Astronomy Network (TWAN) is a proposal to bring together astronomy institutions and groups of scientists working in astronomy and astrophysics (A&A) in Third-World countries (TWCs) under one umbrella, so that they can pool their limited resources and improve their work output through several collaborative ventures. The extraordinary progress of information technology makes it possible to translate this idea into reality through networking with a relatively modest investnment.

What problems will such a network hope to solve? One may identify three main problems faced by astronomers from the Third-World countries:

(1) The field of A&A is advancing very rapidly and it is very difficult to keep pace with it, even under the best of conditions. Isolation from these developments is the main difficulty faced by astronomers from the TWCs. Not knowing where the subject has reached today, their perception is largely based on old literature, with the result that the research problems they undertake to solve are either already solved or are no longer relevant. Similarly, the teaching of A&A in the TWCs can be based on outdated curricula and textbooks. So the students doing Ph.D.s enter the field with inadequate foundations.

(2) The priorities for public spending in the typical TWC are rightly oriented towards providing and upgrading basic amenities like water, food and shelter. Basic research figures on the country's agenda, if at all, very low down the list. And amongst the basic sciences astronomy is not seen as addressing down-to-earth problems. So funding for A&A research, teaching and development (RTD) is either nonexistent or at a very low level. Further, even if there are international pedagogical activities like summer/winter schools, research workshops or conferences presenting reviews of state-of-the-art work, the scientists from TWCs are not able to participate in these for lack of funds.

(3) The lack of access to good observing facilities or usage of good instruments and software for observing and data-analysis deters would-be observers from the TWCs from observational astronomy, orienting them towards abstract theory which demands relatively less expensive and elaborate infrastructure. Very often therefore it is seen that this "by default" recourse to theoretical research produces substandard papers. Given suitable opportunities of observations and/or data analysis, the same people could have produced much better science.

Despite these difficulties many TWCs are members of the IAU and their members make valuable contributions to its activities. It is therefore all the more important that the international community of astronomers comes forward to discuss and resolve these difficulties.

There is no foreseeable change in the economic circumstances of the TWCs, or their attitudes and priorities, to raise expectations of a sea change in the above situation. As the International Astronomical Union is the acknowledged world body for astronomy, it provides the right forum to discuss these issues. Moreover, with its established reputation and human resources, it is the ideal body to initiate developments to improve the situation. Which is why I am raising this proposal at the Special Session organized to discuss the state of astronomy in developing countries.

The bottom line of what I wish to propose is this: *even with limited resources, one can make a beginning by setting up a network linking TWCs to harness the latest available tools of information technology to upgrade the state of A&A in these countries.*

2.   The IT Revolution

The key to resolving the three problems mentioned above lies in the transmission, reception and processing of information. The last decade has seen a remarkable growth of information technology. An example of how this helps can be seen by an episode from my own personal experience, a couple of years ago.

Working in Pune, India, I was then writing a popular book to be entitled *Seven Wonders of the Cosmos* for publication by the Cambridge University Press, Cambridge. My publisher informed me that a photograph that I had sent of the 300-m dish at Arecibo, for inclusion in the book was not of good quality and asked if I could send a better image. I sent an e-mail requesting a friend in the Arecibo Observatory to send a photograph. He readily obliged by sending the picture file to the publisher electronically. The entire process took no longer than twenty four hours. The cost? Hardly anything. In the old days one would have had all this correspondence by air-mail (now called snail-mail!) between Cambridge-Pune-Arecibo-Cambridge, taking at least about a month.

The important thing about this example is that it has by now ceased to become extraordinary. Transmission of manuscripts and pictures is fast becoming routine, the only limitation being access to sufficient bandwidth.

So can we not harness this quiet but sure revolution towards improving the RTD conditions in the TWCs?

3.   Networking

Networking holds the key to this approach. Let me give a few examples, from which we can learn something.

In India libraries of eight A&A institutions have come together under a scheme called FORSA (FOrum for Resource Sharing in Astronomy). These are research institutes and observatories. These libraries share their resources through exchange of information, abstracts of journals, books on inter-library loan, and, more importantly, sharing research journals. Some journals offer special deals for more electronic copies, if one set of hard copy is subscribed to by one library.

My own centre, the Inter-University Centre for Astronomy and Astrophysics (IUCAA) was set up to help upgrade the RTD in Indian universities. We have set up a data centre containing astronomical information, catalogues, etc., a mirror site for the Strasbourg Data Centre, and wide enough bandwidth to access data centres abroad. This facility is made use of by our users, the faculty and students from universities.

In this context the proposal originating in the U.S.A. for a "Virtual Observatory" is very instructive. It is argued in this proposal that archival data from various observations can generate good science through analysis and that they be made available in such a virtual observatory. Thus even without access to world class observatories, one may still hope to do good astronomy from their data.

To show that networking can help also at the other end of the spectrum, for bringing astronomy to schoolchildren, let me narrate another experience at IUCAA. We have set up a link with a 0.4-m telescope at the Mt Wilson Observatory in Southern California. Through this link the telescope can be operated and images taken from Pune, at the other side of the globe. We had sessions of this kind with schoolchildren who had come for a summer programme at IUCAA. Because of the 12.5 hours time difference, these children could operate the telescope in broad daylight, at 2. 30 p.m.!

Perhaps I should mention the pioneering work of the late Professor Abdus Salam of Pakistan, the Nobel prize winning particle physicist, in setting up the International Centre for Theoretical Physics in Trieste more than three decades ago. Long before the IT revolution started, the ICTP had been playing a nodal role in bringing together physicists from TWCs for their individual research as well as for pedagogical activities. The Third-World Academy of Sciences which funds collaborations from TWCs was also set up by Professor Salam.

4. TWAN

With these examples, it is clear that physical separation from the scene of activity is no longer a prime concern; information can be exchanged across physical barriers, and it can be processed to the benefit of astronmy. Using the available technology today (with of course, future improvements) one can develop the idea of TWAN, a Third-World network in astronomy along the following lines.

A few institutions in participating TWCs are identified as nodes of this network. They will ensure that they have certain basic minimum infrastructure which helps promote RTD activities. These could include, for example, fast e-mail and data-transfer facilities, mirror sites of some data centres, electronic access to any virtual observatory, access to astronomy preprint sites, etc. Each nodal institution (NI) will serve the academic community in its zone which may include more than one country to begin with.

Apart from the "passive" mode of making available access to data, there should be the more active role that each NI may be called upon to play. It should organize schools, workshops, teacher-refresher programmes, etc., for students, researchers and teachers in its region. There can be an instrument-making laboratory which encourages scientists from the TWCs to come and build their own modest level instruments under expert guidance, including a telescope making facility for amateur astronomers.

There is a third mode, the "catalytic mode" in which the NI may help a Third-World scientist to submit an observational proposal at a major facility in the world, also providing a travel grant if the proposal is accepted.

I should stress that the IUCAA at Pune is playing all three roles for the Indian university community ever since its foundation in 1989. It is perhaps symptomatic that the IUCAA was set up just at the time that IT was beginning to grow; and so we have been able to take the advantages of the IT revolution as it unfolded.

5. Funds

Now the last but not the least: what funds will be needed to set up the network and to keep it growing? Where can they be generated? Here are some ideas.

The requirements of the network begin with the provision of adequate infrastructure at each nodal institution for data transmission and storage. Thus adequate server capacity and sufficient band-width for data transmission will be required. Some NIs may already have an infrastructure for their own needs. But it may have to be upgraded to take on the additional responsibility of the nodal role. Some project staff may have to be appointed for running and maintaining the facility. Since the job market lures the talented to commercial ventures, it may not be desirable to have permanent appointments, but one may rather concentrate on a floating population of computer-scientists/engineers who may like to take on the job for 2-3 years to pick up experience.

If there are any protocol/royalty charges to be paid for creating the mirror sites of data centres, provision may have to be made for these amounts. The IAU may also request these charges to be waived, considering the use these centres are to be put to.

Some provision may have to be made for travel/subsistence grants for funding guest-observing visits of Third-World astronomers to major observing sites in the world.

The pedagogical activities of each nodal centre will need financial support, mainly in the form of travel of the lecturers and participants. The former may have to be drawn from the advanced countries, but the latter will be limited to the region. It is expected that each centre will have adequate housing facility for hosting such meets. If not, some initial expense will be needed to set up these.

As the information technology advances further, the travel costs will reduce. Thus remote operating of international telescopes, video lecturing with audience participation, e-mail talks between collaborating scientists across the globe will make it easier and easier to improve the interactive role of TWAN. But a beginning has to be made – sooner rather than later.

Having listed the major items requiring funds to run the TWAN, I now consider the likely sources of these funds. These could come from a seed-support from the IAU under the ICSU umbrella, a recurring grant from the UNESCO, some support from the Third-World Academy of Sciences (which may be invited to cosponsor TWAN), recurring commitment from each TWC through its adhering organization to the IAU, and of course, from private foundations.

6. Action Plan

If the idea of TWAN is found attractive, I suggest that the IAU, through its Commission 46 on the teaching of astronomy, may set up a working group to quantify the details and come up with a working budget. This may then be put to the EC of the IAU for further discussion and possible action.

Discussion

Schreuder asked which countries *are* Third-World countries? Kochhar interjected that any country that says it is a Third-World country should be accepted as one. Narlikar said that the Third-World Academy of Sciences has a list which we could adopt.

Astronomy for Developing Countries
IAU Special Session at the 24th General Assembly, 2001
Alan H. Batten, ed.

Security of Equipment

Lesley I. Onuora

*Astronomy Centre, University of Sussex, Falmer, Brighton BN1 9QJ,
U.K. e-mail: lonuora@pact.cpes.susx.ac.uk*

Abstract. Providing adequate security for equipment is probably a
problem in all parts of the world, but needs special consideration in developing countries. Social and economic conditions are very different so
that implementation of advice sought from experts who are unfamiliar
with local conditions can lead to disasters.

The need for adequate security must be taken into account while
considering factors such as type of equipment, the site for the equipment,
budgetary provisions, personnel etc.

Some examples of situations that may arise will be given in order to
provoke discussion and raise the awareness of participants on this important issue.

1. Introduction

It is fairly obvious that when purchasing and installing equipment of any type,
adequate security must be provided. This is true the world over and is not
peculiar to developing countries. However the security problems involved do
vary from country to country. In the context of a foreign expert being asked to
give advice on the installation of equipment it is therefore possible that serious
mistakes may be made. This issue of security is a rather sensitive one (since it
implies that crime is expected to be committed) and is rarely, if ever, discussed
at meetings such as these. However it is hoped that some thought and discussion
may be provoked by this short paper which raises some of the problems that
may be encountered.

2. Siting of Equipment

In choosing the site for an astronomical telescope, the main considerations are
usually scientific e.g. for an optical telescope the quality of the seeing, the
likelihood of clear skies etc. The aim is to site the telescope to obtain the best
possible data. Considerations concerning adequate infrastucture and services are
not usually of prime importance for a developed country. If adequate facilities
are not presently available at the chosen site then provision can generally be
made without any great difficulty.

However for a developing country seeking to install a telescope the situation may be rather different. A major consideration may have to be the ability to maintain a constant power supply. Without power, the whole project will be in jeopardy, not only from the point of view of running the equipment but also from the security aspect. In many developing countries power cuts are common. This may be not just a blackout for an odd hour but could commonly be for days or weeks. Delays in making repairs are often due to lack of resources, either spare parts or money. There is very little incentive to find these resources if the location has little nearby habitation, as is likely for a site chosen purely on its astronomical merits. If the telescope is in an isolated position, lack of illumination leaves the site an open target. Therefore scientific benefits may have to be sacrificed in favour of a site closer to adequate services, such as a university campus. This may be far from ideal but would have the additional benefit of ease of monitoring and quickly dealing with any security problems.

3.   Type of Security and Budgetary Provision

Electronic security equipment is likely to be of little use. A high wall, iron bars on windows and doors, together with armed security officers, possibly with guard dogs, are likely to be far more reliable. Thus a constant flow of funds is required to cover the drain on resources of salaries and possibly also accommodation, since it is common practice to provide on-site accommodation for security guards.

It is therefore of major importance that there be sufficient budgetary provision for any security measures to be maintained after the initial phase of the project. In many developing countries there are no formally organised research councils, or their equivalent, responsible for the funding of scientific projects. Funding may depend on the whim of whoever is in control at either the government or university level and this may change without warning. Therefore it is important to be sure of sufficient funding to pay for security arrangements over the long term before any project is undertaken. For the same reasons it is also important that the funding is not just in the form of an agreement but that it actually exists, for example in a special bank account for the project.

With sufficient funding (which is rarely available) it may be possible to partially solve the power problem by equipping the site with generators. However a constant supply of fuel and maintenance may be problems, particularly if spare parts are needed, often from abroad. The resulting delays may again expose the equipment to risk.

4.   Discussion with Local Scientists

It is vital that the opinions of the local astronomers who will be using the equipment are sought and listened to carefully. Security will almost certainly be uppermost in the minds of the people who are involved in the project but they may be hesitant to raise the issue. One problem here is the common stereotyping that occurs between developed and developing countries. For example an astronomer from a developing country who has never travelled abroad may feel that (a) a foreign expert will be shocked that there is a likelihood that thieves will try to steal the equipment or alternatively (b) that all foreigners think that theft is rampant in developing countries and so mentioning security will only confirm this belief. It is very important therefore that the visitor raises the issue as a common problem that has to be solved everywhere and gives the local astronomers encouragement to express their own views.

Another aspect to be aware of is that the way the equipment is targeted may be very different. Whereas in the West computer components may be the main target, in developing countries everything is recycled, including the power cables. This is in the context of countries which have absolutely no welfare state – if a person has no livelihood there is no safety net. This breeds great ingenuity in order to survive. A visiting expert is therefore unlikely in a brief stay to understand the complexity of the security problem and must rely heavily on local advice.

5. Access to the equipment

One major drawback in providing tight security for the equipment is that it may become counterproductive and this needs to be borne in mind when deciding on the type of security. A very sad situation can arise where the equipment is so surrounded by security measures that its use becomes very resticted. Many of the spin-off benefits of having the telescope, such as teaching students, public education in science etc. can be lost with, at best, a few privileged people using the telescope for research and at worst no-one using it at all. A balance needs to struck between security needs, scientific productivity and outreach activities.

6. Conclusion

Security should be given the highest priority if a project is to be successful. However it is important to realise that there are limitations to the advice which can be offered. The success of a project may depend on giving adequate weight to local experience.

Acknowledgment. LO is a Daphne Jackson Fellow sponsored by the Royal Society.

Discussion

Rijsdijk and Kochhar were concerned that Onuora was generalizing too much from the one country with which she was most familiar. She reminded them that she had emphasized at the beginning that she could talk only from her own experience and it was not her intention to generalize. She admitted that Kochhar's story of wild elephants looking for water and breaking down the campus fencing of Kavalur Observatory illustrated a problem that she had not encountered! Osorio agreed with Onuora on the importance of involving local people. He had worked under very difficult conditions at various remote African sites. Although he had encountered problems with volcanoes, security had not been a big issue, but he had always included local people in the team. Schreuder pointed out that security lighting might contribute to light pollution. He also suggested that perhaps an observatory should have its own power supply. Onuora replied that generators were important but could break down and finding spare parts might be difficult. Ubachukwu remarked that building dormitories for students and living quarters for researchers is a good security measure because the continuous presence of people discourages intruders.

Astronomy for Developing Countries
IAU Special Session at the 24th General Assembly, 2001
Alan H. Batten, ed.

Overcoming the English-Language Barrier

Terry J. Mahoney

Scientific Editorial Service, Research Division, Instituto de Astrofísica de Canarias, E-38205 La Laguna, Tenerife, Spain. e-mail: tjm@ll.iac.es

Abstract. Astronomers from non-English-speaking countries, who form a sizeable proportion of the astronomical research community, are obliged to communicate the results of their investigations in a language that is not their own. Consequently, good science is frequently masked by poor command of English, which can create an unnecessary barrier to the communication of scientific results. A suggested method of surmounting the language barrier is the setting up of scientific editorial services in at least the major astronomical centres. It is further argued that journal editors, rather than scientific referees, should be responsible for judging the linguistic and stylistic quality of articles presented for publication. The peer-review system would then be restricted exclusively to the scientific rather than linguistic content of papers presented.

The scientific Editorial Service of the Instituto de Astrofísica de Canarias, in operation since 1996, is briefly described in this context.

1. Introduction

English has attained an overwhelming hegemony since the Second World War and is now firmly established as the lingua franca of science. Scientists who fail to publish in English run the risk of lack of international recognition of their work with occasionally catastrophic consequences (e.g., Osawa and the prediction of the existence of the C_{60} molecule). The membership list of the IAU reveals that $\sim 60\%$ of working astronomers are non-anglophone (Mahoney 2001), yet all peer-reviewed articles must conform to high standards of written English. Some even claim that there is an "Anglo-Saxon bias" towards work produced by non-anglophone workers (Carter-Sigglow 1997) – especially towards work by those from developing countries (Umakantha 1997).

2. Surmounting the Language Barrier

2.1. The scientific community

The scientific community can help lessen the language burden of non-anglophone researchers by restricting peer review solely to questions of science and subsuming all questions relating to language usage into the editorial process. At the very least, referees who take it upon themselves to comment on the standard of English of a presentation should be prepared to make detailed suggestions on how the language may be improved instead of issuing such useless blanket statements as, 'the English of this paper could do with some improvement.' The question to be addressed is exactly *how* it may be improved, and journal staff are usually better equipped to handle this sort of problem. When a paper is rejected or referred back to the authors for revision, the authors have every right to expect to be told precisely what modifications are required of them in terms of both content and language.

2.2. Publishers

Publishers can (and most do) contribute by providing clear instructions to authors and explaining what happens to an article in the publishing process (a subject on which author knowledge tends to be vague or non-existent).

2.3. Institutions

Research centres need to develop a keen awareness of the language problem and offer courses in communication techniques. Where economically feasible, they could set up in-house editorial services to vet articles for grammar, spelling, adherence to journal styles, etc. Where this is too expensive, such services might be set up by pooling limited resources on a regional, national, or even international basis.

2.4. Author self-help

Above all, non-anglophone authors can help themselves by learning to regard English as an essential working tool, rather than as an objectionable hurdle to be cleared with the minimum effort. They should ensure that their writing possesses a clear logical structure, write with a target journal in mind, *read the journal instructions carefully*, run a spell-checker (with caution!) through their work, learn all they can about editing and publishing and acquire a minimum set of up-to-date standard reference works (dictionaries, etc.) – see poster by Mahoney (this volume, pp. 357-362).

3.   What Sort of Things Can Go Wrong?

It is first of all important to separate language difficulties from inability to think coherently. Language problems tend to occur among scientists at all levels, from doctoral students to heads of research departments, anglophones and non-anglophones, whereas as lack of coherent structure in writing is often associated with lack of writing experience. New researchers often lack an elementary knowledge of how a paper should be structured and how an argument should be developed. Often, their work will have a distinctly classroom air about it, with many references to undergraduate textbooks and no proper insight into current research. Most doctoral students, however, soon pass the juvenilia stage and mature remarkably rapidly; a small number, however, never quite grasp the basics of research writing, and their papers always need drastic overhauling before they are submitted for peer review. For this reason, research centres need to instill the precepts of good research writing in terms of how to structure an article and how to relate their writing effectively to other published work.

Apart from training, it is a good idea for new researchers always to submit their work to some form of internal review—not, I hasten to add, for the purposes of vetting the science, but rather to ensure that the paper is properly structured and well argued, which should be well within the capabilities of any competent researcher.

Common errors not related to language are: incomplete or incorrect references, non-adherence to journal style, poor presentation of illustrations and tables (e.g. unlabelled or illegibly labelled axes, etc.), inclusion of non-English words, phrases, sentences and occasionally entire paragraphs, failure to finish the article, and non-use of spelling checkers to weed out typos and the grosser misspellings. All of these oversights are due to carelessness (and many of them are not restricted to non-anglophone authors!). Authors must learn the need for good presentation at the start of their careers, and research centres can do much in this respect by providing relevant courses in authorship and publishing practices in astronomy. An excellent work that all academic authors should obtain a copy of is Beth Luey's *Handbook for Academic Authors* (Luey 1997), which provides many useful insights into all aspects of academic publishing.

Language difficulties are of an entirely different order and need to be dealt with differently. It is probably unrealistic to expect small non-anglophone research centres to be able to afford the provision of English classes at a suitable level for their research staff, but larger institutions may be able to do something in this respect. Incidentally, given the international nature of astronomical research these days, this also applies to anglophone centres, which often employ staff of many nationalities. Ultimately, however, it is up to individual researchers to strive constantly to improve their mastery of English, which, after all, is as essential a working tool to them as a knowledge of, say, statistics or calculus. While the aim must always be to write to publishable standard, this skill will take considerable time to acquire and is conditional upon such factors as previous language training and innate ability; however, *all* researchers must know sufficient English to avoid outright blunders and at least avoid saying the opposite to what they intend to say (only too easy to do in English, with its inscrutable verbal phrases and innumerable other booby traps for the unwary).

But it is not all just about writing. Non-anglophones are also at a severe disadvantage when it comes to making an oral presentation at a conference. Effective communication in public is a vital skill for all researchers, who need to make a good impression when under fire on the podium. Trainee researchers should take every opportunity of practising this skill in house before venturing to face the music at an international conference (see Gosling 1999 for some useful pointers), and those centres that can afford to provide courses in making public presentations should try to do so.

4. Which English?

From a linguistic point of view, there are many varieties of English, but as far as science is concerned only two need be considered seriously. American journals and publishers invariably require American English; in Britain, Ireland, continental Europe and the Commonwealth countries, British English is the norm. This is not as alarming as it might sound at first sight, since in reality we are mainly talking about questions of spelling and punctuation (it is unwise to try to be idiomatically "bilingual" in both). As far as academic English is concerned, the difference between the two is fairly minimal, full coverage of all the difficulties that are likely to arise being given in the references in the poster by Mahoney (this volume, pp. 357-362).

5.   Essential Working Tools

The basic reference tools for research writing are given in the poster presentation by Mahoney (this volume, just cited). On a purely language level, however, the non-anglophone could usefully browse the Web site of Oxford University Press (http://www.oup.co.uk), which is one of the world's leading English language teaching publishers (its ELT publications, for all proficiency levels, are listed at http://www1.oup.co.uk/elt/). An essential prerequisite for all non-anglophone writers of English is the best affordable bilingual dictionary, of which the latest edition should always be sought. A difficulty here is that such works do not always exist, even for some of the major languages, although the revolution that has taken place in commercial lexicography over recent decades has gradually transformed the situation from what it was, say, thirty years ago. As an example, present-day bilingual Spanish–English/English–Spanish dictionaries are more comprehensive, orders of magnitude more useful and of much higher quality than those published not so long ago.

6.   Scientific Editorial Services

All non-anglophone (and many anglophone) research centres could benefit from setting up—however provisionally—some form of editorial unit dedicated to the task of checking articles and proceedings contributions for English, internal consistency, correct referencing, adherence to journal style, etc. Such scientific editorial services (SESs) should be based on the following concepts:

- Editors should be astronomers, not departmental secretaries (who usually have better things to do with their limited time)

- The decision on whether or not to use the service must be the authors' alone

- All corrections must be indicated either on paper, using standard markup symbols (their use to be explained to authors), or in the electronic text file

- All corrections must be agreed upon between the editor and the authors, the latter always having the final say

- Articles must be corrected, *not* rewritten, and the original wording must be respected (provided it conforms to general standards of idiomatic correctness)

- Articles should be written and corrected with a particular journal in mind

- Authors should undertake to use any feedback from the editorial service to improve their standards of written English and general presentation: the proper role of an SES is to correct articles not to write them for the authors

- With regard to the editing of conference proceedings, editorial services should be used in a strictly advisory capacity: it is the job of the named proceedings editors to edit their own proceedings

7. English at the IAC

The Scientific Editorial Service of the Instituto de Astrofísica de Canarias (IAC) has been in operation since 1996. The SES is manned full time by an astronomer at the IAC's Research Division. The bulk of the work received for correction is in the form of research articles and contributions to conference proceedings, with occasional translation from Spanish into English for the IAC's Web page. Paper copies of double-spaced manuscripts are marked up using standard symbols and the corrections keyed in to the electronic file. The paper copy and electronic file are then returned to the author, who is responsible to submitting the corrected MS to the journal. The typical turnaround time is about two days for a letter and from four days upwards for a main journal article. Letters take priority, followed by main journal articles and proceedings contributions (in that order). Occasionally there are bottlenecks when several MSS arrive over a short period; in such cases it might take weeks to return the corrected article to the author. About half the total published output of the IAC—from main journal articles to proceedings contributions—is pre-edited by the Service before being submitted for publication; to date, over 200 articles and many proceedings contributions have been processed by the SES. IAC authors who use the Service find that many of the usual language problems with referees tend to arise much less frequently or not at all.

8. Conclusions

The English language is seen by many non-anglophone scientists as an obstacle in publishing their work. However, there are many ways in which the language barrier can be minimized by the scientific community (in redefining the role of referees in the peer-review system), by publishers (in explaining to authors what happens to their MSS after submission), by research institutions (by cultivating an awareness of the language problem and providing the necessary training in combating it) and finally by authors themselves (in learning to regard language as an essential working took that needs to be constantly honed to perfection). Scientific editorial services could usefully be set up on institutional, regional or even national levels (according to available funding) to correct articles for English (and perhaps journal style). One such service has been in operation at the IAC since 1996 and is viewed positively by those authors who use it, with the result that many of the common complaints about the language hurdle have either been largely overcome or eradicated altogether.

Acknowledgments.   I thank the IAU for its invitation to attend the General Assembly. My visit was funded by the Instituto de Astrofísica de Canarias.

References

Carter-Sigglow, J. 1997, Nature, 385, 764.

Gosling, P. J. 1999, *Scientist's Guide to Poster Presentations* (New York: Kluwer)

Luey, B. 1997, *Handbook for Academic Authors*, 3rd edn. (Cambridge: Cambridge University Press)

Mahoney, T. J. 2001, in *Organizations and Strategies in Astronomy*, ed. A. Heck (Dordrecht: Kluwer), p. 185.

Umakantha, N. 1997, Nature, 385, 764.

Discussion

Iwanizewska regretted that Mahoney's paper had not been presented at the beginning of the Session. Other speakers might then have not presented their own papers in such an incomprehensible way – with, for example, no abstracts presented on the screen. Speakers from English-speaking countries were particularly guilty. She thanked speakers not from English-speaking countries who had so painstakingly prepared material to be projected – so we could *see* what they were talking about. Please would native English-speakers do the same in future!

Kochhar observed that spell-checkers can be a nuisance: none of them would accept his name! He was worried that in-house checking of papers could lead to scientific censorship. He also recalled that Hidayat had said at the Kyoto General Assembly in 1997 that "Bad English is the international language of science"! Mahoney agreed that spell-checkers should be used with caution but they could weed out the most glaring spelling mistakes and typographical slips. He certainly agreed that there should not be censorship but young researchers sometimes submit immature papers to journals and more experienced workers should check that their papers are up to standard. He agreed with Hidayat, but felt that bad English should be eliminated whenever possible – simply to reduce the delays between submission and publication.

Kosionidis said that the dominance of English made difficulties for visits by astronomers to a non-English-speaking country. In most developing countries it is difficult for local students to benefit fully from visits by astronomers from the U.S., the U.K. or Europe. Mahoney suggested that such countries should take full advantage of any schemes that allow non-anglophone astronomers to visit English-speaking research centres, where they could gradually develop fluency in English.

Pasachoff recalled an incident in "Le Petit Prince" by Antoine de Saint-Euxpéry. St-Exupéry describes an "astronome Turque" who wore native costume while addressing an international meeting of astronomers (perhaps the IAU?). The Turkish astronomer describes the discovery of the asteroid on which the Little Prince lives, but no-one listens to him because of the way he dresses.

Only when he returns later wearing Western dress is his discovery accepted! Pasachoff thought that a reasonable proficiency in English may play the role of non-Western story in this mocking parable. He noted that everyone in the room, although from many different countries, was wearing western dress!

Astronomy for Developing Countries
IAU Special Session at the 24th General Assembly, 2001
Alan H. Batten, ed.

Practical Aid to Libraries in Developing Countries

Peter D. Hingley

Royal Astronomical Society, Burlington House, Piccadilly, London W1V 0NL, U.K. e-mail: pdh@ras.org.uk

Abstract. The problems and rewards of shipping astronomical books to libraries in developing countries are discussed, with particular reference to the author's own experience from his base at the Royal Astronomical Society.

1. Introduction

It is a privilege to prepare this account of one aspect of my work for the benefit of astronomers and librarians all over the world. I am often reminded of a once popular BBC children's radio show "Toytown" in which the principal character, Larry the Lamb, whenever he was assailed by Mr Growser, the Mayor, or Mr Plod the Policeman, would reply "I will do my little best". I, too, am Doing my Little Best!

This will be an uncharacteristically brief paper, no more than a short despatch "from the coal face" on one aspect of aid to developing countries. I believe that many people in institutions in the developed world would vaguely "like to help"; I, myself am in the unusually privileged position of being able to do something practical. I have been able to ship material for libraries to countries and places as diverse as Tanzania, Malta, Brazil, India, the Czech Republic, California, the Canary Islands, Bulgaria, Ireland, Lithuania, Madeira, the Slovak Republic and Hawaii.

2. Sources of Material

One of the privileges of working in a learned Society is that one comes to know, and to like, Fellows of great eminence and learning. Even these people have to die, however, we hope full of years and honours, and the Society's Librarian is often approached by the widow to help to find homes for the late scholar's personal library. It is perhaps well to wait until after the immediate grief before taking action, otherwise the Librarian may find himself acting as a Bereavement Counsellor.

Sometimes a scholar will discard many of his books on retirement and then his (or her) own feelings must be considered; one Fellow said to me, as I carried three-quarters of a ton of books out of his retirement cottage in Cornwall, "That's my whole professional life going out of the door!" Recently, I have assisted in finding homes for books and journals from the libraries of Prof. Sir William McCrea, Prof. Gerald Whitrow and Prof. Keith Runcorn.

At first, I operated by mail, sending out long lists of books and journals and waiting for the replies to come by mail or, later, by fax. Now, e-mail has made things much easier and is a more cost-effective method of disseminating offers of material. I have a series of mailing lists and send offers of material to thise whose names are on the appropriate list. I am willling to add names and addresses to these lists, subject to the caveats mentioned below; if you wish your name to be added, the first step is to send an e-mail to the address given at the head of this paper.

Of course, this work is somewhat outside my principal duties as RAS Librarian and I have to minimise the amount of working time that I spend on it. The attitude of management can vary. I could usually justify the time and petrol I consumed by pointing out that the RAS Library itself had the first pick of anything that was offered and, indeed, some very significant lacunae have been filled in that way. Under a previous management at the RAS, however, I had been told to stop this activity because the janitor was complaining about having to pack up so many parcels!

3. Distribution of the Material

I make little distinction between developing and developed countries; few libraries ever really have as much money as they need and I tend to think only in terms of the need for information. How does one distinguish between a Third-World country and one whose economy has fallen to bits for political reasons?

I should emphasize strongly that material offered through the RAS is very irregular in supply and often quite old. The lack of material in some Third-World countries can be quite humbling. Recently, I was somewhat diffident about offering a list of 24 books on cosmology that seemed a little elderly, having been published over the last 20 years; they were all snapped up within 24 hours. Responses to my messages are often rapid and enthusiastic. I had filled my car with 250 books from Prof. Whitrow's library and every single one has found a home. I find this deeply satisfying; I think all librarians are string hoarders at heart! I try to be even-handed, sending out lists at different times of the day to favour recipients in different time zones. Prof. Whitrow's books were offered to lists of mainly Third-World libraries and then to my personal list of "Fellows, Library Readers, Friends and Miscellaneous Hangers-on". Mrs Whitrow is delighted that her beloved husband's name is commemorated in libraries in many different countries. I always try to inform the recipient of the origin of the journals and books sent and ask them to write a thank-you letter; sadly, such a letter is not always forthcoming. When books are given by a bereaved spouse, it can be a great comfort to know that the loved-one's name is perpetuated in his books in some overseas library. For large consignments, such as the libraries of McCrea and Whitrow, I had a rubber stamp made so that I could insert the source of the donation in all the volumes, before the material left my hands. Other sources of donated books are the PAMNET list operated by the Physics, Astronomy and Mathematics section of the (American) Special Libraries Association and a mailing list for astronomy librarians, called ASTROLIB, which is largely inspired by the LISA conferences. The American Astronomical Society maintains a list of offered material on its Web site.

I feel that donations of back runs of journals can never be an adequate substitue for current subscriptions and, as more past material becomes available on-line through NASA's ADS system administered by Dr Eichhorn, the value of such shippments may diminish. I would welcome discussion of this point. My personal belief is that rumours of the death of the printed book and journal, like those of the death of Mark Twain, have been "greatly exaggerated". A day may come when it will no longer be worthwhile to ship printed material around the world, but I do not think that day has yet come, nor will it for some time.

In a few cases, e.g. the break-up of the former Soviet Union, certain Western bodies have given short-term help, in the form of free subscritptions, to tide observatories in those countries over the immediate difficulties; but this can be only a very temporary expedient. The needs of so many can be rather oppressive and the expectations can be a bit unrealistic. One astronomer, from a country locked in a bloody civil war, whose only observatory had just been completely destroyed by military action, sent a desperate message saying that what they really wanted was a remotely operated 1-m telescope with a full set of CCD detectors! This is not the sort of thing one usually finds lying about in the Library.

4.   Some Closing Thoughts

Sociologically, one might regard astronomy as a form of conspicuous consumption. I wonder how people survive in some countries, let alone do astronomy, and I am lost in admiration for the determination with which that science is pursued in the conditions described by some authors in this volume. I thought I had been rather radical in suggesting that a library is more necessary than a telescope, but I am interested that Hearnshaw (2001) ranked the priority of a library as 3 while a telescope is 6! Telescopes are nice things to have around and no good home should be without one, but much can be done without a professional-level telescope, even if you use a small one to teach the students.

Each participant in the donation process has both needs and obligations:

- The **DONOR** has a responsibility to provide an accurate lsit of what is being offered – all to often supposedly complete runs of journals have numerous issues missing. Donors are making a considerable effort to find homes for the material rather than just discarding it; that effort must be respected and valued. They need to feel that their, or their spouse's, long-cherished journal collections have gone to a good home and are being used. Especially after bereavement, donors wish their spouse's name to be commemorated on a bookplate, or in some similar way. Donors have a right to a proper letter of acknowledgment and the intermediary should always pass on an address for such a letter to be sent to.

- The **INTERMEDIARY**, often a librarian, needs to be prepared to ship material worldwide. This is always a big effort, but I am sad to see messages from American libraries to the effect that they are not willing to ship outside the boundaries of the known universe – which, all too often, seems to be coterminous with the continental U.S.A.! The ideal role for the librarian is as an information broker, connecting the need with the supply, but this ideal, sadly, is infrequently realized. Usually, I end up picking up the material at week-ends in my own transport and then shipping it from Burlington House! The intermediary needs to get responses quickly so that the limited storage space available is not cluttered up indefinitely. I have also had to pay postage and trust the recipient to refund it – and realizing the differences in affluence of different countries, I feel very embarrassed at having to ask a library in a country such as India to refund, say, £100 for postage. There is little point in a recipient asking for the very cheapest form of mailing, usually unregistered sea-mail, and then, as happened recently, expecting packages to be traced, when they are delayed!

- The **RECIPIENT** has obligations to both the other parties – often he must respond very rapidly to obtain any of the material offered. He has an obligation to the intermediary te refund the shipping costs, and to the donor to provide acknowledgment and commemoration.

There are various methods of shipping: using commercial carriers, postal services, and some charities might help. Charities provide services cheaply or even free of charge, but can be very slow; they are mainly suited to general educational material. Things can go wrong with other methods also. Once, 14 large boxes of journals were shipped by commercial carrier to a person building up a library in the western U.S.A.; for some reason they were never delivered and the unfortunate "recipient" had to refund about $300 to his university which had funded the shipping. Some practical measures can be taken, such as fixing prominent labels on each package giving the nature of the contents.

In trendy terms, "Networking" is very important – I prefer to think of it as just making friends. The LISA II and LISA III (Library and Information Science in Astronomy) meetings had been extremely useful in this respect. To my mind, the refusal of the IAU to sponsor the third of these meetings raises questions about the value placed by the astronomical community on librarians. Are we professional advisers, like telescope engineers or software experts, or are we on a level with those who empty the dustbins? I prefer to believe the former, but sometimes I have doubts!

Perhaps the most useful thing a librarian can do is to be a responsive friend on the end of a telephone, fax or e-mail line, but I am painfully aware that the RAS may have fallen short of perfection in this role, especially in the last two years – in the wake of the 1999 eclipse and various other happenings. I have only one assistant and we have only two hands each.

Returning, however, to Larry the Lamb, if we can all cooperate in a friendly manner and "do our little best" together, maybe we can make a difference to the practicability of of practising astroinomy in the less well-off parts of the world.

References

Hearnshaw, J.B. 2001, this volume, pp. 15-28.

Discussion

Percy pointed out that IAU Commission 46 has a program to donate suitable books and journals to astronomically developing countries and he is the "contact person" (jpercy@erin.utoronto.ca). The Commission is looking for suitable material and for ideas about how to ship it securely and inexpensively. The recipients would be institutions targeted by the Commission 46 programs and others the Commission selected. Crawford suggeted that recipients might be encouraged to approach their national airlines to transport material free to their home countries. He emphasized that not all good research is in preprints. There is a wealth of excellent information in old books and journals. Hingley expressed interest in Percy's information and welcomed Crawford's comments. He felt that there is a pressing need for a body with reasonable funds to save us from the embarrassment of asking poor countries to pay the cost of shipping. The recipient country might well be asked, however, to make ad hoc arrangements with its airline – it is probably even more trouble to pack the material and take it to an airport than to mail it! He had recently found that old books (by Eddington and Chandrasekhar) from Whitrow's collection were very much in demand in developing countries.

Astronomy for Developing Countries
IAU Special Session at the 24th General Assembly, 2001
Alan H. Batten, ed.

The Preservation of Library Materials in Developing Countries

Ethleen Lastovica

South African Astronomical Observatory, P.O. Box 9, Observatory, 7935, South Africa. e-mail: ethleen@saao.ac.za

Abstract. The acquisition of books and journals for a scientific library is costly. To ensure that this is a cost-effective process, consideration must be given to the arrangement, care and preservation of the collection. The books should be stored in a designated library area. Items acquired by purchase or donation must be catalogued, and a loan record kept. Besides regular library routines, attention must be given to the protection and preservation of the collection. Among the factors that contribute to the deterioration of library materials are temperature and humidity within the storage area, air pollution, dust, light and pests. Unchecked, they may cause serious damage to library materials at a very fast rate. If they are controlled, deterioration is greatly reduced.

1. Introduction

The acquisition of books and journals for scientific libraries is costly. To ensure that this is a cost-effective process, consideration must be given to the arrangement, care and preservation of the collection so that it does not deteriorate. There are many factors that pose a threat to library collections. For instance, the inherent chemical instability of acidic paper that begins to crumble as it ages; poor environmental conditions in storage areas - too light, too hot, too humid; too wet, too dry; careless handling and storage which causes physical and mechanical damage to books; biological enemies such as birds and animals, e.g. mice, rats, bats and even cats; insects; and micro-organisms which include moulds, fungi and bacteria; and vandalism, including theft.

The cost of professional library conservation is beyond the reach of many, if not most, library budgets. Even non-specialists charged with the responsibility of looking after an informal collection of books, can practice preventative care to establish and maintain better storage conditions in the library. The discussion that follows is not presented with large institutional libraries in mind, but is meant for the small library, which may, or may not, have a professional librarian in charge. Of course, the basics of good care and maintenance can be applied to book collections of any size, whether a small personal collection or a vast national library.

2.   Organizing the Library

It is pointless to apply preservation measures to a jumble of books and documents. Care starts with good management of the collection. If no formal library exists, an area that meets the environmental requirements for storage and is easily accessible to all staff should be used for the collection. It is necessary to know what is in the library, so formal cataloguing of the collection could be done either on cards or by using an electronic database. However, if the collection is small with few users, perhaps all that is needed is a handwritten ledger, or typed list. High-level cataloguing is not essential, but the title, author and date of publication should be recorded for each item. It is useful to assign a running accession number to each book or document. This serves two purposes – firstly it is a unique number that identifies a particular copy, and secondly, accession numbers indicate how many publications have been added to the collection.

In the cataloguing record, the location of the publication should be given. Simple shelf numbers could be used; e.g. section A, shelf 1 (A1), but more conventionally, books are grouped by subject. This is usually based on a standard classification system, such as Dewey.

In order to keep track of the stock, it is essential to maintain loan records. With a small number of users, this could be informal with the borrower writing the details in a loans book, or maybe signing an issue card that is left as a record of a book borrowed. Computerized loan records are being used increasingly in libraries, but care must be taken that the system does not become more sophisticated and time consuming than the size of the collection warrants.

3.   Preservation Assessment and Planning

A clause in the *Code of Ethics for Museums* (American Association of Museums, 1994) could well apply to libraries and archives. It states that the institution must ensure that the "collections in its custody are protected, secure, unencumbered, cared for, and preserved".

Sherelyn Ogden defines preservation planning as a process by which the general and specific needs for the care of collections are determined, priorities are established, and resources for implementation identified (Northeast Document Conservation Center 1999, p.1). She stresses that while the plan must recognize all preservation needs, it should focus on those steps that can be accomplished with existing or obtainable resources.

Both Ogden and Greenfield (1988) outline the need for an assessment survey. In her book on *The Care of Fine Books*, Greenfield includes a two-page "Environmental Survey" form, which can be used for the identification of potential sources of harm to the library collection. It gives consideration to 9 areas of concern relating to the building structure; the climate within the building; fire prevention measures; flood control; security; biological problems from insects, animals and fungus; shelving; housekeeping and, the history of environmental problems.

4. Storage of Books

4.1. Desirable Conditions

Once the collection has been brought into order, and the library area defined, it is essential to ensure that preventative care and maintenance is in place. Basements, attics and outbuildings are unsuitable locations for book storage. At worst, basements may become flooded, but more usually, the dark, damp, humid conditions adversely affect the paper and attract insects and rodents. Attics are too hot and dry, which causes paper to become brittle. Books stored in outbuildings are also likely to be attacked by insects and rodents, and with no protection from the weather, the books may well become brittle or mouldy.

The ideal environment includes controlled temperature and relative humidity, clean air with good circulation, controlled light sources, and freedom from biological infestation. Good housekeeping practices, security controls, and measures to protect collections from fire, water and other hazards complete the range of environmental concerns (IFLA 1998).

4.2. Temperature, Relative Humidity and Light

Ideally, the library should be weatherproof and free of biological infestations. The temperature and relative humidity (RH) of the library area can affect the condition of the books. There is no ideal level for all types of library material but rather a range of values that vary according to the medium of the object being stored. Generally, high temperatures, with very dry air, are more harmful because the paper becomes desiccated. This is why the covers of a book left in the sun, or in front of a fire or heater, become buckled. If the room is constantly damp, and the relative humidity above 65%, books can begin to fall apart, and when the RH rises over 75%, moulds and fungus occur which will severely damage the library material.

There needs to be a careful balance between temperature and relative humidity in the library environment. Paper will retain its chemical stability and physical appearance for longer at a constant, low storage temperature (below 10° C / 50° F) with a relative humidity of 30-40%. For every 10° C (18° F) rise in temperature, the rate of chemical degradation reactions in traditional library and archival material such as paper and books, is doubled. Severe fluctuations or "cycling" of temperature and relative humidity will cause more damage than constantly high readings and so must be avoided (IFLA 1998).

Ultra-violet radiation from sunlight and fluorescent lighting causes irreversible fading and deterioration of paper, ink and bindings. Shelving should not be positioned in direct, or even indirect, sunlight. If necessary windows could be painted, or Venetian blinds or curtains hung to block out the offending light. Another option would be to have UV filtering film applied to windows.

4.3. Pollution

As might be judged by the preceding discussion, library material is susceptible to mould and fungal attack if poorly stored. Pollutants and high moisture content in the room are the likely catalysts for the growth of spores that can be a health hazard, as well as an enemy of library materials. For severe outbreaks of mould, expensive specialist treatment may be required. Fumigation is no longer recommended for mould because fumigants are toxic to people, the residue remains on the object, and it does not prevent the mould from returning. A far more cost-effective measure is to control the environment where the books are stored. This can be done by maintaining an acceptable temperature / relative humidity ratio; by circulating the air; by cleaning regularly with a vacuum cleaner fitted with a high efficiency particulate air (HEPA) filter; by not shelving books on an outside wall; by not having plants in the building; and by ensuring that the building is sound with no leaks or dampness (IFLA 1998).

4.4. Insects and Pests

Insects and pests can be as insidious as fungal spore. The insects that most commonly cause damage in libraries and archives throughout the world are cockroaches, silverfish or firebrats, book-lice, beetles and termites. They feed on the substances that books are made of, and lurk in the dark, damp, poorly ventilated places of the library. Insects cause irreparable physical damage to wood, paper, leather, fabric etc. Their presence is often indicated by tiny holes bored through the covers of books, or a lacework effect on pages caused by silverfish and firebrats that prefer to feast on modern wood-based paper, rather than the rag-based papers of past centuries. Wooden shelving can sometimes provide homes for wood-boring insects that are just as likely to attack books. Dean (1999) gives an account of a library in Vietnam whose wooden shelves were infested with wood-eating beetles; it was only after aluminum shelving was installed that the situation improved.

JICPA, the Joint ICA/IFLA Committee for Preservation in Africa has organized several workshops on the continent to raise awareness on preservation issues among African librarians and archivists. I spoke to some of the conservation specialists who participated in these workshops, and all recommended that non-specialists use simple, non-toxic measures to control pests. Johan Maree, Head of Book Conservation at the University of Cape Town, suggests placing laurel (bay) leaves Lauris nobilis) on the shelves behind the books, or else, placing Epsom salts or borax in small containers at the back of shelves. These are fairly effective means of controlling silverfish and the likes. The salts and borax should not be sprinkled on the shelves because the residue on books may be ingested accidentally by readers after handling the books. Commercial roach traps can also be used but they must be renewed every 3-6 months. The Roodepoort Museum, a cultural history museum in the northern part of South Africa uses Sunlight soap, which is pure, hard, laundry soap, as its only means of pest control. The soap is grated and then tied in a piece of fabric measuring approximately 15 cm^2 (about 6 in^2). They have placed these little bags throughout their collections, including the library, and have found that they provide a very acceptable means of insect contol.

Greenfield (1998) says "termites are not a serious threat because they will presumably be dealt with long before they start on books." The IFLA manual (1998) is more accurate when it says that termites can devastate buildings and collections. Recently a university conservation department in the sub-tropical area of South Africa, moved into their newly built conservation laboratory. It was architect-designed to strict specifications, including cement floors. Some government publications left in the area in cardboard boxes were found to be have been shredded by termites over a weekend. The insects had entered the building through a small crack in the newly completed building.

Mice and rats can also cause devastation in libraries. Not only do they gnaw the books to obtain paper for their nests, but their droppings and urine cause metal shelves to corrode and the room to smell. Besides damaging books, rodents can also spread disease. Good housekeeping goes a long way to controlling rodents – clean floors, no food in the library area and wastebins emptied daily. Mouse and rat traps can be used to catch the intruders. Ultrasonic devices are also suggested as a deterrent, but in my experience, ultrasound has little effect. The least toxic treatment should always be preferred. If the infestation is severe, professional exterminators who understand the environmental requirements of libraries should be called in.

5. Handling of Books

Humans can cause just as much damage to books as environmental conditions and pests do. Greenfield (1988) lists 20 ways to treat books with care, most are self-evident. Whether an item is a printed book or an illuminated manuscript, all Greenfield's points essentially relate to respect for the item. Books should not be left in the sun, or accidentally wet. They should not be dropped, or spines cracked to make the book lie flat. Improper placemarkers (such as a fried egg! as I once saw in a public library exhibit) should not be used, nor should paper clips or pins be used for attaching supplementary notes to the page because they will rust over time.

Do not write in or near a book with an ink pen, and highlighter pens are out of place in a library. Pressure-sensitive tape, such as Sellotape, should not be used for book repair. Archival-quality document repair tape, obtained from specialist suppliers, is an acceptable means of repairing paper that is to be conserved. This tape is non-yellowing, neutral, reversible, transparent and pressure sensitive. Licking a finger to turn pages is unacceptable, this practice weakens the corners of pages, and in addition, if toxic substances have been used in the library for pest control, a person is likely to ingest some of the poison.

6. Good Housekeeping

Good and regular cleaning are essential to ensure the protection of the collection. Clean surroundings discourage fungi, insects and pests. The cleaning programme should include the examination of collections not only to provide early warning of biological or chemical damage, but also to observe conditions throughout the area.

Rooms used for storage should be sound in all respects and easy to clean and inspect. Like all good homes, air should circulate freely. This will ensure good climatic conditions in the areas where library material is stored. At the same time, an even climate should be maintained within the storage areas, with changes occurring as slowly as possible – as was said above, a cool room is better than a warm one because the growth of mould might occur. Floors should be vacuum-cleaned weekly. Library material on the shelves should only be cleaned by specially trained staff. Feather dusters are unsuitable for cleaning in this environment because they simply redistribute the dust; a duster to which particulates adhere should rather be used (Baynes-Cope 1990; IFLA 1998).

7. Disaster Planning

Although preparedness for disaster is a facet of library care and preservation, a detailed discussion of disaster planning is out of the scope of this presentation. Those responsible for preservation of collections need to give serious thought to preparedness for disasters that may be natural (floods, hurricanes etc.), or man-made, for example, acts of war, fires, water and explosions.

IFLA (1998) stresses that it is important to have a plan in place for coping with disasters if they occur. Disaster preparedness requires team effort from start to finish. It can be divided into five phases. A risk assessment needs to be undertaken to determine the potential dangers to buildings and their collections. Measures must be implemented to remove or reduce any danger - for example, light fittings repaired, plumbing, drainpipes and gutters kept in good condition and fire extinguishers serviced. Preparedness requires a written preparedness, response, and recovery plan. To minimize damage, response procedures to follow when disaster strikes must be well rehearsed. And finally there should be a recovery plan for restoring the disaster site and damaged material to a stable and usable condition.

8. Collection Care in Southeast Asia

Although recognized standards for library preservation lay down the foundation for good library management, it should be acknowledged that the approach to preservation should be carefully considered before being applied.

A paper given at the IFLA Conference in Bangkok in 1999 by John Dean, Director of the Department of Preservation and Conservation at Cornell University (www.library.cornell.edu/preservation) presents a new perspective on care and preservation with particular reference to Southeast Asia. Since 1987, Cornell University Department of Preservation has been actively involved in the preservation of collections in the upper regions of Southeast Asia, primarily Burma, Cambodia, Laos, Vietnam and Thailand. Cornell offers a six-month internship course for librarians and archivists from these areas to study achievable solutions to preservation problems, and the development of programs to acquire resources.

According to Dean (1999), the libraries and archives of Southeast Asia operate in a very different time frame, and have a different set of cultural and historical circumstances from those in the West. He draws attention to the fact that solutions to preservation problems are not entirely the same in the East as in the West, and the few attempts to inflict Western standards and practice, unaltered by locale, have been unsuccessful. He gives as an example the stock response to high levels of temperature and relative humidity by Westerners is to call for air conditioning systems to be installed. This can be a costly mistake in tropical regions, especially when books and manuscripts are removed from the controlled library environment, or the untrustworthy electricity supplies fail. In these circumstances, the drastic increase in temperature and relative humidity causes condensation on colder materials and interior walls and the consequent rapid development of mould.

9.   Care of Computer Disks

Information stored on computer disk can be as priceless as the printed page. A brief mention needs to be made on the care of disks. An extensive search on the web located a site created by Skidmore College (2000) that is loaded with practical advice.

Never touch the disk surface; store disks in cool, dry places and always in a box of some kind; keep disks away from things that generate magnetic fields such as monitors, electric motors, telephones, and other electrical devices; don't bend disks - mail or transport them in rigid packaging; use a felt-tip pen to write on disk labels; make frequent backups of important information and store the backup away from your workplace; don't spill liquid on disks and don't use them as coasters for your drinks; and finally, never remove a disk while the drive light is on since this frequently results in deleting everything off the disk.

10.   Sources of Information on Preservation

Two organizations, IFLA and NEDCC, are fully committed to encouraging worldwide cooperation for the preservation of library materials, and both provide outstanding resources on the web.

IFLA - the International Federation of Library Associations and Institutions, based in Paris has established a Core Programme for Preservation and Conservation (PAC) with regional centres in Asia (HQ - Tokyo), Latin America and the Caribbean (HQ - Caracas, Venezuela), United States/Canada (HQ - Washington, D.C.), Eastern Europe and the Commonwealth of Independent States (HQ - Moscow), and Western Europe, Africa and the Middle East (HQ - Paris, France). An excellent 74-page manual, *IFLA Principles for the Care and Handling of Library Material* is available free of charge on two websites, http://ifla.inist.fr/VI/4/pac.htm *and* http://www.clir.org.

The document is a general introduction to the care and handling of library material for individuals and institutions with little or no preservation knowledge. It does not provide detailed methods and practices, but gives basic information to assist libraries in establishing a responsible attitude to looking after their collections.

The NEDCC - Northeast Document Conservation Center, a non-profit American organization devoted to the preservation of library and archival materials, has had a strong commitment to the dissemination of information since its founding. Their 410-page manual on *Preservation of Library and Archival Materials* edited by Sherelyn Ogden (1999) can be freely downloaded from the web in English, Russian or Spanish (http://www.nedcc.org). It is also available in book form at a very reasonable cost.

Two additional inexpensive books that present practical advice on the care of books and documents in a concise, readable way are *Caring for Books and Documents* by A.D. Baynes-Cope (1990) and *The Care of Fine Books* by Jane

Greenfield (1988). The cartoons illustrating the first book attract the reader to the informative text that was originally compiled by the British Library in response to the many queries they receive annually. The other book deals not only with the care, but also the repair of books, and provides very clear diagrams on how to do the various procedures.

Of course the importance of networking should not be overlooked. There is a strong network of astronomy librarians around the world who are always willing to provide assistance or advice to others if at all possible. The *Directory of Astronomy Librarians and Libraries* maintained by the library of the European Southern Observatory on the web at: http://www.eso.org/genfac/libraries/astro-addresses.html can help locate other librarians within one's geographical area or elsewhere.

A forum for electronic communication between astronomy librarians is provided by Astrolib, a project based at the National Radio Astronomy Observatory (library@nrao.edu). As a final thought, there is what Kathleen Robertson (1998) described as "A Clearinghouse for Astronomy Librarians: the PAM Web Site" (http://pantheon.yale.edu/ dstern/pamtop.html). This is the web site of the Physics-Astronomy-Mathematics Division of the Special Libraries Association. PAMnet, the electronic discussion list of the division, provides a forum for the discussion of library and information resource issues relevant to the fields of physics, astronomy and mathematics. It is open to subscribers who need not be PAM division members.

It may seem that I have strayed from my theme of preservation of library materials in developing countries, but the responsibility for preserving information for future generations is great, so therefore, the more we can share and discuss the challenges that face us, the more likely we are to succeed.

References

American Association of Museums 1994. *Code of Ethics for Museums*, rev. & adapted 1993 (Washington, D.C.: AAM), 8.

Baynes-Cope, A.D. 1990. *Caring for Books and Documents*, 2nd ed. (New York: New Amsterdam Books).

Dean, J. 1999. *Collection care and preservation of Southeast Asian materials.* Int Preserv News, 20, 10 (http://ifla.inist.fr/VI/4/pac.htm).

Greenfield, J. 1988. *The Care of Fine Books* (New York: Lyons Press).

IFLA (International Federation of Library Associations and Institutions) 1998. *IFLA Principles for the Care and Handling of Library Material*, ed. E.P. Adcock, M.-T. Varlamoff, V. Kremp (Paris: IFLA). Electronic version: http://ifla.inist.fr/VI/4/pac.htm .

International Preservation News: a Newsletter of the IFLA Core Programme on Preservation and Conservation, no. 14, May 1997 + Electronic version: http://ifla.inist.fr/VI/4/pac.htm .

Northeast Document Conservation Center 1999. *Preservation of Library & Archival Materials: a Manual*, 3rd ed., ed. S. Ogden (Andover, MA.: NEDCC). Electronic version: http://www.nedcc.org .

Robertson, A.K. 1998. A clearing house for astronomy librarians: the PAM web site. *Library and Information Services in Astronomy III, ASP Conf. Ser. 153*, eds U. Grothkopf et al., p.244. (San Francisco, ASP).

Skidmore College, Center for Information Technology Services, Saratoga Springs, NY 2000. Computing help@skidmore. http://www.skidmore.edu/help .

Appendix: Useful Addresses

IFLA Core Programme for Preservation and Conservation (PAC), International Focal Point, Bibliothque Nationale de France, T3 N4 97, Quai Franois Mauriac, 75706, Paris cedex 13, France. fax: +33-1-53-79-59-80; http://ifla.inist.fr/VI/4/pac.htm

Northeast Document Conservation Center, 100 Brickstone Square, Andover, MA 01810-1494, USA. fax: (978) 475-6021; http://www.nedcc.org

Astronomy for Developing Countries
IAU Special Session at the 24th General Assembly, 2001
Alan H. Batten, ed.

Electronic Access to Journals

Helmut A. Abt

Kitt Peak National Observatory, Box 26732, Tucson, AZ 85726-6732, U.S.A. e-mail: abt@noao.edu

Abstract. The use made of electronic access to journals by astronomers in some developing countries is estimated and compared with the use made of it by astronomers in developed countries.

1. Introduction

Most astronomical journals are now published in two editions: a printed version and an on-line version. The latter has at least five advantages in addition to being a convenient form for computer-literate users: (1) unlimited free access by all faculty, employees, and students at any institution subscribing to the journal, (2) ability to click on references and to bring them directly to the screen in most cases, (3) unpaginated editions of the letters a month or two before the printed edition appears, (4) virtually unlimited space for tables and illustrative material that would cost too much to reproduce in print, and (5) the on-line versions have the full tables and illustrations, rather than samples.

We ask here whether scientists in developing countries are taking advantage of these services in this rapidly-changing medium. To do so, we asked the University of Chicago Press staff to count the number of visits ("hits") to the *Astrophysical Journal* (ApJ) in April 2000 by people in seven developed countries and four developing countries. The numbers are given in Table 1 for the North American site and its two mirror sites in France and Japan. A user is not limited to the nearest site, e.g. a Japanese may find that the North American site is less busy during his daylight hours than the Japanese mirror site.

The numbers in Table 1 are impressively large except for China, where Internet charges are significant and access is not encouraged. How can we meaningfully compare the numbers for the developed and developing countries? We will use three comparisons: with the gross national domestic product, with IAU membership, and with estimates of the national numbers of papers published.

2. Comparison with Gross Domestic Product

We obtained gross domestic product (GDP) values from the 2000 edition of the *World Almanac Book of Facts*; the amounts are 1997 estimates and are expressed

354

Table 1.   Visits to the ApJ in April 2000

Country	N. America	Europe	Japan	Total
(Developed)				
France	10,767	26,515	713	37,995
Germany	24,883	14,636	582	40,101
Italy	24,246	13,404	160	37,810
Japan	38,217	552	47,101	85,870
Spain	9,393	11,583	57	21,033
U.K.	47,335	10,350	164	57,849
U.S.A.	529,089	13,237	4,788	547,144
(Developing)				
Argentina	6,497	14	0	6,511
China	15	33	0	48
India	10,767	147	113	11,027
Russia	11,560	881	45	12,486

in billions of U.S. \$. Table 2 gives the GDP numbers and then the visits per GDP (in 10^9 U.S. \$). The average for the seven developed countries is 36.8 ± 15.5 ($1\ \sigma$) visits per billion US\$, while for three of the four developing countries it is 13.3 ± 5.8 ($1\ \sigma$). The difference, a ratio of 2.8, is not significant because countries devote different fractions of their GDP to R & D.

3.   Comparison with IAU membership

The national IAU membership numbers are taken from the 1997 *Transactions of the International Astronomical Union* and are tabulated in the fourth column of Table 2. The fifth column gives the number of visits per IAU member. The average number of visits per IAU member is 126 ± 66 ($1\ \sigma$) for the developed countries and 52 ± 18 ($1\ \sigma$) for the three developing countries (China excluded). The difference, a ratio of 2.4, is probably not significant because the fraction of the IAU members active in research may differ widely from country to country.

4.   Comparison with papers published

We counted the number of papers published in the January 2000 issues of *Astronomy and Astrophysics* (and its *Supplement*, the *Astronomical Journal* the *Astrophysical Journal* (and its *Supplement*), *Icarus*, *Monthly Notices of the Royal Astronomical Society* and *Solar Physics*. For each paper we determined the fraction of authors coming from each of the 11 selected countries, plus other countries. The numbers of January 2000 papers from each country are given in the sixth column of Table 2; the seventh column gives the number of visits per published paper. The average for the seven developed countries is 1820 ± 400 (dispersion in the mean) while for the three developing countries, excluding China, it is 1500 ± 177 (dispersion in the mean). Therefore the difference is not significant.

Table 2. Comparison of April 2000 On-line Visits to the ApJ with GDP, IAU Memebership, and Papers Published in January, 2000.

Country	GDP	Visits/ GDP	IAU Members	Visits/ Member	Papers	Visits/ Paper
(Developed) France	1,320	28.8	611	62	28.3	1,340
Germany	1,740	23.0	489	82	45.1	890
Italy	1,240	30.5	410	92	36.4	1,040
Japan	3,080	27.9	448	192	25.6	3,360
Spain	642	32.7	204	103	13.6	1,540
U.K.	1,242	46.6	535	108	37.6	1,540
U.S.A.	8,080	67.7	2,250	243	179.9	3,040
Mean		36.8		126		1,820
(Developing)						
Argentina	348	18.7	90	72	3.7	1,750
China	4,250	(0.0)	367	(0)	6.2	(8)
India	1,534	7.2	228	48	8.8	1,250
Russia	892	14.0	345	36	8.3	1,500
Mean		13.3		52		1,500

We conclude that most of the developing countries are making use of the advantages of on-line publications.

Astronomy for Developing Countries
IAU Special Session at the 24th General Assembly, 2001
Alan H. Batten, ed.

Making the Most of Publishing Software

Terry J. Mahoney

Instituto de Astrofísica de Canarias, E-38205 La Laguna, Tenerife, Spain. e-mail: tjm@ll.iac.es

Abstract. Astronomers nowadays have a wealth of sophisticated publishing software at their fingertips, but are they achieving acceptable standards of presentation?

1. Introduction

Increased commercial pressure on academic publishers to reduce overheads (see Mitton 1992; Sharpe & Gunther 1994) and the ever-increasing numbers of papers being published each year (Abt 1998) have resulted in more responsibility for quality of presentation being placed on the shoulders of authors and editors of proceedings.

The new technology enables publishers to provide sophisticated macros that obviate the need for traditional typesetting (Heck 1997); authors are now frequently handling this expensive aspect of production. The respective benefits to publishers and authors are much reduced overheads and increased authorial control over the appearance of the final product.

The downside is a general unawareness of good typesetting practice among authors, who receive little or no training in the principles of publishing practice. So much is evident from even a cursory examination of many volumes of conference proceedings, where cost usually forbids much direct publisher intervention in presentation beyond the provision of standardized macros. For more formal products (monographs, textbooks, etc.), publishers are often obliged to redo much of what the author or volume editor has done badly, and this can be extremely expensive (Mitton 1992) since it requires the services of a professional typesetter.

2. Publishing Practices

Astronomers need to:

- Read publishers' instructions

- Have at least a survival grasp of LaTeX

- Know basic typesetting and page makeup

- Learn to copy-edit

- Understand what happens between sending an article or proceedings MS to a publisher and its final publication

- Be aware of the problems of presentation

3.   Universality of LaTeX in Astronomy

Most research journals and science publishing houses now provide LaTeX (Lamport 1986; Goossens, Mittelbach & Samarin 1994) or TeX (Knuth 1986) macros for authors and volume editors. New researchers should acquire a sound knowledge of LaTeX.

4.   Improving Standards

Apart from writing research articles and contributions to proceedings, a researcher may also occasionally have to edit an entire volume of proceedings. Today's researcher, therefore, needs a basic understanding of typesetting; an editor must also have some idea of page makeup (the arrangement of text and illustration on the page). This knowledge is usually picked up haphazardly. Formal training, both at undergraduate and postgraduate level and including relevant aspects of authorship, editorship and publishing, should be provided by research centres.

5.   Common Errors Committed by Authors

- Failure to read publishers' instructions

- Redesigning the page layout by altering fount sizes, indentations, line and paragraph spacings, etc. (often done to squeeze too much content within the set page limit)

- Wrong use of founts

- Poor mathematical typesetting

- Ignoring guidelines for tables

- Unawareness of conventions on referencing and bibliographical lists

6.   Common Errors Committed by Editors of Proceedings

- Failure to read publishers' instructions

- Poor or non-existent copy-editing

- Careless placement of tables and figures

- Failure to provide indices

- Failure to impose series style of presentation on the contributions

- Non-checking of cross-references within the volume

- Failure to update reference lists before closing the edition

- Failure to meet delivery deadlines

Many of these errors could be avoided through suitable training. A short course in publishing practices might include:

- A description of the publishing process

- The basics of copy-editing

- The elements of typography, correct use of founts, etc.

- Mathematical typesetting

- Typesetting of tables

- Preparation and labelling of illustrations

- Bibliographical referencing

- Page makeup

- Introduction to LATEX for publishing

- The rudiments of book production

7. Understanding the Publishing Process

Publishers and journals can help improve the input for authors and volume editors by:

- Providing explicit and complete instructions to authors and editors

- Explaining to authors and editors exactly what happens to a MS once it is delivered

Many journals and publishing houses do a good job in providing author/editor instructions, but there is generally insufficient information on the publishing process itself (*Nature* has shown what can be achieved in this respect (e.g., Nature 1999).

Many universities and research centres offer courses in writing and the techniques of communication. Other courses dedicated to publishing skills could be added to these in order to provide a fully rounded training in the presentation of scientific results.

8. How Can the IAU Help?

The problems of quality of presentation could usefully be addressed by Commissions 5 (Documentation and Astronomical Data) and 46 (Teaching of Astronomy).

The shift in the editorial responsibilities of authors could well justify the setting up of an entirely separate commission or WG.

Moves to standardize certain aspects of presentation (referencing styles, etc.) need to be accelerated. The IAU has a crucial role to play here.

Now is a good time to produce an updated *Astronomer's Handbook* (Pecker 1966; Wilkins 1989). This could be made relevant to all astronomy writers (not just to authors of IAU publications) and could recommend, in agreement with the main peer-review journals, the standardization of referencing styles, physical units, mathematical typesetting, etc. While the nomenclature of astronomical objects, planetary features, etc., is now well in hand through the various nomenclature committees of the IAU, astronomical terminology – which deals with the meanings of the terms used in all branches of astronomy and its related sciences – is by its nature (as part of the living English language) uncontrollable from "above" and is better studied than legislated (Mahoney 1998).

9. An Editorial Survival Kit

All astronomy writers should have easy access to the following publications:

Dictionaries:

- The *Concise Oxford Dictionary*, 10th edn. (Pearsall 1999; for British English, but now with American spellings included; a very close approximation to the house styles of the main British scientific publishers)

- *Webster's Third New International Dictionary* (Merriam-Webster 1976; for American English)

- All non-anglophones should have the most up to date version of a relevant bilingual dictionary

Style Guides:

- *The Chicago Manual of Style*, 14th edn. (University of Chicago 1993) the last word in all aspects of book production)

- *Hart's Rules for Compositors and Readers*, 39th edn. (Oxford University Press 1983; a compact and highly informative guide to OUP house style— soon to be updated)

- *The Oxford Dictionary for Writers and Editors*, 2nd edn. (Ritter 2000; an excellent trouble-shooter; both British and American spellings)

Editorial Manuals:

- *Copy-Editing*, 3rd edn. (Butcher 1992; an authoritative guide to British editorial practices)

- *Editing Fact and Fiction* (Sharpe & Gunther 1994; a general guide to book editing the American way) *The Chicago Manual of Style* also gives and exhaustive treatment of American editorial practices.

English Grammar, Punctuation, Style, etc.:

- *The Elements of Style*, 4th edn. (Strunk & White 2000; a masterpiece of conciseness—for American English)

- *The Right Word at the Right Time* (Reader's Digest Association 1985; a superb usage guide for British English with encyclopaedic coverage of the many kinds of English)

If funds will permit another extremely useful guide to academic publishing in all its aspects is the *Handbook for Academic Authors* (Luey 1997).

10. Conclusions

1. Pressure on publishers to reduce overheads and the continuing increase in the number of publications, combined with availability of typesetting software, have shifted a considerable degree of editorial responsibility on to research authors.

2. Astronomical centres should train their research staff in writing and editorial techniques.

3. Astronomers can help themselves by studying the available literature.

4. The IAU could usefully examine the question of the presentational quality of publications.

References

Abt, H. A. 1998, PASP, 110, 210.

Butcher, J. 1992, *Copy-editing*, 3rd edn. (Cambridge: Cambridge University Press).

Goossens, M., Mittelbach, F. & Samarin, A. 1994, *The L^AT_EX Companion* (Reading, Mass.: Addison-Wesley).

Heck, A. (ed.) 1997, *Electronic Publishing for Physics and Astronomy* (ApSS, 247, 1).

Knuth, D. E. 1986, *The T_EX Book* (Reading, Mass.: Addison-Wesley).

Lamport, L. 1986, *L^AT_EX: a Document Preparation System* (Reading, Mass.: Addison-Wesley).

Luey, B. 1997, *Handbook for Academic Authors*, 3rd edn. (Cambridge: Cambridge University Press).

Mahoney, T. J. 1998, in ASP Conf. Ser., Vol. 153, *Library and Information Services in Astronomy III* (LISA III), ed. U. Grothkopf, H. Andernach, S. Stevens-Rayburn, & M. Gomez (San Francisco: ASP), p. 218.

Merriam-Webster 1976, *Webster's Third New International Dictionary* (Springfield, Mass.: Merriam-Webster).

Mitton, S. 1992, in *Desktop Publishing in Astronomy and Space Sciences*, ed. A. Heck (Singapore: World Scientific), p. 67.

Nature 1999, *Cómo se publica en nature: una guía*, Nature, 402 (suppl.).

Oxford University Press 1983, *Hart's Rules for Compositors and Readers*, 39th edn. (Oxford: Oxford University Press).

Pearsall, J. (ed.) 1999, *Concise Oxford Dictionary*, 10th edn. (Oxford: Oxford University Press).

Pecker, J.-C. 1966, Trans. IAU, XIIC, *Astronomer's Handbook* (London: Academic Press).

Reader's Digest Association 1985, *The Right Word at the Right Time* (London: Reader's Digest Association).

Ritter, R. M. (ed.) 2000, *The Oxford Dictionary for Writers and Editors*, 2nd edn. (Oxford: Oxford University Press).

Sharpe, L. T. & Gunther, I. 1994, *Editing Fact and Fiction* (Cambridge: Cambridge University Press).

Strunk Jr., W. & White, E. B. 2000, *The Elements of Style*, 4th edn. (Boston: Allyn & Bacon).

University of Chicago Press 1993, *The Chicago Manual of Style*, 14th edn. (Chicago: University of Chicago Press).

Wilkins, G. A. 1989, *IAU Style Manual* (Dordrecht: Kluwer).

Abstracts of Poster Papers

Abstracts of poster papers relevant to this section of the Special Session are presented below.

The Physics-Astronomy-Mathematics Asia-Pacific Forum: A Network for Librarians and Information Specialists

Jeanette Regan et al., Australian National University, Private Bag, Weston Creek PO, Canberra ACT 2601, Australia

Non-commercial astronomical publications have been freely distributed worldwide to relevant institutions for more than a century. Astronomers have considered collaboration a very important aspect of their work and, to assist them, librarians have established a supportive international network.

In 1988 the first Library and Information Services in Astronomy meeting was held in Washington, USA. From this meeting, a discussion list was established. Shortly afterwards, the lack of resources for astronomers in the former Eastern Block countries was realised and the twinning of libraries in this region with sister institutions in developed countries was introduced. These resources were not available for libraries in the Asia Pacific Region.

The Physics Astronomy, Mathematics Division (PAM) of the Special Libraries Association, working with the Australian Library and Information Association founded the PAM-Asia Pacific Forum (PAM-APF) in 1998.

This discusses the aims of this network, the advantages offered by publishers to participating libraries, achievements so far and proposed developments.

(Co-authors are M.C.Koch, and M.E.Gomez, of Santiago, Chile, C.Louis, Bangalore, India, and B.G. Corbin Washington, D,C., U.S.A.)

ALA: Astronomy in Latin America

Philippe Eenens, Department of Astronomy, University of Guanajuato, Apartado Postal 144, Guanajuato, CP 36000, México.

An electronic newsletter was launched in Latin America with the purpose of fostering communication and collaboration between the national astronomical communities. The unique characteristics of this newsletter are presented, together with the first results and plans for the future.

The Working Group on Space Sciences in Africa

Peter Martinez et al., South African Astronomical Observatory, P.O. Box 9, Observatory 7935, South Africa.

IAU membership is a good indicator of a nationally organized astronomical community. Although IAU membership statistics for Africa continue to be very poor, other indicators (such as publications) suggest that there are many individual scientists in Africa who are attempting research or promoting education in astronomy. The Working Group on Space Sciences in Africa seeks to support these individuals through various means. This poster provides an overview of astronomy in Africa and the activities of this Working Group.

Co-author is François R. Querci, France.

Pollution-Free Road Lighting

Duco A. Schreuder, Spechtlaan 303, 2261 BH Leidschendam, The Netherlands

The beneficial effects of road lighting are often seen as very important. They relate to reducing road accidents and some forms of crime but also enhance the social safety of residents and pedestrians and the amenity for residents. Road traffic in developing countries is much more hazardous than in industrialized countries. Accident rates in 'low' income countries may be as much as 35 times higher than in 'high' income countries. Thus, it might be much more cost-effective to light roads in the developing world than in the industrialized world. Fighting light pollution is more pressing in developing countries as most of the major high-class astronomical observatories are there. Astronomical observations are disturbed by light from outdoor lighting installations, part of which is scattered in the atmosphere to form 'sky glow'. The International Lighting Commission CIE has published a Technical Report giving general guidance for lighting designers and policy makers on the reduction of the sky glow.

Lighting improves visibility, essential for almost all human activity. However, light that hits the road contributes to visibility only if it is reflected. In poorly designed lighting equipment much of the lumen output of the lamps is sent directly upwards. This can be avoided by properly defined light fittings. The light output of fittings is determined by their optical quality and by the installation maintenance factor. Open fittings are to be preferred. If mounted horizontally, they make street lighting with the least light pollution.

Zimbabwe – The Place for Astronomy at the next Total Solar Eclipse

Francis Podmore, Dept. of Physics, University of Zimbabwe, Harare, Zimbabwe

With clear skies most of the year, low levels of light and industrial pollution and location (we can see 96% of the celestial sphere) Zimbabwe is an excellent place for astronomy. For nearly 100 years a small but dedicated and talented band of amateur astronomers have been making hundreds of observations of occultations and variable stars, and contibuted 10% of the global total of reports to the International Halley Watch. The Astronomical Society of Southern Africa (Harare Centre) is 25 years old and the largest telescopes (mostly 'home-made') in the country are owned by members. Active preparations for the next two solar eclipses include site selection, coordination or safaris and free distribution of information packs and over 100 000 eclipse viewers to all schools. If the economy doesn't collapse, good government and respect for law and order return, the planes keep flying and fuel shortages end, we look forward to welcoming hundreds of eclipse watchers to a dramatic 3 minute spectacle on 21 June 2001.

Panel Discussion

Summary of Panel Discussion

Abstract. The Special Session closed with a general discussion led by a panel consisting of R.K. Kochhar (India, moderator), M. Gerbaldi (France), A. Ubachukwu (Nigeria) and M.C. Pineda de Carias (Honduras) and then opened to general participation. The account given here has been prepared by the Editor from summaries provided by the four panellists, notes taken during the discussion and records of the questions and answers provided by the other participants.

Kochhar pointed out that astronomy is unique among modern scientific disciplines in the sense that it depends on international pooling of efforts. It is important to interpret the history of astronomy in the same spirit, that is, by emphasizing scientific content and continuity rather than the geographical and denominational compartments. By presenting astronomical developments through the ages as a cultural continuum, we will be emphasizing astronomy as a global heritage which must therefore be globally enriched and appreciated. The West must give up its fixation with ancient Greece. It should recognize that the Greek contributions had their antecedents and that time did not stand still between Ptolemy and Copernicus. On their part, cultures with memories of past contributions should use tradition as a source of inspiration and then selectively break away from it to enhance modern science.

Gerbaldi spoke of the need for organizing documentation for astronomy education. This point had come up several times (see, for example, p. 100). Particularly at school level there seemed to be a lack of documentation or, if it exists, people do not know where to find it. Perhaps a Centre should be set up to collect such documentation.

She emphasized the cultural differences in the ways of learning and teaching, that must be taken into consideration by those using the sort of documentation she had just described. Particular courses are always conceived for some specific environment, for teachers with prerequisites about the sort of activities they could undertake with their pupils.

Besides the Centre she proposed, and perhaps before it is set up, there should be a "tutorial service". Besides documentation, teachers need tuition or advice –but always as a dialogue with someone who has had face-to-face contact with pupils. These tutorials are needed to help in the acquisition of the necessary astronomical background, in the selection of the best documentation, and to provide a means of feedback from it, etc. Ways of learning are so different that no one procedure will fit all situations; but there is a wide variety of experience on which to draw. Indeed, from Gerbaldi's point of view, there is no lack of documentation – she is flooded by it – but the opportunities for "lonely teachers" to have personal contact with tutors, who will share their experience and not just provide documentation, are severely limited.

Gerbaldi also drew attention to the many small telescopes around the world, with apertures less than 0.4 m or 0.5 m. How can we organize connections between them, in order to coordinate observations, to initiate observing programmes and to provide tutorials for the "lost observers" who could use them?

Ubachukwu stressed the problems of communication. There had been much talk at the Session on what to do with telephones, the Internet and libraries. There had been no talk on what to do without these. There is a danger in judging other developing countries by the experience of India. In Nigeria, they have to use their salaries to do any scientific work. There are no telescopes, no Internet (they have to pay for e-mail) and they can only receive faxes. There are no personal computers: they have to pay people to type, to analyze and to draw graphs. It takes a minimum of two years to get an article published. Ubachukwu still has not seen copies of his last two papers. They have to select journals that do not levy page charges, but they never see even them and cannot cite their own articles properly. There is no-one around to advise and criticize (compare the acknowledgments at the end of many Western papers!).

Recently, at the University of Nigeria, Nsukka, the National Space Research and Development Agency (NASRDA), formed in 1998, has established a Centre for Atmospheric Sciences and Astronomy (CASA) with Prof. P.N. Okeke as Director. NASRDA is improving commnuications among Nigerian scientists and sponsors meetings and workshops that motivate space scientists and increase the awareness of the relevance of space science and astronomy to national developments. The Astrophysics Group of the University of Nigeria has so far been largely self-supporting through international collaborations. The IAU has provided most of the travel grants and Jodrell Bank has provided us with preprints as astrophysical journals. The Group has never had any specific financial support either from the University or from the government. The establishment of CASA thus marks the beginning of funding by the Nigerian government for academic research in astronomy and will, undoubtedly, help to solve the long-standing problem that has crippled the development of astronomy in the country.

Pineda de Carias asked a question posed before in the Session (see p. 328) "What is a developing country?". Her answer: a country that needs help, that feels the need to raise its level. Independently of how one defines a developing country there is a common factor that defines the problem: we want our country or region to contribute significantly to the worldwide effort. How? There is not one recipe that suits all cases. One has to think about a strategy, to work out projects, to state policies. When you are the astronomer, the "lonely one" in a small country, you have to think what to do, how to do it and when to do it. But you need resources: money, equipment, staff, well-trained astronomers. How do you begin? We are all inhabitants of the same planet –the only inhabited planet we know; but language, culture and tradition separate us and present natural cores around which we can develop. We must transcend these divisions.

She supported Narlikar's concept of Third-World networking and agreed with Batten's suggestion of links between institutions: let us work out these ideas.

From the audience, Aguilar spoke of the problems that she faces in Peru. Although "how to learn science" is a worldwide problem, each country has its own specific difficulties. In Peru, the Faculty of Education emphasizes the methodology of education rather than the knowledge content of courses. The current globalization of industry leads to stress on production; basic science is seen as unnecessary by government officials. We have to use our imagination to convince them. People are concerned about the contamination of the environment by technology and we need to relate these concerns to astronomy. Astronomy is multidisciplinary and interdisciplinary.

Cooperation between countries is very important, especially within regions. Each Latin-American country is preoccupied with its own problems and they all forget that cooperation within the region would help them all, in a very short time. There is not enough interaction within the region.

There are educational reforms under way in Peru and two courses *Naturaleza y Ambiente* and *Ciencias de la Tierra* have been written from the point of view of astronomers. For the first time there is astronomy in the high-school curriculum –about one- third of a first-year course. Previously, astronomical topics were scattered in several different courses encountered by the students over a period of five years.

Taking up the theme of the relation between astronomy and environmental concerns, Crawford remarked that light pollution is a problem for all countries. He urged developing countries not to make the same mistake that the West has made. Light pollution is an issue with many social and environmental implications. Start campaigning in your own country now! Use the International Dark-Sky Association as a resource (www.darksky.org). (*Editor's note*: see also Schreuder's abstract, p. 364).

Martinez also stressed the need for cooperation between developed and developing countries: it is not taking place now at the right level. He spoke about the World Space Observatory (WSO), a concept that had emerged from the UN/ESA Workshops. The UN Committee on the Peaceful Uses of Outer Space was concerned with non-commercial and non-military uses of space for all humanity –i.e. astronomy. The mission concept is of an orbiting telescope outside the Earth's shadow and in a low-radiation zone. No new technology needs to be developed for it. Science Operations Centres on each continent (or even within each country), located in already existing space-science centres, could become mission-operations centres and provide personal- computer services. The WSO would promote cooperation between developed and developing countries and help to retain the best talent in each developing country. There are complex management issues and global cooperation is needed. He urged all those interested to contact himself or Willem Wamsteker.

Mattei commented that if the WSO were built there would be a need for workshops to educate those interested in using it. Such workshops should receive high priority because, without that sort of education, the satellite would not be used by astronomers from developing countries but by a few from developed countries. Adelman added that, although the WSO was a good idea, ground-based observatories could be provided much more cheaply. He drew attention again to the availability of data-bases through the Internet. NASA has also put all its educational materials on the Web.

Much of the general discussion centred around the issues of education and networking. Rijsdijk felt that the IAU should form a Working Group for the development of educational resources which could be made available to organizations such as UNESCO, ICASE, ICSU, etc. Such a group could take cognizance of local culture and needs and make accurate information available. (*Editor's note*: Commission 46 already does much of this sort of work.) Ros also stressed the importance of contact between teachers in developing and developed countries. The European Summer Schools invite some participants from non-European countries (see p. 98) and those participants can organize courses in their own countries, based on what they have learned in the Summer School. She offered the opportunity to some teachers from developing countries to participate in the Summer Schools.

Bhatia voiced his support of Narlikar's scheme for networking, also supported by Pineda de Carias. He hoped that a request would go from this Session to the IAU to organize a group to develop the idea. Such a network could also help with education in astronomy, in schools, at higher levels and for the lay public. For example, it could identify groups of scientists able and willing to help and which could be approached by those wanting to use their services. Tancredi pointed to the example of networking given by the International Centre of Theoretical Physics in Trieste. It might be wise to ask them to broaden their interests to include research in astronomy and astrophysics.

Metaxa thanked Kochhar for his remarks about the Greek influence in astronomy but felt that this topic would be more appropriately discussed in Commission 41. Tancredi said that the IAU should be requested to encourage reflection of the diverse human cultures in the naming of celestial objects (e.g. minor planets) which, he believed, still had a strong tradition of using the names of Greek gods. Kozai pointed out that the discoverers of minor planets have the right to propose names for their own discoveries and many of them are quite happy to receive suggestions. Adelman agreed that it was important to be sensitive to the different cultures of the world and regretted that, in his opinion, few astronomers knew much about any culture other than their own. Chamcham felt that one should be cautious in making use of local traditions.

A few other isolated points were made. Adelman pointed out that many U.S. universities could donate equipment only if there was a cooperative agreement with the country concerned. Ratnatunga appealed to everyone with access to the Internet to publish on it so that they broadcast their results worldwide. Publication on the Internet is much less expensive than printing and can reach many more readers. Fierro commented that astronomers in developing countries need to spend time learning about fund-raising and how to encourage donations.

Index of Countries

Albania 26,
Algeria 26, 171ff, 227, 230,
Angola 101,
Argentina 26, 65, 66, 72, 75, 77, 115, 313.
Armenia 26, 115, 354-5.
Australia 20, 26, 102, 106, 322.
Austria 26, 97, 98, 115,
Azerbaijan 26, 167,

Bahrain 26,
Belgium 23, 26, 97, 98, 115,
Bolivia 26, 65, 230,
Botswana 101,
Brazil 26, 65, 66, 72, 75, 77, 98, 102, 115, 340.
Brunei 200.
Bulgaria 26, 102, 115, 230, 340.
Burma (Myanmar) 313, 350.

Cambodia 350.
Canada 4ff, 19, 26, 115, 126, 253.
Chile 9, 18, 26, 33, 98, 115, 272, 313.
China (Beijing) 9, 26, 80ff, 115, 133, 210ff, 245, 319ff, 355-6.
    (Taipei) 23, 26, 115, 163, 219, 320ff.
Colombia 26, 115.
Costa Rica 70, 72, 78.
Croatia 26, 115.
Cuba 23, 26, 115, 163.
Czech Republic 340.

Denmark 26, 97, 98, 115.

Ecuador 26, 233.
Egypt 26, 157ff, 167, 179ff.
El Salvador 70.
Estonia 23, 25, 26, 97, 115.

Finland 26, 97, 98, 115.
France 26, 45, 56, 59, 72, 97, 98, 133, 176, 178, 276-7, 354-5.

Georgia (Rep. of) 26, 97, 98.
Germany 26, 97, 98, 115, 181, 222, 272, 354-5.
Greece 26, 97, 98, 115, 367.

Honduras 26, 69ff.
Hungary 26, 102, 115.

Iceland 23.
India 3, 9, 14, 26, 86, 115, 133, 164, 165, 258, 272, 303ff, 313, 322, 326, 340, 354-5, 369.
Indonesia 9, 26, 115, 133, 197ff, 233, 313.
Iran 26, 115, 188.
Iraq 26, 157.
Ireland 26, 340.
Israel 26.
Italy 26, 61, 97, 98, 115, 354-5.

Japan 11, 18, 21, 25, 26, 47, 56ff, 67, 115, 166, 184, 250, 266, 271, 312-3, 319ff, 354-5.
Jordan 160, 163, 313.

Kazakhstan 26.
Korea (Dem. Rep.) 23, 26.
Korea (Rep. of) 9, 21, 27, 115, 133, 319ff.

Laos 350.
Latvia 26, 97, 98.
Lithuania 26, 115, 340.
Luxembourg 97, 98.

Macedonia (Rep. of) 232.
Madagascar 101.
Malaysia 26, 200, 313.
Malta 340.
Mauritius 26, 227.
México 26, 66, 107ff, 115, 164, 165, 167, 178, 208.
Morocco 26, 59ff, 178, 227.
Mozanbique 101.

Netherlands 26, 39, 97, 98, 115, 126, 257, 265.

New Zealand 9, 11, 18ff, 26, 115, 116, 126, 130, 163, 208, 222, 269-70, 322.
Nigeria 19, 26, 115, 227, 370.
Norway 23, 26, 97.

Pakistan 26.
Paraguay 27, 65ff, 313, 315.
Peru 27, 115, 368-9.
Philippines 27, 49ff, 313.
Poland 27, 115, 222.
Portugal 27, 64, 97, 115.

Romania 27, 102, 115, 164, 232.
Russia 27, 97, 98, 115, 167, 188, 230, 355-6.

Saudi Arabia 23, 160.
Singapore 21, 27, 271, 313.
Slovakia 27, 340.
Slovenia 27, 115.
Somalia 191.
South Africa 27, 86, 101, 115, 117ff, 133, 152, 221ff, 348-9.
Spain 27, 64, 72, 78, 97, 98, 115, 133, 159, 166, 176, 277, 337, 354-5.
Sri Lanka 27, 115, 266ff, 279-80, 285ff, 313.
Sweden 27, 97, 98, 115, 191.
Switzerland 23, 27, 97, 115.

Tajikistan 27, 187ff.
Tanzania 166, 340.
Thailand 27, 56, 133, 165, 243ff, 313, 350.
Turkey 27, 133, 188.

Ukraine 25, 27, 115.
United Kingdom 20, 27, 64, 97, 106, 115, 131ff, 176, 179, 180, 222, 338, 334-5.
United States of America 20ff, 27, 48, 54, 85, 110, 115, 139, 142, 152, 222-3, 271, 276, 285, 287, 313, 326, 338, 355-6, 372.
Uruguay 27, 65, 115, 313.
Uzbekistan 27, 166, 230, 232, 313.

Vatican City State 8, 23, 110ff.
Venezuela 27, 114, 115, 205ff.
Vietnam 27, 46ff, 92, 115, 262ff, 313, 348, 350.

Yugoslavia 23, 27.

Zambia 39ff, 101.
Zimbabwe 101, 364.

Subject Index

AAVSO 4, 47, 63, 84, 89ff, 252ff.
Amateur astronomers 4, 30, 32, 60, 84, 89ff, 160, 163, 171, 231, 287.
Archaeoastronomy 75, 157.
Astrometry and positional astronomy 146-7, 173-4, 184, 192.
Astronomical Research Index 15, 16, 21ff.
Astronomy curricula 61, 72ff, 81ff, 141ff, 194-5.
Astrophysics 46, 58, 60ff, 69ff, 89ff, 110ff, 141-2.
Atmospheric research 173ff, 189ff, 252.

Bureaucracy 7, 61, 238, 272, 287.

Calendars (religious) 51, 63, 116, 171, 178, 200, 232, 271.
Carte du Ciel 19, 172, 178.
Catalogues, astronomical 230, 291ff.
CCD cameras 180ff, 47-8, 149, 199, 207, 224, 231, 237, 241ff, 252ff, 269. 287, 303, 306, 342.
   photometry 61, 232, 199, 252, 306.
   spectroscopy 184, 245.
Celestial mechanics 142, 148.
Climate effects
   on books 347, 350.
   on telescopes 247.
Cooperation
   International 42, 62, 176, 287.
   Regional 42, 69ff, 319ff.
Comets and meteors 53, 58, 163, 166, 179, 188ff, 252, 364.
Computer availability 16, 62.
Crises, economic and political 164, 193-4, 197ff, 208.

Data bases 193, 231, 232.
Dependence 31, 290.
Distance learning 131ff.

Eclipses (solar) 45, 51, 52, 84, 101ff, 171, 281, 363.

Education 3, 31ff, 40ff, 59ff, 69ff, 80ff, 107ff, 117ff, 131ff, 141ff, 152ff, 159-60.
Electromagnetic radiation 143ff.
Employment opportunities in astronomy 20, 40, 85-6, 151, 265, 283.
English language, see Language barriers.

Gravity 143, 145-6.
GDP 5, 353ff.
GNAT 237ff, 271.
GNP 4ff.

Hands-on Astrophysics 47, 61, 89ff, 252.
HIPPARCOS 92, 93, 142, 144, 146-7, 300.
HST 271, 280, 284-5.

IAU 3-4, 8-9, 88, 18, 23, 25, 39, 47, 49, 55, 57, 59, 61, 62, 65ff, 78, 101, 191, 250, 253-4, 271, 322, 328, 333, 343-4, 354, 360-1, 364, 371-2.
ICSU 4, 328, 371.
ICTP (Trieste) 327. Interferometry 257ff, 310.
Internet xii, 61, 92, 137, 167, 240, 276ff, 279ff, 324ff, 354ff.
Isolation 9, 19, 161-2, 325.
IUCAA 165, 324ff.

Journals
   Electronic access to 20, 277, 280, 284, 354ff, 363.
   Page charges 19.

Language barriers 63, 66, 79, 99-100, 112-3, 118, 333ff.
Leonids, see Comets and meteors
Libraries 16-17, 32, 74, 161, 200, 326, 340ff, 345ff.
Light Pollution 242, 269, 280, 331, 364, 371.

Minor Planets 231, 253.
Meteors, see Comets and meteors.

Near-Earth Objects 75, 163, 191, 195, 218.

Networking 11, 227, 287, 324ff.

OAS 71, 78.
Observatories
    AAO 271.
    Arecibo 326.
    Auckland 269-70.
    Bosscha 197ff, 233.
    Bouzaréah 172ff.
    Boyden 127.
    Cagigal 205ff.
    Calern 173-4, 178.
    Cerro Tololo 271-2.
    Chinese National 211ff.
    ESO 9, 221, 271-2.
    Gissar 188ff.
    GMRT 257-8, 265.
    GONG 271.
    Hartebeesthoek 126.
    Helwan 179ff.
    Jodrell Bank 370
    Las Campanas 272.
    Manila 49ff.
    Mauna Kea 271.
    McDonald 152ff.
    Mérida 206ff.
    Pamir 188ff.
    Paris 265.
    Pulkovo 230, 232.
    Roma 61-2.
    SAAO 41, 42, 120ff, 153ff, 221-2, 228, 271-2.
    Sanglokh 188ff.
    Santa Ana 230.
    Sirindhorn 243ff.
    Suyapa (Central America) 69ff.
    Tamanrasset 173-4, 178.
    Trieste 62.
    Yunnan 245.
    Westerbork 257, 265.
    World Space 371.

Photometry 198, 241, 244, 252.
Planetaria 48, 49ff, 60, 66, 159, 201, 221ff, 230, 312-3.
Planetary systems 293.
Public appeal and outreach 31, 34, 47, 51ff, 60, 61ff, 84, 101ff, 152ff, 160, 197ff, 226.

Publishing 19, 357ff.

Radio astronomy 48, 255ff, 307ff.
Religion and science 63, 116, 271, 285.

SALT 20, 23, 128, 152ff.
Security 7, 61, 251, 267, 329ff.
Site testing 175, 188, 320.
Solar astronomy 59, 60, 81, 173-4, 212, 230-1, 252, 262, 308.
Solar-Terrestrial Relations 175-6, 218.
Space Science 5, 30, 40ff. 60, 70, 200ff.
Spectrographs 179-80, 184, 188, 198, 224.
Stars (binary) 108, 198-9, 232, 244, 252, 291ff.
Stars (variable) 89ff, 174, 188ff, 251.
Summer schools
    European 95ff, 166, 372.
    ISYA 59, 65, 66.
    Vatican 110ff.

TAD 46ff, 62, 78, 274.
Telescopes
    Schmidt 198, 207, 212, 273, 303.
    Small xii, 8, 30, 90, 231, 237ff, 243ff, 250ff, 279-80, 303ff, 312ff.
Time measurement 51, 56, 142, 146-7, 206, 230.
TV 106, 107ff, 134, 160, 163.
TWAN 11, 324ff, 368, 370, 372.

UN xiii, 8, 33, 44, 48, 55, 57, 60, 61, 69, 78, 266, 268, 287.
UNESCO 198, 328, 3671.

Wealth (disparity between nations) 5, 23, 313.
Web access, see Internet
WGSSA 9, 42, 43, 228, 364.
WSO, see Observatories.

INTERNATIONAL ASTRONOMICAL UNION (IAU)
VOLUMES

Published by

Astronomical Society of the Pacific

PUBLISHED: 1999

Vol. No. 190   NEW VIEWS OF THE MAGELLANIC CLOUDS
eds. You-Hua Chu, Nicholas B. Suntzeff, James E. Hesser, and David A. Bohlender
ISBN: 1-58381-021-8

Vol. No. 191   ASYMPTOTIC GIANT BRANCH STARS
eds. T. Le Bertre, A. Lèbre, and C. Waelkens
ISBN: 1-886733-90-2

Vol. No. 192   THE STELLAR CONTENT OF LOCAL GROUP GALAXIES
eds. Patricia Whitelock and Russell Cannon
ISBN: 1-886733-82-1

Vol. No. 193   WOLF-RAYET PHENOMENA IN MASSIVE STARS AND STARBURST GALAXIES
eds. Karel A. van der Hucht, Gloria Koenigsberger, and Philippe R. J. Eenens
ISBN: 1-58381-004-8

Vol. No. 194   ACTIVITY IN GALAXIES AND RELATED PHENOMENA
eds. Yervant Terzian, Daniel Weedman, and Edward Khachikian
ISBN: 1-58381-008-0

PUBLISHED: 2000

Vol. XXIV A   TRANSACTIONS OF THE INTERNATIONAL ASTRONOMICAL UNION
REPORTS ON ASTRONOMY 1996-1999
ed. Johannes Andersen
ISBN: 1-58381-035-8

Vol. No. 195   HIGHLY ENERGETIC PHYSICAL PROCESSES AND MECHANISMS FOR
EMISSION FROM ASTROPHYSICAL PLASMAS
eds. P. C. H. Martens, S. Tsuruta, and M. A. Weber
ISBN: 1-58381-038-2

Vol. No. 197   ASTROCHEMISTRY: FROM MOLECULAR CLOUDS TO PLANETARY SYSTEMS
eds. Y. C. Minh and E. F. van Dishoeck
ISBN: 1-58381-034-X

Vol. No. 198   THE LIGHT ELEMENTS AND THEIR EVOLUTION
eds. L. da Silva, M. Spite, and R. de Medeiros
ISBN: 1-58381-048-X

PUBLISHED: 2001

Vol. No. 204   THE EXTRAGALACTIC INFRARED BACKGROUND AND ITS COSMOLOGICAL
IMPLICATIONS
eds. Martin Harwit and Michael G. Hauser
ISBN: 1-58381-062-5

IAU SPS   ASTRONOMY FOR DEVELOPING COUNTRIES
Special Session of the XXIV General Assembly of the IAU
ed. Alan H. Batten
ISBN: 1-58381-067-6

INTERNATIONAL ASTRONOMICAL UNION (IAU)
SYMPOSIA
Published by the Astronomical Society of the Pacific

Complete lists of proceedings of past IAU Meetings are maintained at the
IAU Web site at the URL: http://www.iau.org/publicat.html

Volumes 32 - 189 in the IAU Symposia Series may be ordered from
Kluwer Academic Publishers
P. O. Box 117
NL 3300 AA Dordrecht
The Netherlands

All other book orders or inquiries concerning volumes listed should be directed to the:

Astronomical Society of the Pacific Conference Series
390 Ashton Avenue
San Francisco CA 94112-1722 USA

Phone:	415-337-2126
Fax:	415-337-5205
E-mail:	catalog@aspsky.org
Web Site:	http://www.aspsky.org